THE ROUGH GUIDE to

The Universe

2nd edition

John Scalzi

www.roughguides.com

Credits

The Rough Guide to The Universe

Editing: Joe Staines & Matthew Milton
Cartography: Maxine Repath, Ed Wright & Katie Lloyd-Jones
Illustration: Peter Buckley
Proofreading: Diane Margolis
Production: Rebecca Short

Rough Guides Reference

Director: Andrew Lockett
Editors: Peter Buckley, Tracy Hopkins, Sean Mahoney, Matthew Milton, Joe Staines, Ruth Tidball

Publishing Information

This second edition published April 2008 by
Rough Guides Ltd, 80 Strand, London WC2R 0RL
345 Hudson St, 4th Floor, New York 10014, USA
Email: mail@roughguides.com

Distributed by the Penguin Group

Penguin Books Ltd, 80 Strand, London WC2R 0RL
Penguin Putnam, Inc., 375 Hudson Street, NY 10014, USA
Penguin Group (Australia), 250 Camberwell Road, Camberwell, Victoria 3124, Australia
Penguin Books Canada Ltd, 90 Eglinton Avenue East, Toronto, Ontario, Canada M4P 2YE
Penguin Group (New Zealand), Cnr Rosedale and Airborne Roads, Albany, Auckland, NZ

Printed in China

Typeset in DIN, Myriad and Minion

The publishers and authors have done their best to ensure the accuracy and currency of all information in *The Rough Guide to The Universe*; however, they can accept no responsibility for any loss or inconvenience sustained by any reader as a result of its information or advice.

No part of this book may be reproduced in any form without permission from the publisher except for the quotation of brief passages in reviews.

© John Scalzi, 2003, 2008

416 pages; includes index

A catalogue record for this book is available from the British Library

ISBN 9-781-84353-800-4

1 3 5 7 9 8 6 4 2

Contents

Part 3: Resources

List of Star Charts

Introduction

"We are made of starstuff"

Carl Sagan

Once upon a time, billions of years ago, a star exploded in the endless night. It exploded at the end of its life cycle, after first imploding as its core collapsed. When the core could collapse no further, the inward-falling gas bounced off the core and started back up, causing the star to explode violently in a blast that outshone its entire galaxy.

Some of this material became part of a nebula which served as a "stellar nursery", a place where new stars form from dust and gas. It was here that our own Sun was born out of the elements torn from the exploding star. Other parts fell into the disc of dust and gas that formed around the Sun, from which the Sun's family – its planets – would later be born. Among these planets was our own Earth, on which a unique combination of these star-born elements arose in the form of life: algae, trilobites, early fish and amphibians, giant dinosaurs, mammals and eventually humans.

We are all made from stars, as is everything you've ever known. Is it any wonder, therefore, that from the first moment humans had the idea of looking up into the night sky, we've felt a connection to those far-distant points of light? One of the aims of *The Rough Guide to the Universe* is to get you a little better acquainted with these long-lost "ancestors" of ours.

Knowledge of the universe is expanding – nearly at the same rate as the universe itself! In terms of astronomy, there has never been a more exciting time to be alive and to take part in its exploration. I hope that this book captures some of the excitement of the moment, sharing much of what is known, and suggesting what's left to discover.

People are often intimidated by science because they assume it is always complicated. But the universe is not as hard to understand as is often thought, and there's a tremendous amount that can be learned about it without having to drag in obscure astronomer-speak. *The Rough Guide to the Universe* is aimed at all those people who have ever looked up into the night sky and wondered what the heck is going on up there, and how did it get there in the first place? After all, it's easy enough to see and enjoy the night sky, so it shouldn't be that difficult to explain what's behind some of its main attractions.

As well as giving a general overview of the current state of our understanding of the universe, *The Rough Guide to the Universe* is also a practical guide for the enthusiastic home astronomer. It covers all the highlights of the night sky that can be seen using the naked eye, binoculars or home telescopes. It also provides context, explaining why you see what you see. To assist observation of the stars, charts are provided for every constellation with information on what you'll see there, as well as the stories behind the constellations. This book won't answer every question or explore every astronomical mystery, but it covers a lot of ground, and will enable you to find your way around the darkness of the night sky.

How this book works

The main part of the book comprises a section, entitled "The Universe", made up of fourteen chapters. These begin with a quick tour through the life of the entire universe (including how it began and how it might end), followed by some basic information on what you need to know when observing the night sky, plus a survey of useful equipment like binoculars and telescopes. Chapters 3 to 11 concentrate on our own solar system and how best to view the objects within it.

Each chapter has the same organizational structure – information on the chapter subject comes first, and then a guide on observing that object in the night sky (with the exception of the Sun, which for obvious reasons is observed in the daytime). For example, the chapter about our Moon begins with the birth of the Moon before describing how it is now, and what you might experience if you were to walk on its surface yourself. The second part of the chapter focuses on how to view the Moon, highlighting the major features that are visible with the naked eye, and with binoculars or a telescope.

Chapters 12 to 14 cover stars, galaxies and some of the other, stranger, phenomena that exist in our universe. Little of the really weird material – such as black holes, dark matter and extra-solar planets – can be seen through home telescopes, but they are, nevertheless, fascinating to read about.

Details on how to view stars and galaxies that are observable appear in the "Star Charts" section of the book which comes next. After a brief introduction, explaining how to read a star chart, come the charts themselves: beginning with two showing all the constellations in the northern and southern hemispheres, followed by 68 charts covering every constellation in alphabetical order. For each chart there is accompanying text providing historical and scientific information, and noting the major objects that are worth a closer look through binoculars or a telescope.

Finally, the "Resources" section gives tips on how to find astronomy information both online and off. This includes a short list of books for further reading as well as a substantial website directory listing information on telescopes, planetaria, observatories, space agencies, image galleries and much else besides.

About the author

John Scalzi is the author of three Rough Guide titles: *The Rough Guide to the Universe* (you're reading it now), *The Rough Guide to Sci-Fi Movies* and *The Rough Guide to Money Online*. He's also an award-winning science fiction novelist, whose books include *Old Man's War*, *The Android's Dream* and *Zoe's Tale*. He lives in rural Ohio, USA, where he can look up and see the Milky Way any night of the year without cloud cover. Having grown up in Los Angeles, with all its night-time light pollution, he thinks that's pretty cool.

Acknowledgements

This book is dedicated to Kristine and Athena Scalzi with all the love in the world, in the stars, and beyond.

God is said to have created the universe in a week (plus breaks); it took me fifteen months to finish this one small book. Indeed, I might still be writing it were it not for the support, help and inspiration I received from friends, editors and others. So I'd like to take this opportunity to recognize the following people and offer them my thanks.

First, to Robert Shepard, an excellent agent and friend who always checked in to make sure I wasn't getting lost in the cosmos. Every writer should have an agent like him, and I'm sorry for those who don't. At Rough Guides my editor Joe Staines did yeoman's work letting me know when I was going over people's heads; he's the man to thank for the consistently accessible tone. Also many thanks to Helen Prior for the book's clear design and layout, and to Maxine Repath and the cartography department, in particular Katie Lloyd-Jones, for handling the star charts with such competence and efficiency. Peter Buckley likewise created several neat illustrations that help to illuminate the text.

Thanks to Barry Gandelman at Space.com, and to all the people who make the Starry Night software, which was instrumental in creating the star charts in the book; they made it far easier than it would have been otherwise. Over at NASA, Bert Ulrich heard me declare several times that he was my hero for the day for helping to clear up permissions questions I had.

Dr Joan Rohrbach and Dr Matthew Carlson were instrumental in fact-checking early chapters of the book; any subsequent errors are attributable to me, and not to them. On the other side of things, my friend Stephanie Lynn read the book to make sure the concepts and ideas were accessible to everyone. Thanks also to the members of the Miami Valley Astronomical Society who offered to give the book a read-through.

I'd also like to acknowledge astronomy's great communicator, Carl Sagan, whose *Cosmos* book and TV series inspired me as a kid, and whose mission to popularize science (without dumbing it down) is an ideal that I have aspired to in my own writing.

Finally, thanks to Kristine Scalzi, my wife, for just about everything. And to my daughter Athena Scalzi, who asks me to get out my telescope so she can look at the planets, and who knows Saturn from Jupiter, which is pretty cool considering she's not quite four. I'd give you both the entire universe if I could; here's a book about the universe instead.

THE UNIVERSE

The first 15 billion years

in 15 minutes

In the beginning God created the heaven and the earth.
And the earth was without form, and void; and darkness was upon the face of the deep. And the Spirit of God moved upon the face of the waters.
And God said, Let there be light: and there was light.
Genesis, Chapter One, Verses 1–3

In the beginning

The scientific account of the creation of the universe, which occurred some fourteen billion years ago, is even shorter than the biblical version. For all intents and purposes, it can be summed up in two sentences. In the beginning there wasn't anything. And then there was.

Was *what*? Literally, everything – every bit of matter and energy that would ever be in our universe was created in the unimaginable crucible of heat and light that has been loosely termed the **Big Bang**. Just for the record, it was neither big (in the beginning our universe would have fit onto the head of a pin), nor was there a bang.

It's not even accurate to call it an "explosion", because the phenomenon marked the creation of everything and hence occurred everywhere in the universe, all at once. In fact the term Big Bang was something of a slur, coined derivatively in 1950 by physicist **Fred Hoyle**, who

Hundreds of galaxies captured by the Hubble telescope at various stages of their development. Some are over 11 billion light years away, from when the universe was just one-sixth the size it is today (see p.8).
Hulton-Deutsch/CORBIS

thought the theory was bunk. But Big Bang is what we all know it by, and it's too late to change it now.

Planck's Time

What caused the Big Bang? We don't know. Also, at this point, we *can't* know. Our current understanding of physics allows scientists to trace events back to a mere fraction of a second after the creation of the universe, to an instant called **Planck's Time** – that is 10^{-43} of a second after the creation of everything (or, if you'd like to see that numerically, .0000000000000000000000000000000000000000001 seconds after the creation of everything). In order to have an idea of what happened before that moment, physicists would need a theory of **unified force**: a model of the universe to tie all four universal forces – **gravity**, **electromagnetism** and **strong and weak nuclear force** – together. Grand Unified Theories (or GUTS) currently suggest that at the very high temperatures and densities in the early stages of the Big Bang all four fundamental forces of nature were unified. The forces separated as the universe expanded and cooled. Unfortunately, no one has come up with a model for it. Yet.

The universe at Planck's Time was hot and crowded. Hot because the temperature of the universe was 10,000,000,000,000,000,000,000,000,000,000°C; crowded because the entire universe was crammed into a space 1.6×10^{-35} metres long, about 1/10,000,000,000,000,000,000th the size of a **proton** (a subatomic particle which is substantially smaller than the entire atom itself).

Evidence for the Big Bang

The theory of the Big Bang isn't perfect, and over time additions and modifications have been made to it in order to make it more consistent with the universe as it appears to us. One big modification was the inclusion of the inflationary era, which allowed scientists to explain variations in the structure of the universe that were otherwise inexplicable. Be that as it may, much of what we see in the universe today conforms to what we would expect to see under the Big Bang hypothesis. The examples include:

▶ **The Expanding Universe** The Big Bang hypothesis dictates that the universe would continue to expand after its initial fiery sprint into existence. It is expanding and shows no signs of stopping its expansion, even as gravity tugs at everything in the universe in an attempt to bring it all together.

▶ **The Abundance of Helium** According to the Big Bang theory, roughly 25 percent of the mass of the universe should be in the form of helium created in the first seconds of the universe, and observations show that this is indeed almost exactly the case. Helium is also created in stars through nuclear fusion, but not in the amounts required to explain the abundance of helium in the universe. A tiny proportion of the universe is made of deuterium (a hydrogen atom with one neutron); since deuterium is only infrequently made in stars, nearly all of it was made during the Big Bang.

▶ **Background Radiation** The Big Bang theory posited the existence of cosmic background radiation, which is the 'residue' of the early expansion of the universe. Discovered (almost entirely by accident) by astronomers Arno Penzias and Robert Wilson in 1965, one of the notable aspects of background radiation is that it is almost entirely 'smooth' and functions like a black body (a physical entity that is a perfect radiator of energy, one which absorbs all electromagnetic radiation that falls onto it, reflecting none), which can be explained by the Big Bang theory. Later observations of the cosmic background radiation by NASA's COBE satellite showed tiny variations in the radiation, fluctuations which are believed to be the 'seeds' of galaxies and the stars within them.

The next few seconds

Physicists have discerned the presence of **gravity** at Planck's Time, a momentous event in the history of this young, hot universe. In the next few seconds, the universe continued to expand and cool rapidly, and that allowed for some other very interesting events to occur.

At about 10^{-35} seconds after the Big Bang, a **strong nuclear force** broke off into a separate force, and many physicists believe the universe entered a brief but spectacular **inflationary period**, in which it expanded from the size of an atom to something the size of a cantaloupe melon – rather like a grape expanding to fill up the observable universe in less time than it takes you to blink. This might seem improbable, but such an inflationary period would explain much of what came later, including how galaxies formed. This inflationary period shut off at around 10^{-32} seconds.

In this suddenly roomier universe, at about 10^{-6} seconds after the Big Bang, **nucleons** formed. These are a class of particles that includes protons and neutrons, both of which we will later find in the nuclei of atoms (but not yet – it's still too hot for atoms to form). **Leptons**, which include electrons, photons and neutrinos, will be "born" shortly after.

At one hundred seconds after the Big Bang, the universe had cooled down enough (to a mere 1,000,000,000°C) for protons and neutrons to begin clumping together in atomic nuclei – **deuterium** (one proton and one neutron), and **helium** (two protons and two neutrons). The abundance of both deuterium and helium in the universe is in fact one of the best bits of evidence we have that the Big Bang theory is largely correct. Elements heavier than hydrogen are usually produced in stars, but deu-

Sir Fred Hoyle believed in an unchanging, "steady-state" universe; his disparaging description of a competing theory gave the "Big Bang" its name.
Hulton-Deutsch/CORBIS

Alternative theories to the Big Bang

Current scientific thinking considers the Big Bang creation theory to be the one that best explains the origins of the universe, but that is not to say that there haven't been other attempts to explain how we got to where we are today. Here are some of them:

▶ **Steady State** This theory was proposed in 1948 by Fred Hoyle, Thomas Gold and Hermann Bondi. Their view of the universe held that at the largest scale, the universe looks the same at all times and all places, and for this to be possible, the universe could not have a real "beginning" or "end". It was a theory that did not hold up to close scrutiny, however. For instance, the discovery of quasars in the 1960s showed that the universe really wasn't the same in all times, a fact that was reinforced by the discovery of the background radiation at the edge of the observable universe.

▶**No-Boundary Universe** This interpretation of the universe, presented by Stephen Hawking and Jim Hartle in 1983, gets rid of the need for the infinitely hot and dense singularity from which our universe presumably expanded. It does so by dispensing with the conventional definitions of space and time. Instead, it employs an abstract, mathematical "imaginary time", which has no beginning or end. In a practical sense, this doesn't have much effect on the observable universe, since Hawking and Hartle's universe would appear to behave in the same way as one that had started from a singularity – it just gets rid of the *need* for the singularity. Sound complicated? It is.

▶**Ekpyrotic Universe/Cyclical Model** This wild new theory, initially advanced in 2001 by Paul Steinhardt, Burt Ovrut, Justin Khory and Neil Turok, supposes that the Big Bang was not a superdense singularity but the result of a five-dimensional collision between our universe – which until this time had been a featureless expanse – with another universe. The collision knocked loose the energy required for the Big Bang, and began the formation of the galaxies that pepper our universe today. The theory is still being revised, notably by Turok and Steinhardt, who suggest that the brane collision that created our universe wasn't the first – and won't be the last.

terium is only rarely produced in them, and the amount of helium in the universe is too great to be accounted for simply by stellar formation. They had to come from somewhere else, or more accurately some *time* else – the first few seconds after the Big Bang.

The recombination era

After the first few seconds in the life of the universe, the big events began to space themselves out a little more. The next really big step in the evolution of the universe happened roughly 300,000 years after the Big Bang: a point in time in which our universe was 1000 times smaller than it is today, and about 1000 times hotter, with an average temperature of about 5000°F/2700°C. Before this point, the temperature in the universe was too hot to allow electrons to orbit the protons and neutrons that had clumped together. But now it was just cool enough, and suddenly there were **atoms** everywhere. (Hydrogen and helium atoms, that is; heavier elements would come much later.)

Thus began what is known as the **recombination** or **decoupling** era. One of the side effects of the appearance of atoms was that the universe, previously opaque (thanks to the tremendous amounts of energy banging around) became transparent. We can still see the residue of this era at the borders of the observable universe. It's a reminder that the Big Bang didn't happen in one place in the universe, it *was* the universe.

Next comes a big gap in our knowledge covering nearly a billion years – a chunk of time for which we have no observable artefacts of universal

Seeing the past

No matter where you look in the universe, you are always looking into the past. The reason for this is that light, while the fastest thing in the universe, still travels at a finite speed – about 186,000 miles or 300,000 km per second. So when you look at any object, you're actually seeing the object as it was at the instant light began to move from that object to your eye.

For your daily life, this hardly matters. The distance between your eye and this book, for example, is so small that the time lag is trivial. As distances increase, however, the lag becomes greater. The light you see from a full Moon has travelled almost 250,000 miles to you, which means that you're seeing the Moon as it was just over a second ago. The light from the Sun finds its way to you from a distance of 93 million miles, so the sunshine you feel on your face (remember never to look directly into the Sun) took eight minutes to get from the Sun to you. The light from the nearest star system to ours, which harbours Alpha Centauri, took over four years to get here, so we say the system is four light years away – the distance of a light year being just under 5.9 trillion miles.

The finite nature of the speed of light allows astronomers not only to look out into the distances of the universe, but back into the depths of time, thus allowing them to see the universe at various stages of its formation. The Andromeda galaxy, our nearest galactic companion, is two million light years away, an amount of time that doesn't add up to much in terms of the evolution of the universe. But as we look further back we see stranger and stranger objects: radio galaxies, quasars and more, back to the first observable galaxies more than ten billion light years distant. Back beyond all of this is the wash of background radiation which signals the frontier of the visually observable universe. In order to "see" the universe any further than this (where telescopes are of no use), we are only equipped with physical theories and our own imaginations.

Located 2.7 million light years away, the nebula NGC 605 is a hotbed for new star formation with hundreds blazing away at its centre. Our own Sun got its start in a similar nebula of gas and dust more than five billion years ago.
NASA

evolution. The first things we see after the visual barrier of the recombination era are the very **first galaxies**, forming one billion years after the Big Bang. It was in these galaxies that many of the **first stars** were forming – nuclear furnaces which, in processing the abundant hydrogen and helium floating around, created the heavier elements (everything from lithium to uranium) that occur naturally in the universe today. Some of these very old stars are still around: the oldest star we know of, the elegantly named HE 1523-0901, is about 13.2 billion years old.

The birth of the solar system

At one billion years after the Big Bang, the universe was well on its way to becoming what we observe today: filled with galaxies, stars, planets and all other sorts of strange and exotic things which we will explore in later chapters. For now, let's fast-forward to a time about four and a half billion years ago, to a place which we now call the **Milky Way galaxy** – the galaxy in which we live. Within the galaxy was a nebula, or cloud, of stellar gas and dust, which was not doing very much until something happened that caused the nebula to collapse on itself.

What caused the collapse is a matter of speculation – the best idea at the moment is that a nearby star went supernova, and the resulting shockwave caused the nebula to compress and contract – but collapse it did, and in doing so, it bulged in the centre and flattened around the edges, and began to rotate. That bulge in the centre was what became our Sun.

The entire universe radiates energy in infrared frequencies, as shown in this 1990 survey from NASA's COBE spacecraft. The bright images in the first two pictures are the infrared energy of our own galaxy and the stars within it; the final image subtracts them to show the dim, residual glow of energy absorbed and re-emitted since the days of the Big Bang.
NASA

The birth of the Sun

It took some time, however. For a very long time, the rotating ball of gas and dust in the centre of the disc of material pulled in by gravity (called an **accretion disc)** was just that – a ball of gas and dust. But as the ball contracted and gathered more gas and dust from the disc, the pressures at the centre of the ball became greater, and the temperature grew hotter correspondingly. When the temperature reached 18 million°F/10 million°C, a self-sustaining thermonuclear reaction came into being. A star was born, our star, **Sol** (otherwise known as the **Sun**).

The Sun was not alone at its moment of birth. Out in the swirling disc that surrounded the newborn star, clumps of material were busily growing and collapsing as well. These clumps of matter were far too small to become stars in their own right; even the largest of them, which would become Jupiter (the second most massive entity in the solar system) was not nearly big enough to create a thermonuclear reaction in its core. But they were still large enough to suck in more and more material, and in time caused smaller chunks of matter to orbit them, as they themselves orbited the Sun. Within ten million years of the birth of our Sun the **planets** were born. The hard, rocky, smaller planets (Mercury, Venus,

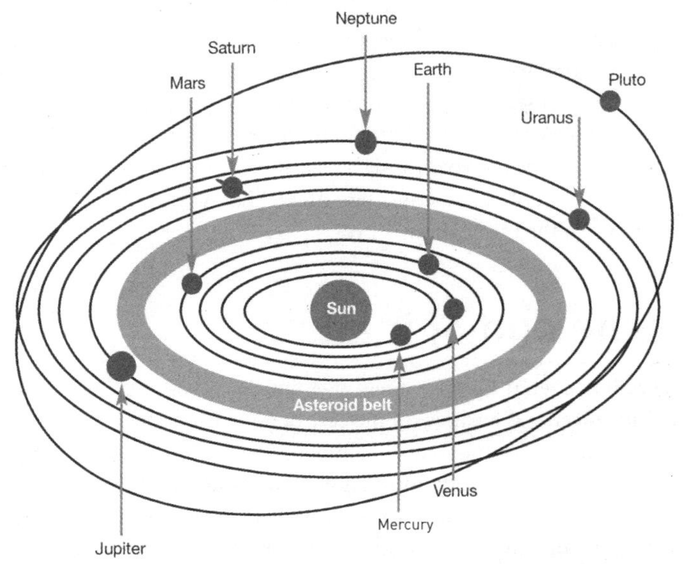

The solar system

Earth and Mars) huddled close to the Sun while the larger gas giants (Jupiter, Saturn, Uranus and Neptune) staked their places in the colder, further reaches of the system.

The solar system's bit players

The Sun and the planets may be the major characters in the drama of celestial birth, but they are not the entire cast. Created along with the Sun and the planets are the solar system's bit players: the **satellites**, captured by the larger planets (or, as is suspected in the case of our Moon, ripped bodily from the planet itself); the **asteroids** and **planetesimals**, small failed planets that float mostly between Mars and Jupiter, or beyond Neptune, with a few careening uncomfortably close to Earth from time to time; and **comets**, far-flung bits of ice and rock, which tumble down the gravity well towards the Sun from the furthest edges of the solar sytem, streaming tails of volatilized gas as they fall, before (most of the time) being launched back into the depths of space from which they came.

These bit players are not unimportant to astronomers. When the planets formed, the heat of their creation may not have been enough to turn them to stars, but it was more than enough to melt metal and rock, and in doing so, to obliterate the record of what it was like way back when our solar system was young. The Earth, the most geologically active of all the planets, is particularly unhelpful because very few of its native rocks date back more than a billion years. To see what it was like when the solar system was young, you need to consult these bit players, some of which (or parts of which) fall to the Earth in the form of meteorites. Radioactive elements in these **meteorites**, set when they were formed, give us the date of the birth of our solar system. At 4.5 billion years, the Sun and its planets are in the prime of their lives.

At the end of it all

How will it all end? Because it *will* end, one day – both our solar system and the universe as a whole. In the case of our solar system, the end will come sooner rather than later – although "sooner" is strictly a relative term. The Sun is currently in the middle of its normal life span, fusing hydrogen into helium in its nuclear furnace. In another five billion years, however, it will have worked through its hydrogen and will start burning helium instead. When this happens, it will swell into a **red giant**,

The right universe

If the laws of physics were even slightly different, we wouldn't be here at all. Take **strong nuclear force**, for example, the force that holds the nucleus of an atom together against the electrical repulsion of the protons. If strong nuclear force were just a little bit weaker, fusion couldn't happen – no fusion, no stars, no us. If it were just a little stronger, protons would always stick to each other, meaning that hydrogen couldn't exist – no hydrogen, no water, no us. The universe we live in seems to be precisely calibrated so that we can exist in it. Indeed, the relationship between the conditions in our universe and our existence in it is so strong that one school of thought places life at the centre of universal evolution.

This idea is known as the **Anthropic Principle**, and it comes in two versions, Weak and Strong. **Weak**, posited in the early 1960s by **Robert Dicke**, allows for the existence of many sorts of universes but stipulates that life can only exist in universes very much like our own. **Strong**, posited by **Brandon Cater** later in the same decade, states that there can only be one type of universe – that in which we can exist. The Strong Anthropic Principle seems a little wild, although those who adhere to it point out that some interpretations of quantum physics seem to require observation from intelligent beings (not necessarily humans) in order for the laws of physics to snap into place. At this point, a lot of people (not just scientists) start getting a little edgy.

bloating so enormously that it will swallow up the planets Mercury and Venus outright. The Earth's oceans and atmosphere will boil away, rendering the planet utterly uninhabitable. The Sun might consume the Earth as well – it's impossible to know.

The Sun will then remain as a red giant for a few billion years more before destabilizing and ejecting its outer levels of gas into space. These expanding shells of gas will become a **planetary nebula** (so named because they looked like planets to early astronomers). The Sun will then be reduced to a small, dense corpse known as a **white dwarf**. Nearly as massive as the Sun is today but compressed into a space barely larger than the Earth, this white dwarf will slowly cool down over billions of years.

The Big Crunch

As for the end of the universe – well, what ultimately happens to the universe depends on how dense it is. Currently the universe is expanding, continuing the process that began with the Big Bang. But if the universe is dense enough, if it has enough matter within it, that expansion will one day slow, then stop, and then reverse itself. The universe may then fall back on itself, culminating in a **Big Crunch**. This will proceed almost exactly like the Big Bang, except in reverse, with mat-

A white dwarf, the burnt-out cinder of a star, sits in the centre of a nebula comprised of its former outer layers. Over time, it will cool and fade into darkness. A similar fate awaits our Sun – eventually.
NASA

ter losing its grip a few minutes before the crunch, and all the forces in the universe consolidating again into a single unified field the very instant before the crunch is completed. What happens after that? We can't say.

But the Big Crunch isn't a very likely scenario. There simply isn't enough observable mass in the universe to trigger the slowing of universal expansion. Beyond the observable mass in the universe is the **dark matter** – material that exists in the universe but which is nearly impossible for us to detect directly. It's estimated that there is ten times more dark matter in the universe than observable matter. But in terms of universal density, that would still be only a fifth of what would be needed to make the universe eventually contract. A recent and exotic theory has postulated the existence of **dark energy**, a type of energy that might explain recent observations that seem to show the universe

The universe is ending!

The fact of the matter is that either scenario of universal death isn't going to be pleasant. If it's a **Big Crunch**, then the entire universe will compress in on itself, raising temperatures into the millions, billions and trillions of degrees, causing all matter to fly together in a blistering hot implosion into a singularity. If it's the **Slow Death**, then all the matter in the universe will get sucked into black holes that will then slowly evaporate into complete nothingness.

If you're wondering if there's anything you can do about the end of the universe (or, more locally, the end of our solar system) in the next five billion years, you should probably just relax. Yes, the end is coming, but not for a very, very, very long time. The end of the solar system, is slated for five billion years from now (more or less).

To put this in perspective, the history of life on our planet only goes back about a billion years – one fifth of the time that our planet has left before the sun swells and possibly swallows it whole. The species *Homo sapiens* is only about 100,000 years old, and most of what we understand as history has taken place in the last 10,000 years. In other words, the end of the world is going to happen in a timespan that is 500,000 times longer than the history of civilization – and the end of the universe will happen in a time frame that is far longer. That's a whole lot of time.

is expanding at *increasing* rates, not just a constant one. That is to say, it expands faster the more it flies apart.

Expanding – up to a point

So what would that mean for the universe? It means that it would keep expanding, and within galaxies, stars would continue to form from the residue of previous stars and from the hydrogen and helium produced at the Big Bang. But this resource won't last forever. One trillion years after the Big Bang, all the hydrogen and helium will run out, and the stars and galaxies will begin to die, flickering out and becoming **black holes**. These are superdense piles of matter with gravitational pulls so strong that not even light can escape. And even black holes have expiration dates: trillions upon trillions of years from now, even the most massive black holes will evaporate into nothingness. In about 10^{106} years, what will be left of the universe will closely resemble what it was in the beginning – nothing at all.

Getting around the sky

This chapter introduces you to the night sky, what you can see in it and how to find your way around it by using celestial coordinates. It also assesses the relative delights and advantages of viewing it with your own eyes, with binoculars or by using a telescope.

Charting the celestial globe

The universe is vast and its true dimensions are not known to us, but when we look at the night sky, the shape and scope of the cosmos is clearly apparent. It looks like a giant globe or sphere viewed from the inside. More specifically for us on the ground, it appears to be a bowl, or hemisphere, bounded on all sides by the horizon of the Earth and arcing up to a top, or **zenith**.

If you look at a globe of the Earth, you'll notice that it has been criss-crossed with a grid of coordinate lines – lines of **latitude** that mark positions north and south, and lines of **longitude** that do the same for east and west. Lines of latitude are labelled by their distance north or south of the equator, which girds the globe around the middle. Lines of latitude are all parallel to each other (no line of latitude ever meets up with another). Lines of longitude are labelled according to their distance east or west of the Prime Meridian, a line of longitude that runs through Greenwich, England. Lines of longitude aren't parallel to each other – they all meet at two points, the north and the south poles. Both longitude and latitude are divided by degrees, which are further divided into minutes and seconds. If you know the latitude and longitude of a place, you can pinpoint exactly where it is on the surface of the Earth: 151° 30' N, 0° 7' W, for example, are the geographic coordinates of London; 33° 54.996' S, 151° 17' E gets you to Sydney, Australia.

Declination and right ascension

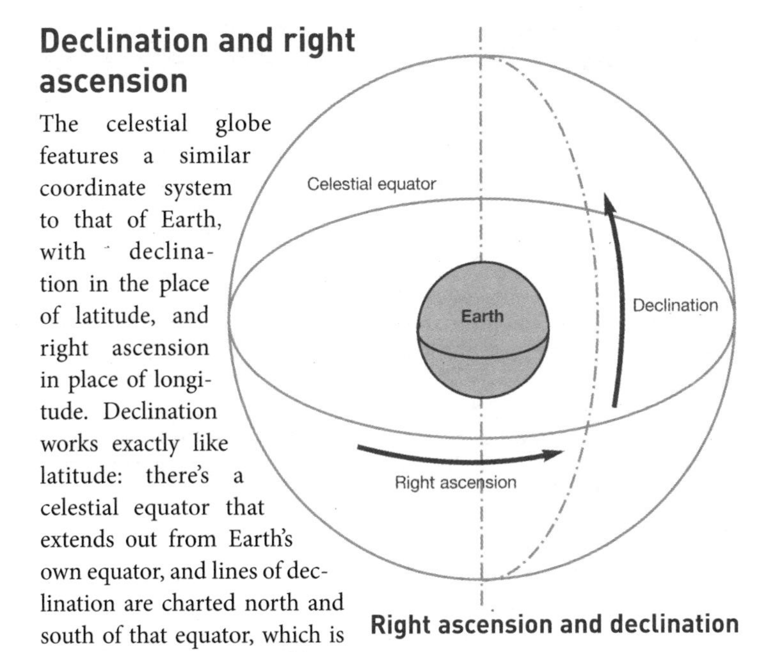

Right ascension and declination

The celestial globe features a similar coordinate system to that of Earth, with declination in the place of latitude, and right ascension in place of longitude. Declination works exactly like latitude: there's a celestial equator that extends out from Earth's own equator, and lines of declination are charted north and south of that equator, which is given the designation of having zero degrees. The degrees go up to +90 degrees in the north (directly above the Earth's north pole) and down to -90 degrees in the south (above the south pole). Degrees for declination, like for latitude, are once again further divided into minutes and seconds for precision.

Right ascension works slightly differently to longitude, however. Rather than making the distinction of "east" and "west", right ascension carves up the celestial globe in hours, with the zero-hour line of right ascension running through the point of the **vernal equinox** (that is, where in the sky the Sun is located the moment it becomes spring in Earth's northern hemisphere). Logically enough, there are 24 hours of right ascension, going counterclockwise around the celestial equator. These are further broken down into minutes and seconds.

Everything that appears within the celestial globe can be located by its coordinates. For example, going to coordinates RA: 00 42 44.17 (that is, right ascension of zero hours, 42 minutes and 44.17 seconds) dec: +41 16 25.74 (declination 41 degrees, 16 minutes and 25.74 seconds north of the celestial equator) will present you with the dazzling **Andromeda galaxy**, the Milky Way's closest full-sized galactic neighbour.

Just an arcminute!

Objects in the celestial sphere are located by using right ascension and declination, but to describe distances from one point on the celestial globe to another, or the size of objects on the globe, we use degrees of arc. There are 360 degrees in a circle; likewise, one degree of arc describes one three hundred and sixtieth of a span across the celestial globe. Each degree is further sliced into sixty arcminutes, which are themselves sliced into sixty arcseconds.

Need a further sense of scale? Look at the Moon. It takes up half a degree of arc (or, alternatively, about 30 arcminutes). You'd need about 720 of them to circle the sky. Coincidentally, the Sun also takes up half a degree of arc, but the various planets in the sky are much smaller – no more than about 45 arcseconds large at their largest (less than one fortieth the size of the Moon), and usually rather smaller than that.

Bear in mind that the size of an object in the night sky has no bearing on its true size in the cosmos; the Sun and the Moon take up the same number of arcminutes in our sky but in reality the Sun is more than a million times larger. The Moon just happens to be rather closer to Earth.

One important thing to know about celestial coordinates is that they change slightly over time – not because the stars move enough in their positions for you to notice (they don't), but because we do, through procession of the Earth's axis. Astronomers deal with this by putting their star maps in epochs, giving the coordinates of stars and other objects relative to where they were on a particular date. For much of the twentieth century, star atlases and charts used the 1950.0 epoch, while more recent atlases and charts are tuned to the 2000.0 epoch.

The Sun's path

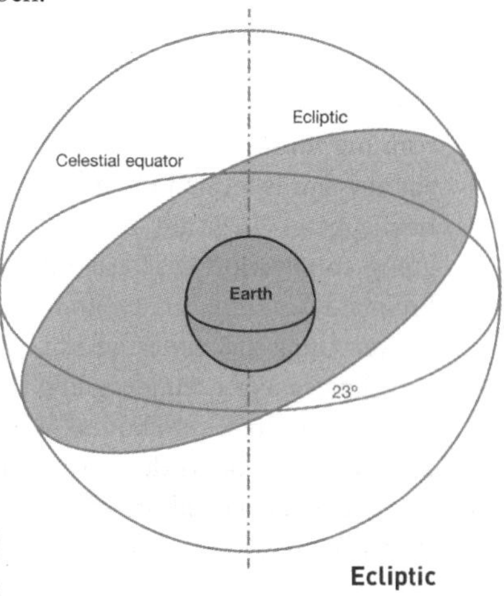

Ecliptic

Aside from the coordinate system, there's one other feature of the cosmic globe you'll want to know about – the **ecliptic**. The ecliptic describes the path the Sun takes as it makes its way across the sky over the course of the year. Also, and not coincidentally, it's the plane in which nearly all the planets orbit, within a few degrees. It's also the plane in which you'll find all the constellations of the **Zodiac**

– again, this is not a coincidence, since the definition of the Zodiac is that it is comprised of constellations in which the Sun moves over the course of the year. The ecliptic is not the same as the celestial equator – it rises and dips just over 23 degrees above and below the equator. The reason for this is pretty simple – the Earth's axis is tilted by this amount. It's our planet that's actually skewed, not the ecliptic.

What's out there

What are you going to find waiting for you in the night sky? Here's a quick rundown of the main performers.

- **Stars** There are roughly 6000 stars in the sky that humans are capable of seeing with their eyes, from Sirius, the brightest star in the sky with an apparent **magnitude** of -1.5 (the smaller your magnitude number, the brighter you are), down to 6th-magnitude stars more than 600 times dimmer. Many more are of course visible with binoculars and telescopes. How many of these stars (and everything else in the sky) you'll see depends on a number of factors, which will be examined a little later.
- **Constellations** Humans like to keep things compartmentalized, and this includes the night sky in which hundreds of stars are linked together in connect-the-dot pictures called constellations. Some of these are famous and easily recognizable, such as Orion or Ursa Major, while others, like Vulpecula and Lynx, are completely obscure. There are 88 internationally recognized constellations.
- **Nebulae** These are the clouds of gas and dust that populate the heavens. Only a few can be seen with the naked eye, but with the help of a telescope, these clouds can reveal a delicate beauty.
- **Star clusters and Galaxies** Star clusters are conglomerations of thousands of stars, galaxies of billions. Like nebulae, these best reveal their magic when viewed through a telescope.
- **Planets** Our ancestors saw them as wandering stars, moving against the fixed background of the night sky. We now know them as our neighbours in our solar system, all with distinct characteristics. The night sky also holds asteroids, tiny leftovers from the earliest times of the solar system, which occasionally make a close call on the larger planets.
- **Comets** Occasional visitors to the inner solar system, these travellers stream glowing tails of gas and dust behind them as they swing toward and then away from the Sun. Most are too faint to be appreciated by the naked eye, but now and then one brightens enough to make for an amazing sight in the sky.
- **Meteors** These are shooting stars – debris from space that has a brief, hot and dramatic encounter with the Earth's atmosphere, which are best appreciated by the naked eye alone.
- **The Moon** The Earth's closest neighbour in space. The most familiar and reassuring presence in the night sky, and the brightest object visible from Earth after the Sun.

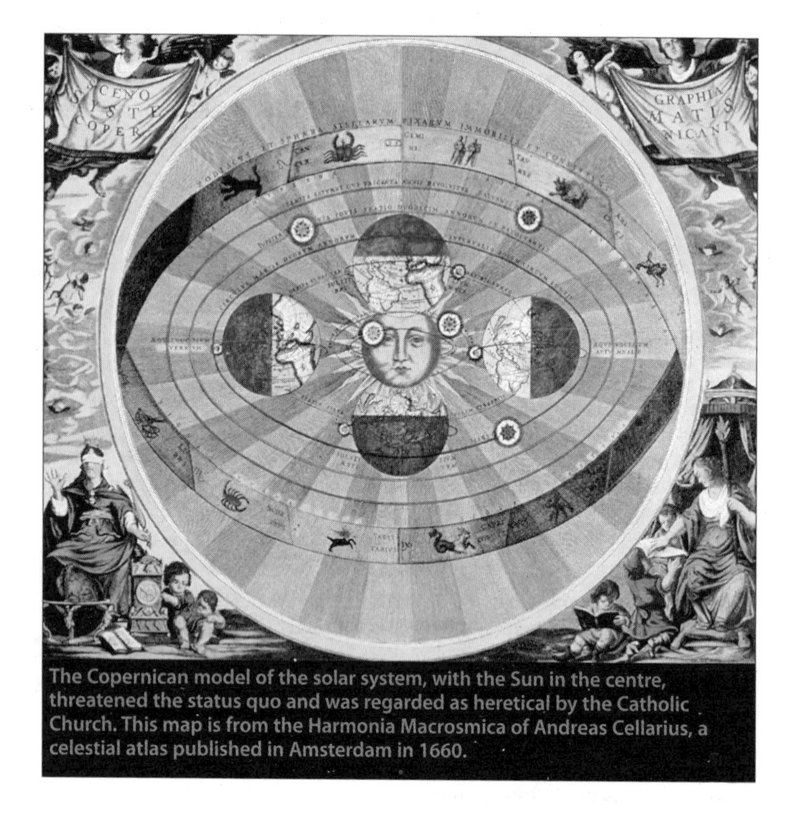

The Copernican model of the solar system, with the Sun in the centre, threatened the status quo and was regarded as heretical by the Catholic Church. This map is from the Harmonia Macrosmica of Andreas Cellarius, a celestial atlas published in Amsterdam in 1660.

What moves and what doesn't

The night sky moves, and is continually changing, offering new sights to enjoy and to excite the stargazer. But some objects move more than others. The Sun and the Moon, of course, travel round the sky in a day; planets change their positions over weeks, months and years, while stars (from a human point of view) hardly move over an entire lifetime.

Stars and other "fixed" objects

Stars do move, but not enough for you to notice over the course of your life (or indeed, over the course of hundreds of years). From our perspective, the stars are fixed, maintaining their positions relative to other stars in the night sky; likewise, **galaxies**, **clusters** and **nebulae**. However, which of these objects we see in the night sky changes over the course of the year. While it takes the Sun 24 hours to get back to the exact position it held in the sky the day before, the stars and other fixed objects make it back to that same position some four minutes earlier. This is because the Earth is moving in its orbit, and that motion shortens the amount of time the Earth needs to rotate to get back to the same position it was the day before, relative to the stars.

The amount of time it takes to rotate relative to the stars is called a **sidereal day**, the time to rotate relative to the Sun is called the **solar day**. Because of the discrepancy between the solar and sidereal days, the stars appear to move in the sky. Test this for yourself by going outside each night at the same hour and watching the stars on the horizon. In the east,

Obscuring the view – annoyances to sky watching

As breathtaking as the night skies can be, today's amateur sky-watcher has a number of things that can hinder enjoyment of the stars and planets. Here are some of them and what can be done to overcome them.

▶ **Light Pollution** Many inhabitants of large cities can't see the Milky Way from where they live, because of the brightness caused from lights in urban areas expending much of their energy – often uselessly – up into the sky. This isn't just a problem for amateur astronomers; observatories near urban areas are increasingly in danger of having their instruments made useless thanks to the glare from cities. For amateur astronomers, there's the additional nuisance of bright, glaring lights interfering with the eye's ability to adjust to the darkness – even in an area that is otherwise free of light pollution, a neighbour's all-night garage light can create problems. The long-term solution for light pollution includes smarter lighting in urban areas, including the shielding of lights so that their wattage is not wasted upwards. In the short term, however, the best solution is often simply to get out of town, and to find areas near you where light pollution is low.

▶ **Your Own Eyes** The human eye is a wonderful instrument, but it can take a very long time to adjust to darkness. For maximum night-time sensitivity, this can be anything up to a half-hour, but for most stargazing 10 to 20 minutes will do the job. While you're acclimatizing your eyes, avoid bright lights of all sorts – even penlights or car lights can set you back several minutes. Some stargazers place red filters on their flashlights in order to keep their eyes sensitized while looking at star maps (red light doesn't affect your sensitivity to darkness as white light would), and the latest stargazing software often comes with a dimmer, redder "night viewing" option to protect sensitized eyes.

▶ **Weather** Some things you can't control, and weather is one of them. Not every cloudy night is automatically bad for stargazing, but nights with cirrus clouds (wispy, feather-like clouds of ice particles in the stratosphere) aren't good for observation. Keep an eye on the weather reports for your area. Warm weather and windy nights can also trouble stargazing by providing a turbulent atmosphere.

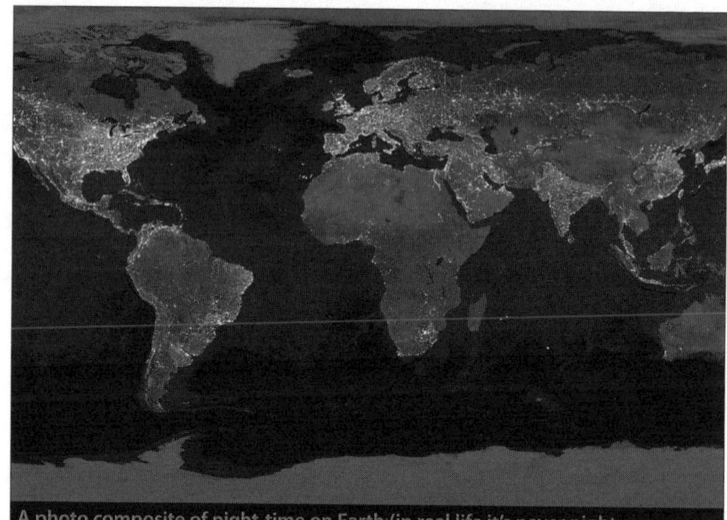

A photo composite of night-time on Earth (in real life it's never night everywhere on the planet, of course). The brightness of the world's cities is bad news for stargazers; some estimates suggest that "light pollution" prevents 90% of Americans and a similar proportion of Britons from seeing the Milky Way.

you see stars slowly climbing up, while in the west they slowly slide under the horizon. No matter where you are on the globe, some stars will always be visible in the night sky. Which stars these are depends on your latitude. If you were living at the north pole, the stars you see would never set, they would merely trace slow circles in the sky. At 45 degrees north latitude, only those stars with a declination of +45 or higher would be in your sky all the time; those stars below the declination would spend some time hidden from you by the horizon.

Only someone living directly on the **equator** will see all the stars in the sky rise and set at some point during the year. The flip side of having some stars never set in your sky is that there are some stars that will never rise. For example, a citizen of Minneapolis will never see a star in the sky whose declination is less than -45 (so much for the constellations Calum, Centaurus and Corona Australis).

Planets and other wanderers

While the stars appear fixed to human eyes, the **planets** noticeably change their positions over time (the word planet derives from the ancient Greek *planasthai*, meaning "to wander"). Some planets seem to do rather more than move in the night sky. Mars, Jupiter and Saturn (all visible to the naked eye) appear to reverse their course in the night sky, changing their eastward motions to travel west for a short time before reversing and resuming their eastward drift. Uranus, Neptune and Pluto can also be seen exhibiting this reversing movement when seen through a telescope.

The planets don't actually reverse in their orbits, of course. Their apparent backwards motion is due to the fact that the earth's orbit is nearer to the Sun than that of these planets, so in their passage around the Sun, the Earth overtakes them and moves ahead of them, and in doing so, their position against the stars seems to reverse from our perspective. Once the Earth has moved ahead of their positions, these planets appear to move forward again. Mercury and Venus, whose orbits are inside our own, are seen to wander as well, always close to the Sun, which they forever appear to be leaping out of and then back into.

Also moving out there in the blackness are **comets**, which sweep down in parabolic orbits from the outer reaches of the solar system. **Asteroids** also move, mostly in orbits between Mars and Jupiter but with a few operating a little closer to home. And of course there's the Moon, changing phases in its graceful monthly orbit.

Tools of the trade

Every job is done better when you use the right equipment, and viewing the night sky is no different. A range of optical tools exists to help you get the best results, each with its own distinct advantages (and occasional disadvantages).

Your eyes

When properly acclimatized, your **eyes** can pick out stars down to the 6th magnitude, as well as a number of faint nebulae and galaxies. They are also the best instruments you have to look at constellations (which can fill out vast stretches of the sky) and meteor showers (which happen over a broad area).

You can also use your eyes to estimate the approximate size of objects in the night sky. For example, the average adult index finger, held at arm's length, can cover the Moon or the Sun. Both of these objects are half a degree wide in the sky – therefore, an index-finger-width at arm's length equals about half a degree. Your entire hand held out at arm's length is about ten degrees; splay your fingers wide, and you'll cover about fifteen degrees of sky. This will give you a sense of scale when you use this book or other guides while looking at the night sky.

Binoculars

You might initially think of **binoculars** as a poor substitute for a telescope, but in fact binoculars are a fine way of getting a better look at the night sky. Binoculars open up the night sky, showing many more stars than can be seen by the naked eye, while still allowing for a wide enough field of viewing to get the most out of many celestial objects (like some nebulae, clusters and comets). Binoculars can also be used in conjunction with telescopes; in particular they can help you locate an object you're looking for before positioning your telescope to zero in for a closer view.

You can work out the **magnification** and the **light-gathering** power of binoculars from the numbers used to describe them. A 7x35 pair of binoculars, for example, has a magnification power of 7 (meaning objects appear 7 times larger than they do to the naked eye), and an objective lens (the lens at the front of the binoculars) with a diameter of 35 millimetres. The size of the objective lens gives you an idea of the binoculars' ability to gather light. As a comparison, the average dilated pupil on a human being is about 7mm, so the light gathering capacity of a pair of binoculars is a great deal larger than the naked eye.

When it comes to binoculars, larger isn't always better. You'll find that binoculars with a **magnification level** greater than 10x are dif-

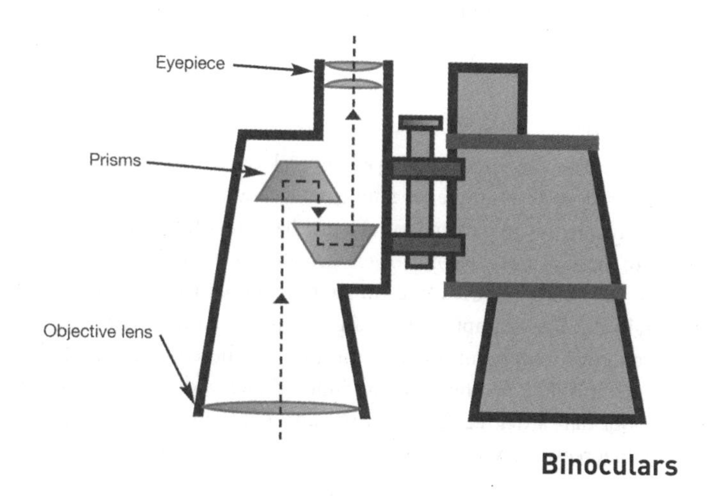

Binoculars

ficult to use in your hands because the image in the eyepiece won't stay steady. For magnifications of 10x or greater, you'll probably want to get yourself a **tripod**.

You also want to make sure that the image coming out of your binoculars is a good one. First, consider the **exit pupil** – the size of the image coming out of your binoculars' eyepiece. In this case, a large exit pupil means a brighter image for you. To determine the size of the exit pupil, divide the size of your binoculars' objective lens by its magnification. For a 7x35 pair of binoculars, for example, the exit pupil would be 5mm. Remember that a dilated pupil can be 7mm across.

Binoculars feature **prisms** inside them, so be sure that these are of reasonably high quality. Hold the eyepieces of the binoculars at a distance and look at them; if the exit pupil image is squared off, the prism is too small for the binoculars, and your image quality will suffer. Also look at the edges of the field of view in your binoculars; too much **chromatic aberration** (red and blue outlines) is a sign of inferior workmanship. Finally, the image you see in the binoculars should be clear and sharp, as opposed to grainy, filmy or low contrast. Be sure to shop around – the most expensive binoculars aren't necessarily the best.

Telescopes

The **telescope** has come a long way since Galileo made his own and pointed it in the direction of Jupiter. Today's amateurs have the sort of power in home telescopes that would have made Galileo weep – and a range of telescope types to fit a stargazer's personal needs for observation.

Refractors

The **refracting telescope** was invented in 1608 by Dutch lens-maker **Hans Lippershey** (originally for military purposes), and the following year **Galileo Galilei** became the first astronomer to use it to observe the skies. In a refracting telescope lenses are employed to collect and focus light coming from the outside. Light enters through the **objective lens** (the big lens at the end of the scope) and focuses down the length of the primary telescope tube. At the end of the telescope, an eyepiece lens then magnifies the resulting image. Refractors are relatively uncomplicated mechanisms compared to some other types of telescope, which makes them a good choice for beginners (although they can be expensive compared to telescopes with similar capabilities). Refractors have what are known as long **focal ratios**. Focal ratios are

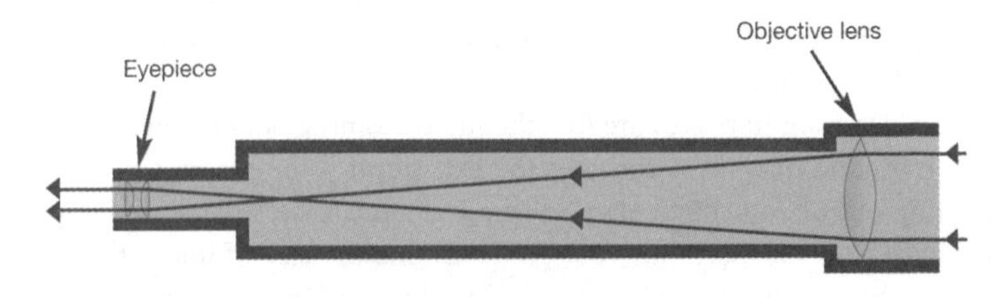

Refractor telescope

determined by dividing the diameter of the objective lens (or mirror for other telescope types) by the distance it takes the lens or mirror to form an image. Longer focal ratios (f/10 and above) create sharper contrasts, while shorter focal ratios, like f/4 or f/5, allow for wider fields of view and more sensitivity to light. Since refractors usually have longer focal ratios, they are generally regarded as better telescopes for tasks such as observing the Moon, the planets, or certain types of stars, like double star (which need sharp contrasts to separate the two stars in your scope).

Reflectors

Instead of lenses, **reflectors** use mirrors to collect and focus light. Light enters the telescope at the front and hits a curved, parabolic mirror at the back of the telescope, which sends it back towards a second, smaller flat mirror close to the front of the scope. This secondary mirror then sends the light to the eyepiece lens, which then magnifies the resulting image. The first reflector-telescope design was by **Sir Isaac Newton** and most reflectors used by amateurs are **Newtonians**.

Reflectors generally have focal ratios of between f/4 and f/8, which make them good telescopes for looking at faint or distant objects like clusters, galaxies or nebulae. They're also relatively inexpensive compared to other types of scope. Some of the drawbacks include the fact that they're not very useful for looking at anything other than the sky, and the fact that, thanks to an open tube, reflectors are liable to gather dust and imperfections over time, which will compromise image quality.

Catadioptrics

Catadioptrics use both lenses and mirrors to collect and focus light; two of the most popular variations of this type of telescope are **Schmidt-Cassegrains** and **Maksutov-Cassegrains** (these two types differ in the

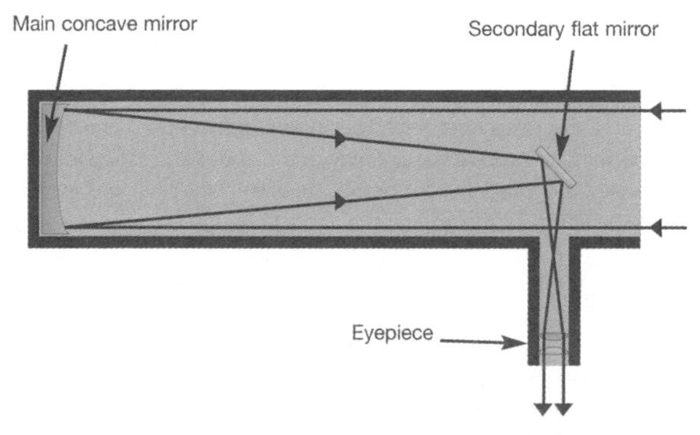

Reflector telescope

type of corrector lens used at the front of the scope to increase the field of view). These telescopes inhabit a middle ground between reflectors and refractors in terms of price and focal ratio, and most offer the additional advantage of being relatively compact, and therefore, somewhat easier to lug from place to place.

Choosing your telescope

Which of these telescopes is best for you will depend on what you want out of your night-sky experience. Those who like looking at planets should go for a refractor, while those fascinated with clusters and deep-sky objects will be better off with a reflector. In the absence of a preference one

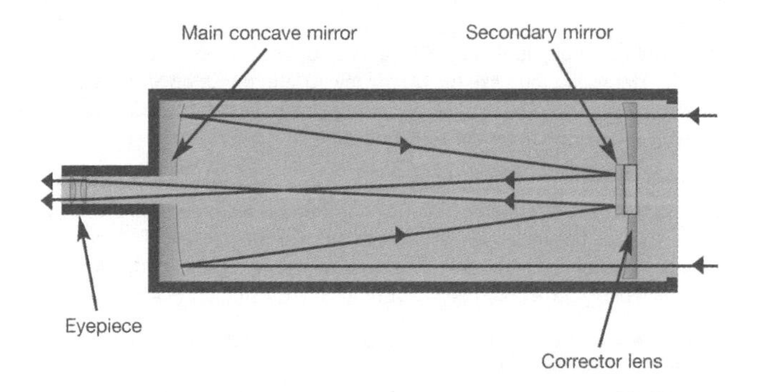

Catadioptric telescope

way or another, a catadioptric telescope will provide you with the greatest flexibility. Some other considerations to bear in mind when choosing a telescope are:

- **Size** Most amateur telescopes have apertures of between 2.5 and 10 inches – **aperture** being the size of the opening at the end of the scope for reflectors (and the size of the mirror inside), or the size of the objective lens in refractors. The larger the aperture, the greater the light-gathering power; a telescope with an 8-inch aperture, for example, can gather almost 850 times more light than the naked eye. The more light, the more distant or faint objects you'll be able to see. If you want to find Pluto, for example, you'll need to have a telescope that can pick up objects down to the 14th magnitude – that's a telescope with at least ten inches of aperture. Make sure your telescope can find what you want to find; but also make sure that the added expense of a large scope is worth it to you. Almost all the objects we'll be pointing out in this book should be viewable with a telescope with an aperture of four inches, which will resolve objects down to the 12th magnitude in optimal viewing conditions.
- **Eyepieces** that come with telescopes and which handle magnification are usually switchable, allowing you to fit your telescope with eyepieces that offer greater or lesser magnification depending on what it is you want to look at. There's a natural tendency to think that more magnification is always better, but in general the best magnification for your telescope is a multiple of the size of

Computerized telescopes

In recent years, amateur telescopes have been produced with optional **computerized mountings** that will locate stars, planets and up to thousands of other celestial sights with just the push of a button. All you have to do is set up the telescope and feed it some basic geographical information, and the scope does all the rest, taking the guesswork out of finding many of the most intriguing objects in the night sky. Many of these systems can be updated with a PC, which can pipe the latest information into the telescope's onboard tracking system. They'll also slowly move the telescope to keep the object you're looking at in your field of view. And with the advent of the now-ubiquitous GPS (Global Positioning System), there has come a new class of stargazing tools like the Meade mySKY and the Orion SkyScout, where all you do is point at an object in the sky and the instrument rattles off information on what you're looking at – no skill required.

Many fervent starwatchers look askance at these computerized telescopes, the argument being that part of the enjoyment of stargazing is learning to find one's way through the sky to favourite viewing destinations. For them, using a computerized object-finding system is rather like using a motor scooter in the London Marathon – it gets you where you're going quicker, but it's also missing the point.

But computerized systems open the night sky to people who might not otherwise get to appreciate its pleasures. Those who want to can still learn how to navigate the sky on their own; everyone else can use the computer assist to dive right in and see the night sky in a whole new way. There's no downside.

the aperture – 3 to 4 times the aperture for the lowest magnification, and 25 times the aperture for the highest resolution. Anything above that range and you start to degrade the quality of the image you get out of your telescope, and the time and effort it takes to get a good image goes up quite a bit. Go for quality rather than pure magnification – and remember that not every object you'll look at requires (or even looks good with) a maximum magnification.

- **Filters** Telescopes can be fitted with filters that perform various functions. **Polarized filters** cut down glare from light pollution or from bright objects like the Moon (and even the Sun), and **coloured filters** help highlight features on planets. We'll look at some of these filters in further chapters.
- **Mounts** While you're looking at the night sky through your telescope, the Earth is rotating – and that means you'll need to adjust your telescope from time to time to keep the objects you're looking at in view of your scope. How you do this depends on your telescope's mounting; an **alt-azimuth mount** requires you to make adjustments both horizontally and vertically, while an **equatorial** mount allows you to align your telescope with a celestial pole, so all you have to do is turn the telescope around the polar axis. Equatorial mounts are more expensive.
- **Inversion** Objects in your telescope's eyepiece are probably inverted (unless you have a corrective lens). As a practical matter for stargazing, this won't matter a whole lot, but it's something to keep in mind when you're using this book during its discussions of where things are on a planet or in the sky.

Here comes the Sun

For human beings, the greatest astronomical phenomenon we experience is not visible in the night sky. The Sun is the single non-terrestrial object we regularly see in the daytime sky, with the exception of the Moon which, of course, shines only because of the sunlight that reflects from it.

Every aspect of human existence revolves around the Sun. Most obviously, our planet circles its celestial parent. Our various ancestors worshipped it as a deity, even as modern sun-worshippers bathe in its light. Sunlight and heat enable plants to grow, which feed us. Oil is the residue of former life on the planet, nourished by the Sun's rays. The Sun's heat generates wind, for windmills both ancient and modern. We owe hydroelectric power to an orbital position that helps water keep its liquid form. Solar power is self-evident – nuclear power is merely a flawed simulacrum of the processes that power the Sun itself.

Warming up to the Sun

This chapter begins with a warning: never look directly at the Sun. Everybody was told this as a child, but a surprising number of people do it anyway. Looking directly at the Sun with your unprotected eyes – even during an eclipse – can cause permanent vision damage; you can literally fry your retinas. Looking at the Sun through an unfiltered telescope, which focuses and magnifies light, is just plain idiotic. If you want to know what focused magnified light will do to your retina, think what light focused

FACT FILE

Distance from Earth 92 million miles (147 million km)

Diameter 865,000 miles (1.39 million km)

Rotation About 25 days at the equator and 36 days at the poles

Mass 4.39×10^{24} lb (1.99×10^{24} kg)

Surface Gravity 899 ft/s^2 (274 m/s^2)

Surface Temperature 10,000°F (5700°C)

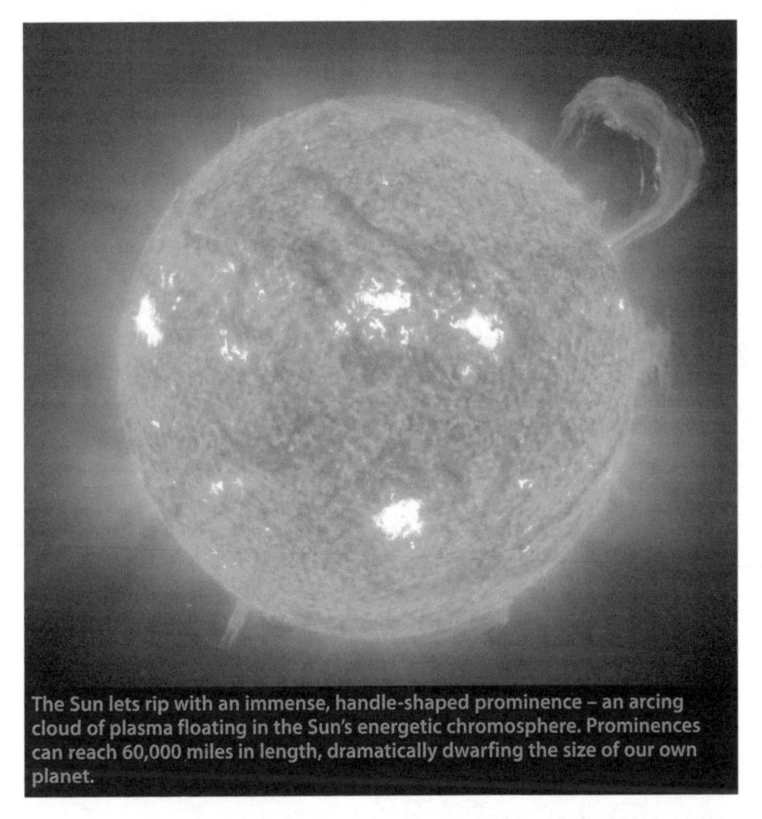

The Sun lets rip with an immense, handle-shaped prominence – an arcing cloud of plasma floating in the Sun's energetic chromosphere. Prominences can reach 60,000 miles in length, dramatically dwarfing the size of our own planet.

through a magnifying lens will do to an ant. This would be much worse, so never, ever try it.

What is the Sun? Simply put, it's a **star** – which is (to put it even more simply) a ball of gas that releases immense amounts of energy thanks to a continuous **nuclear reaction** at its core. What makes the Sun of especial interest to us is that it is our star, the vast glowing ball that ultimately powers every aspect of life on Earth. It's huge, large enough in volume to swallow the Earth more than one million times over. ninety-nine percent of all matter in our solar system is contained within the Sun. When it's in the sky, its brightness blots out every other astronomical object save the Moon. Alpha Centauri, the next closest star and one of the brightest objects in the night sky, is 40 billion times fainter.

Not so remarkable?

Despite the indisputable might and power of our Sun, celestially speaking it's relatively unremarkable. For a start, at about 865,000 miles in diameter and with a mass of 4.4 x 10^{24} lb, it's not an especially big star, coming between the **red dwarfs** (small stars with a lower temperature than the Sun and only a fifth of its mass), and the **blue supergiants** (which can be up to sixty times as massive). In fact, the Sun is actually a bit on the smallish side, as stars go, but this means that it burns its nuclear fuel at a pace that allowed life to evolve at leisure. Bigger is not better as far as life is concerned.

Technically speaking, the Sun is a **"G" type star**, which means that it is a yellow star whose surface temperature is around 10,000°F/5700°C. More precisely, it's a "G2" star, which means it's slightly hotter than most other G type stars. G type stars are unimpressively common in the universe; there are millions in our own Milky Way galaxy alone. Swing your telescope in any direction and look out for the yellow stars.

The Sun is roughly 4.5 billion years old, which, for a star of its size and mass, means that it's about halfway through its normal life-span. The Sun burns **hydrogen** in its stellar core, and because of this it is said to be on the **main sequence**, a stellar continuum that accounts for the size and luminosity of nearly all stars (see p.170). In about five billion years, the Sun will have used up its hydrogen and begin burning **helium**, at which point it will fall off the main sequence, swell up to a red giant and start the slow and occasionally violent process of dying. The Earth will be obliterated as a result.

Distinguishing features

There are really only two things that distinguish the Sun from millions of other stars that are almost exactly like it. The first is that it is a **single star** – it does not share its system with another star (or stars) as so many others do (nearby Alpha Centauri, for example, is part of a system with two other stars). Indeed, it could be said that the Sun is nothing but a single, middle-aged ball of gas; something we should be pleased about since the presence of a second star in our system might have destabilized the orbit of the Sun's planets, including Earth's. And it is this stability which provides our Sun with its second distinguishing characteristic, the fact that – as far as we know – it is the only star which harbours **life** within its system.

A tour of the Sun

Most of the Sun is made up of just two elements: **hydrogen** (73 percent of the total mass), which serves as the fuel for the nuclear furnace at the Sun's core, and **helium** (25 percent), which is the product of the hydrogen burning in the core. The other two percent is composed of **carbon**, **nitrogen**, **oxygen**, **iron** and other elements that exist in the solar system and in the universe as a whole. But while these facts are relatively simple, the hydrogen and helium reside in a stellar system that is incredibly complicated – from the unimaginable heat and violence of the stellar core, to the roiling, sunspot-pocked solar surface, to the blazingly brilliant corona.

The Sun's core

The 10,000°F/5700°C temperature of the Sun's surface is as nothing compared to the heat of its **core**, which is some 27,000,000°F/15,000,000°C. In addition to this immense heat, there is huge pressure, of the order of 9400 pounds per

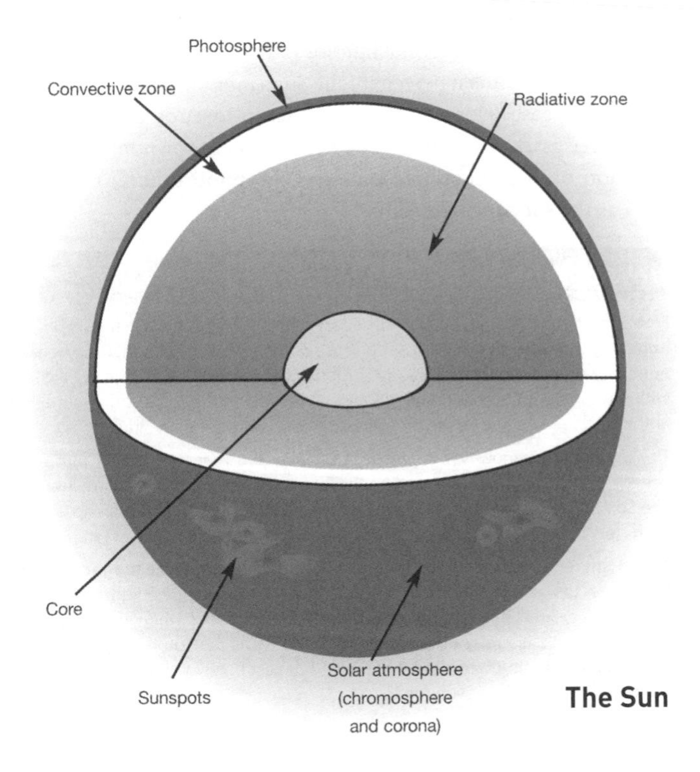

The Sun

cubic foot (150,000kg per cubic metre). What's actually happening in the core is **nuclear fusion**: because of the immense heat and pressure, hydrogen atoms are stripped of their electrons, and their nuclei are fused together to create helium nuclei (four hydrogen nuclei for each helium nucleus).

The resulting **helium nucleus** is slightly less massive than the combined mass of the four original hydrogen nuclei. Where did that missing mass go? It was converted into energy when the hydrogen nuclei fused together – energy that ultimately pours out of the Sun and reaches Earth, mostly in the form of light and heat. The Sun's fusion process is happening on an immense, massive scale: each second, more than 600 million tons of hydrogen slams together in the core to form helium and four million tons of energy.

The energy created in the core can't simply burst directly out of the Sun – things are far too dense and chaotic for that. Outside the core is a **radiative zone**, in which energy in the form of **photons** bounces back and forth and slowly makes its way towards the surface of the Sun. A photon released in the core can spend hundreds of thousands of years in the radiative zone, being slammed to and fro by other particles that are also slowly making their way to the surface. This radiative zone, along with the core, takes up 85 percent of the interior of the Sun.

The uppermost fifteen percent of the Sun's interior is a **convective zone**, in which hot gases closer to the core rise up towards the Sun's surface, while cooler gases at the surface are plunged back down into the interior. At the top of this convective zone lies the **photosphere**, more simply known as the surface of the Sun.

How bright is the Sun?

From where we're positioned, the Sun is easily the brightest object in the sky, with an **apparent magnitude** of -26.74 (in the case of magnitude, the higher the negative number, the brighter the object). By contrast, the brightest star in the sky is **Sirius**, with an apparent magnitude of -1.5. Since the magnitude scale is geometric, not arithmetic, the Sun appears to be billions of times brighter than Sirius.

The key word here, however, is "appears". In reality, the larger star in the Sirius binary system is both bigger and hotter than the Sun and is about 23 times more luminous. The Sun appears brighter simply because we are closer to it – if we were to examine both stars from an equal distance, Sirius would easily outshine our own star.

Astronomers gauge how bright a star really is by using **absolute magnitude** – the brightness of a star viewed from a distance of ten parsecs (32.6 light years). From this distance, Sirius has a magnitude of 1.4. Not as bright as when viewed from the Earth's sky, but still pretty bright and easily visible. The Sun, on the other hand, has an absolute magnitude of just 4.8 – it would be just barely visible to the eye in the night sky. Perspective is everything.

On the surface and above

Our perception of the Sun's surface is of something featureless and bright. In fact, it isn't uniform at all but mottled by gas bubbles (some as large as 600 miles/965km across) rising to the surface of the Sun. This process, known as **granulation**, is dwarfed by **supergranulation** – huge convective cells into which one could tuck two or three Earths and still have room to spare. These supergranules are connected to the massive magnetic fields on the surface of the Sun.

The Sun's **magnetic fields** get twisted because its surface doesn't rotate at a uniform rate (as a huge ball of gas there's nothing solid to help it rotate at one speed), so that eventually they become so twisted that magnetic "loops" pop out of the Sun's surface. At the base of these loops are **sunspots**, dark regions the size of the Earth (or larger). These have two parts to them: the **penumbra**, which is slightly darker than the rest of the Sun's surface, and the **umbra**, which looks darker still.

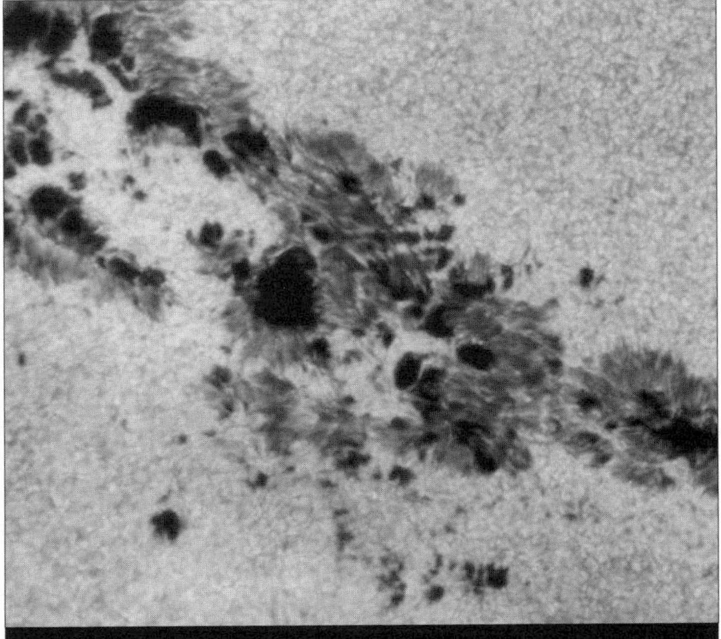

Sunspots are caused by the twisting of the sun's magnetic field and appear dark because they are slightly less hot than the surrounding solar surface. They can be large – spots the width of the Earth are not uncommon.

Two coronal mass ejections, spewing billions of tons of the Sun's gas into space. When such ejections are let loose in the direction of the Earth, their energies create dazzling auroral displays.
ESA/NASA

Sunspots

Sunspots look dark because the temperature of a sunspot (5660°F/3125°C) is less than the surrounding surface, which makes them less luminous. However, the Sun is actually brighter the more sunspots it has, thanks to faculae – intensely bright spots on the Sun whose presence is associated with sunspots. The duration of a sunspot depends on its size: as little as a few hours if they're small, or several months if they're larger.

Sunspots generally travel in pairs: a **leading** sunspot with a **trailing** sunspot ("leading" and "trailing", relative to the Sun's rotation). These have differing **magnetic polarities**; if the leading spot is positive, the trailing spot will be negative. What's more, all the leading sunspots on the Sun's northern hemisphere will have the same magnetic polarity, while all the leading sunspots in the southern hemisphere will have the opposite polarity. So that when all the leading sunspots in the northern hemisphere are positive, all the leading sunspots in the southern hemisphere are negative (trailing sunspots in these hemispheres would be negative and positive, respectively).

How a solar flare messes up your radio reception

When the Sun sends up a **solar flare** or a **coronal mass ejection**, we're only seeing half of the story; it's what we don't see that affects us the most. In addition to sending us energy we can see, solar flares also let fly incredible amounts of other **radiation** – ultraviolet radiation, X-rays, and other, slower particles that launch themselves outwards and, on occasion, towards Earth.

It's these slower, but still powerful, particles that cause the trouble. When such particles collide with the Earth's upper atmosphere they can interact with the Earth's magnetic field. If the particles hit in a certain way, they may well cause a **magnetic storm**, which in turn can create severe disruptions in power grids, electronics and radio reception. In March of 1989, for example, a solar flare caused six million people in Quebec to lose power. These massive doses of solar particles are also able to fry orbiting satellites and pose a danger to whatever humans may be orbiting the planet at the time.

The silver lining to all this magnetic and electrical chaos is that magnetic storms also create shimmering and breathtaking **auroral displays**. These dazzling Northern and Southern lights are created when **ionized solar particles** are trapped in our magnetic field. These particles excite gases in the atmopshere, and there you have it – aurorae. The intensity of the displays varies with the cycle of sunspots.

An aurora over Canada. Auroras are caused by solar particles interacting with the Earth's magnetic field.

Michael DeYoung/CORBIS

Why make a big issue about sunspot polarity? Because it's tied into the **22-year sunspot cycle**. At the beginning of a sunspot cycle, a small number of sunspots will appear in the mid-latitudes between the Sun's equator and its poles. As the cycle continues, more spots appear, closer to the equator, until the cycle reaches its peak, at which point the spots begin to peter out. This takes eleven years. As the second portion of the cycle begins, sunspots appear in the same way but this time with their polarities reversed from the first half of the cycle.

Chromosphere and corona

Above the photosphere is the **chromosphere**, which is hotter than the Sun's surface (it can reach 17,000°F/9725°C) and is the home of some of the most captivating and violent of solar displays. Chief among these are **solar prominences**. These whirls and loops of gas can reach 60,000 miles/96,000km in length and attain altitudes of 25,000 miles/40,000km above the solar surface, and can last for weeks. **Solar flares** are even more violent, searing solar eruptions with temperatures up to 180,000,000°F/100,000,000°C.

Beyond the chromosphere is the **corona**, an intensely hot (1.8 million°F/1 million°C) field of electrified gas whose shape is determined by the magnetic fields on the Sun. Every now and again, the Sun experiences a coronal mass ejection, when billions of tons of coronal gases launch into space. These ejections can create dazzling displays on Earth in the form of **aurorae** but can also create severe radio disruptions and in some cases can even knock out power-grids.

Eclipses of the Sun

Without a doubt, the most dramatic solar event humans can actually witness is an eclipse. Throughout human history **solar eclipses** have been the objects of fear and wonder. A solar eclipse in Greece in 585 BC was

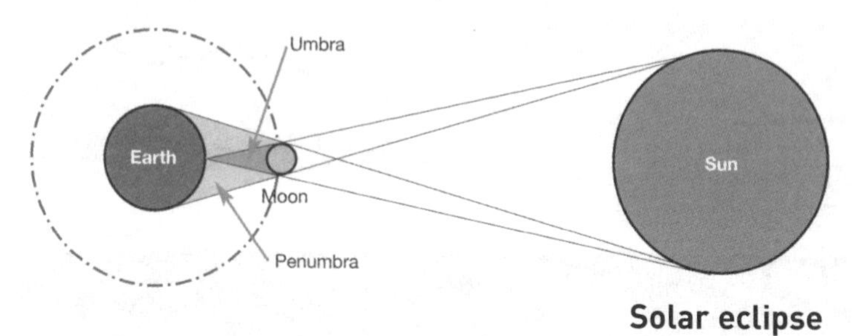

Solar eclipse

wondrous enough to cause warring Medes and Lydians to declare a peace (the eclipse had been predicted by pre-Socratic philosopher Thales). In 1504, **Christopher Columbus** used a predicted solar eclipse to impress the not-so-friendly natives of Jamaica. In 1919, a solar eclipse over Africa and South America helped verify Einstein's **theory of relativity** by allowing scientists to see light bent by the gravity of the Sun, just as Einstein's theories predicted. These days, astronomers, enthusiasts and the idle rich take eclipse vacations, travelling to far-flung portions of the globe in the hopes of seeing a solar eclipse with their own eyes.

For all the fear and wonder, eclipses happen for a very simple reason: the Earth's Moon lines up in the sky with the Sun, blocking direct light from our star. In one of those strange coincidences, the size of the Moon in the sky is almost exactly the size of the Sun in the sky, a fact that allows us to see parts of the Sun during an eclipse (such as the solar corona) that are normally lost in the glare of the Sun's photosphere and in the daytime sky.

Types of solar eclipse

A **total solar eclipse** occurs when the Moon passes directly in front of the Sun and blocks its disc entirely. An **annular eclipse** is when the Moon passes directly in front of the Sun but doesn't block out the entire disc, leaving a "ring". This is due to the fact that the Moon's orbit is not perfectly circular, and sometimes the Moon is further away, and thus smaller in our sky.

A **partial solar eclipse** occurs when the Moon only partially obscures the Sun, taking out a chunk but not the whole thing. This happens in tandem with total and annular eclipses, in latitudes north and south of a total eclipse, in which the Moon is not perfectly aligned with the Sun. These areas are said to be in the **penumbra** of an eclipse, whereas the areas that see the total eclipse are in the **umbral path**. The umbral path of a total eclipse is relatively small, so that the chance of a total eclipse coming near you is reasonably slim. For example, the next total eclipse visible in the US won't happen until 2017, while the UK has only seen four total eclipses in the last thousand years, with the next one due in 2090.

The predicting of eclipses has gone on for millennia, with varying degrees of success (in 2136 BC, the Chinese emperor Chung K'ong had his royal astronomers put to death for their failure to predict an eclipse). Today we know that eclipses follow a predictable pattern that repeats on a regular basis (although not at the same location). There's even a name for the period of time in which an eclipse cycle repeats itself: a **saros** takes 18 years, 11 days and 8 hours. The 8 hours explains why eclipses don't happen in the same places on the planet each time – those eight hours allow the planet to turn an additional 120 degrees, placing the eclipse a third of the globe away from where it was in the previous cycle.

Using your telescope to view the Sun

Never look directly at the Sun through an unfiltered telescope: doing so will cause permanent damage to your vision. Also, depending on the sort of telescope you own, pointing an unfiltered telescope at the Sun is a fine way to fry your telescope's inner workings. The warnings apply during solar eclipses as well as during normal sunny days; the Sun is still powerfully bright, even during an eclipse. The good news is that, provided you use proper safety precautions, you can use your telescope to explore the Sun. There are two ways to do it: with **projection**, and using **front-end filters**.

Viewing through projection

The simplest and least expensive way to view the Sun through a telescope is by using **projection**. With projection, instead of looking through the eyepiece, you allow the Sun's image to project through the telescope onto a piece of cardboard (or other viewing surface) held at a sufficient distance from the eyepiece. This is the safest way of making solar observations, although you'll still need to take precautions.

Projection should be done only with simple telescopes, such as **refractors** or **Newtonian reflectors**. More complex telescopes, such as Schmidt-Cassegrain models, use mirrors and lenses within the telescope tube, and the heat generated by the solar image can severely damage its inner workings. If you're not sure what sort of telescope you have, it's better to be safe than sorry. Likewise, it's best to use simple eyepieces for observing the Sun; complex eyepieces with multiple elements also run the risk of damage from the heat generated by the unfiltered solar rays.

Once you've ascertained that your telescope can be used for projection, prepare your telescope by placing it on a tripod and installing the lowest power eyepiece you have. Cover your **finder scope** (it can focus light too, remember), then (without looking at the Sun) point your telescope in the direction of the Sun. Use the **shadow** of the telescope to align it with the Sun; when the shadow is at its smallest, it'll be pointing at the Sun.

Now that you have the Sun in your sights, take a piece of cardboard or stiff paper and slowly move it towards your eyepiece. An image of the Sun will appear. Move the cardboard or paper until the Sun is focused to your satisfaction. Ambient sunlight (the sunlight not coming through your telescope) can actually make the image of the Sun harder to see, so you may want to use another piece of cardboard as a shade: simply hold this piece in front of the cardboard or paper on which the Sun's image is displayed (but not in the path of the projection), and there it is.

Remember to be careful. As you are getting the Sun's image to focus on

A panoply of Sun gods

The power of the Sun is reflected in the appearance of a Sun god (or gods) in virtually every mythology and religion. Here are some of the most notable.

- **Amaterasu** The Japanese Sun goddess whose full name, Amaterasu Omikami, means Great Divinity Illuminating Heaven. The Japanese imperial family claims direct lineage from her, and her emblem, the rising Sun, appears on the national flag. In her most famous story, Amaterasu hid in a cave after a family quarrel, thus plunging the world into darkness. She only reappeared after catching a glimpse of her own magnificence in a mirror.
- **Helios** The Greek Sun god who in Roman mythology is called Sol and is identified with Apollo. Helios's task was to drive his horse-drawn chariot across the sky each day. When he allowed his son Phaeton to have a go, the young man was unable to control the horses and came so close to Earth that he burned parts of it. To save the day, Zeus struck him with a thunderbolt and Phaeton tumbled into the River Eridanus.
- **Huitzilopochtli** The principal Aztec god is both the Sun god and the god of war. His name means Blue Hummingbird on the Left, and he is usually depicted either as a hummingbird or as an armed blue man with a headdress of feathers. The Aztecs believed that Huitzilopochtli needed to be fed human hearts on a daily basis, and would tear out the hearts of captured warriors or slaves.
- **Lugh** This Celtic Sun god, whose name means brilliant one, was skilled in all the arts and crafts. In the Irish version of his myth, he killed his evil grandfather Balor (who tried to have him drowned) by throwing a magic stone at his eye in battle. Lugh became the leader of the Tuathu De Danaan (descendants of the goddess Danu). His festival, Lughnasadh, celebrates the beginning of the harvest.
- **Ra** Also known as Re-Horakhty, this Egyptian Sun god was believed to traverse the sky in a boat by day, before being swallowed by the goddess Nut to be reborn each morning as a scarab beetle. His symbol was the winged Sun disc, but he was also depicted in human form with a falcon head. Ra was associated with the pharaohs who saw themselves as his sons. He was also believed to have been the creator of the world, first creating himself and then the other gods.

The Egyptian Sun god Ra, also known as Re-Horakhty, travelling across the sky in his solar boat. Archivo Iconografico, S.A./CORBIS

your paper, the closer you get to the eyepiece, the more intense the image will be. At a point very close to the eyepiece the image will be concentrated enough to start burning your cardboard. The beam of light coming out of your telescope will also burn pretty much anything else if the object is close enough to the eyepiece, including you or anyone else who may be observing the image. Keep yourself and those viewing with you out of the path of the image at all times. Do not let children (or anybody inexperienced) operate the telescope, as it's easy to get hurt.

Viewing through filters

Your second option for viewing the Sun through your telescope is to use **optical filters** that block out nearly all the light from the Sun, leaving only a small (and visually harmless) amount to go through your telescope and to your eyes.

The only type of filters you should even consider for the task of viewing the Sun are ones that go on the front end of your telescope. Sun-blocking filters that are near or that replace the eyepiece simply are not safe – either for you or your telescope. These filters allow unfiltered Sun rays to enter the telescope, and this can cause serious damage, especially to complex scopes. And if the rear-end filter fails in any way, your eyes will receive the brunt of focused sunbeams, which can cause severe damage to your eye. **Front-end filters** keep light from entering your telescope in the first place, thereby keeping your eyes and your telescope's inner workings safe.

There are two types of solar telescope filter to consider using, depending on the sort of telescope you have. Smaller telescopes (4-inch apertures or smaller) can use full **aperture filters** which cover the entire telescope diameter; larger telescopes will require an **off-axis filter** which limits the amount of sunlight coming to the telescope. Solar filter prices range from about £20/$30 for a basic filter for small refractors, up to £2000/$3000 for the kind of high-end **H-alpha filters** which offer a great deal of detail.

What you will see

On good days, the two phenomena you'll observe most clearly are **granulation** and **sunspots**. Depending on viewing conditions, granulation can be hard to see (and on bad days it is hardly observable at all). When you do get to view it, it really does look like granules, with the Sun taking on the appearance of a grainy photograph.

Granules are the appetizer for the sungazer's main course, sunspots. Sunspots range in size from small spots, called **pores**, to full-sized beauties which show you **penumbrae**. The structures within penumbrae contain darker areas called **filaments**, bright spots known as **grains**, and spoke-like structures radiating out from umbrae called **fibrils**.

As well as sunspots there are bright spots known as **faculae**. These can be difficult to see and are best spotted at the edge of the Sun's disc, where some of the natural darkening that occurs can bring faculae out in contrast. Observers with H-alpha filters will be able to observe **prominences**, immense arcs of gas that lift off from the Sun.

Those fortunate enough to observe the Sun during a **total solar eclipse** will see parts of the Sun not normally visible, the most spectacular of which is the **corona** – the immense halo of electrified gas which encloses the Sun. This is an amazing sight, but one that varies from eclipse to eclipse, since the shape of the corona changes depending on the current state of sunspot activity. Those with high-end lenses will also be able to see **solar prominences** and **flares**. The interplay of sunlight on the surface of the Moon also creates some striking effects, most notably **Baily's Beads** – flaring spots of light caused by light peeking over mountains and crater rims on the Moon.

Meet the Moon

The Moon is the Earth's next-door neighbour, and it is the only place in space humans have visited. Between July 20, 1969, and December 19, 1972, twelve men landed on the surface of the Moon, to collect samples, make observations and (on at least one occasion) hit golf balls in the Moon's forgivingly light gravity.

But it is more than merely the first (and so far only) step in humanity's climb out of our planet's gravity-well. Throughout history, the Moon has enchanted and intrigued us, its calm, silvery glow balancing the Sun's fiery brilliance, its waxing and waning charting a space of time – a month – whose name still holds the Moon's influence within it. It pervades our culture in numerous ways, from legends about werewolves to romantic songs about blue moons. Its pull influences the tides, and even, according to some, our states of mind.

FACT FILE

Average Distance from Earth 239,000 miles (384,400km)

Diameter 2159 miles (3476km)

Rotation 27.3 days

Orbital Period 27.3 days

Average Temperature -4°F (-20°C)

Gravity 5.3ft/s^2 (1.6m/s^2)

The Moon's birth and early life

Four and a half billion years ago the Sun's nuclear furnace had only recently ignited, and the planets were little more than roiling balls of heat, dust and molten lava, coursing through a young solar system still jammed with **planetesimals** (see p.12). The Earth was still young, only 50 to 100 million years old, and without its junior partner, the Moon. All of that changed forever when the Earth was violently hit by a wandering planetesimal, a huge ball of rock and metal roughly the size of the planet Mars.

The **collision** caused an immense amount of damage to the newborn Earth, peeling off a substantial chunk of the planet's rocky skin, which,

The Moon is "tidally locked", meaning that its orbital period and rotation period are the same – this means it's always showing the same face towards us. Moon features noted here: 1: Mare Crisium; 2: Mare Tranquillitatus; 3: Mare Imbrium; 4: Crater Tycho; 5: Crater Copernicus; 6: Crater Plato; 7: Crater Aristarchus; 8: Crater Grimaldi.
NASA

along with fragments of the shattered planetesimal, was hurled into orbit around the wounded planet. Over time this ejected material came together to form the Moon.

Despite alternative theories, this version of how the Moon was formed best fits the reality of the Earth–Moon system. The composition of the Moon, for example, is rocky but iron poor; it is much less dense than the Earth, but has very near the density of the Earth's rocky mantle; and the high angular momentum the Moon possesses was the result of the speed of the planetesimal's impact with the Earth.

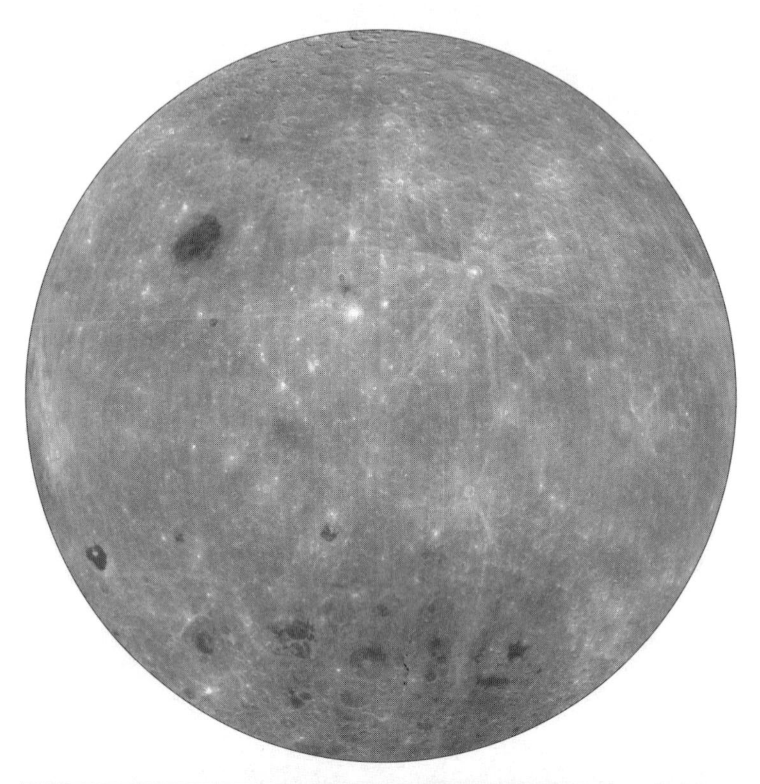

We knew almost nothing about the "dark side" of the Moon until spacecraft photographed it in the 1960s. Though always facing away from us, this side receives as much sunlight as the one facing us.
NASA

If the Moon's birth was violent and messy, its childhood was no less so. The first 200 million years of its life saw the Moon battered by impacts that heated its surface, causing it to melt and reform, then melt and reform again. We can still see some of the later impacts of this era on the face of the Moon in the **craters** that are the primary feature of the Moon's surface.

Impacts heated the surface of the Moon; and beneath the surface, the Moon retained heat in its interior until three billion years ago. During this period, whenever a particularly powerful impact cracked the Moon's surface, **magma** would flow out, covering vast areas in the thick, dense, dark rock that had migrated into the Moon's interior. Billions of years later, Galileo would call these areas **maria**, or seas, even though even in those

The Moon's origins – alternative theories

The current theory of lunar creation was suggested by William K. Hartman and Donald R. Davis in 1975, and has withstood various challenges since that time. If it doesn't work for you, here are some alternative theories to ponder:

▶ **Sibling Rivalry** Rather than having the Moon born from the Earth, this theory (proposed by astronomer Edouard Roche) suggests that both the Earth and the Moon formed simultaneously from the same dust cloud, with no impacting planetesimal required. The drawback is that co-creation would suggest the Moon and the Earth are very similar in overall composition, and they're not – the Earth has much more metal.

▶ **The Tilt-a-Whirl** This theory (championed by Charles Darwin's son George) suggests that the Moon was derived from the Earth, but without the help of a violent planetesimal impact. Instead, what happened was that the primordial Earth was spinning so quickly (say, once every couple of hours), that some of its outer layers literally spun right off, to produce the Moon. The glitch here is that such a "birth" would make it difficult to have the new moon twirl into a stable orbit.

▶ **Come Here, Little Planet** The Moon already existed, fully formed somewhere else in the early solar system, and then got too close to the Earth and was captured by our planet's gravity field. The same theory is used to explain other moons in our system (Neptune is thought to have captured its moon Triton this way). The snag is that such a capture would involve exacting physics, because the Moon and the Earth are so close in size (whereas Neptune is massively larger than Triton). Another drawback is the number of compositional similarities between the Earth and the Moon that are better explained by the Moon having once been part of the Earth.

early stages of astronomical exploration it was clear that there was no water in these lunar "oceans". Any lingering, romantic doubts about this were dispelled a few hundred years later when Neil Armstrong and Buzz Aldrin landed **Apollo 11** at Mare Tranquillitatis (the Sea of Tranquillity), and stepped into Moon dust rather than Moon surf.

Once the interior of the Moon cooled and the inner solar system was largely swept clean of obstruction, the Moon's appearance took on the form we recognize today, with craters and maria seemingly frozen in time. This is altered only occasionally, by increasingly rare large impacts and a slow, undramatic accretion of dust from smaller bits of material caught in the Moon's gravitational pull.

Unlike Earth, the Moon has no atmosphere and no water to wear down its features, and it no longer has any volcanic activity. Barring further dramatic hits from space, the Moon will appear largely as it does now for as long as we care to look at it. The human footprints set in the Moon's dust will last for millions upon millions of years – possibly longer than the human race itself.

An astronaut's footprint from the Apollo 11 mission.
NASA

Locked together but drifting apart

While some planets have larger moons than Earth (and others have more), only the recently demoted Pluto (see p.152) has a moon that is larger than our own relative to the size of the planet it orbits. At 2160 miles, the Moon has a quarter of the diameter of Earth and 1/80th of its mass, and orbits the planet from a mere 240,000 miles – in astronomical terms, it's a stone's throw (or rocket launch) away.

Because the Earth and Moon are so close in size, composition and distance, many astronomers regard them as a **double planet**, more intimate and interlinked than most planets and their satellites. They used to be even closer together in their younger days. Computer simulations of the impact on Earth by that Mars-sized planetesimal show a proto-Moon condensing into an initial orbit with a radius equal to about ten Earth radii – just 40,000 miles away. Over time, however, the Earth and the Moon have drifted apart, and they are still doing so.

This drift is happening because of the way the Earth and the Moon tug at each other. The Earth's tug on the Moon is what keeps our satellite in orbit, while the Moon's tug on the Earth helps to create the **tides**, the twice daily rise and fall of the planet's oceans. It also slightly deforms the crust of the Earth, making the planet ovoid rather than spherical, with the smaller end of the oval pointing towards the Moon. The tides create friction that slows down the rotation of the Earth by an incredibly small amount: 2.3 milliseconds every hundred years. So, days really are getting longer. At the time when the planetesimal rammed into Earth, a day was a mere six hours long – meaning that there were over 1200 "days" in a year. The Earth exerts tidal forces on the Moon, too, and over time slowed down its rotation to the extent that the Moon's rotational period is now exactly the same as its orbital period. This is why the Moon always shows the same face to the Earth.

The Moon will return the favour. Billions of years from now, as the Earth's rotation slows, the Moon and Earth will lock rotations and present the same faces to each other forever. Assuming you're on the side of the Earth where you can see it, the Moon will appear to hang immobile in the sky, and the length of the "day" will then be over 1100 hours long.

As the Earth slows down, it loses angular momentum. The Moon responds by speeding up, which it does by increasing the size of its orbit, adding a distance of 1.5 inches every year. It's not much, taken a year at a time, but over millions and billions of years, it adds up. Slowly but surely, the Moon is pulling away from us, shrinking in the Earth's sky. One obvious side effect of this will be the fact that millions of years from now, there will be no such thing as a total solar eclipse: the Moon and the Sun will no longer be the same apparent size.

Walking on the Moon

On May 25, 1961, US president **John F. Kennedy** stood before a joint session of Congress and outlined a risky and daring plan: the US would send a man to the Moon by the end of the decade. The proposal had more to do with geopolitics than it did with the desire for exploration (in 1961, the Soviet Union had the jump on the US in space exploration), but whatever the motivation, it worked. On July 20, 1969, **Neil Armstrong** set foot on the Moon and uttered his famous if grammatically confused line: "That's one small step for man, one giant leap for mankind" (what he meant to say was: "That's one small step for *a* man, one giant leap for mankind").

No atmosphere

Owing to its low mass relative to the Earth, the Moon's **gravitational pull** is only about one-sixth that of our home planet's. This means that if you weighed 180 pounds on Earth, you'd be a mere 30 pounds on the Moon.

The Apollo 11 mission of July 1969: astronaut "Buzz" Aldrin is photographed by the first man on the Moon, Neil Armstrong. The gap between the last manned Moon mission (Apollo 17, 1972) and the present day is longer than that between the flight of the first jet aircraft (1939) and the first Moon mission.

Your muscles, designed to carry you along in Earth's gravity, would propel you six times further when you walked, and launch you six times higher when you jumped. You'd have to be more careful, however, because while your weight would be less, your mass would be the same as on Earth. You'd still be carrying those 180 pounds, no matter what it felt like.

You'd also need to keep your suit on – particularly your helmet. The Moon has **no atmosphere** of any consequence; its gravitational well is not nearly deep enough to keep one from leaking off into space. No atmosphere also means no protection from harsh cosmic rays and surface temperatures are extreme: over 240°F/150°C during the two-week-long day, and around -270°F during the equally long night. The good news is that you probably wouldn't freeze or fry to death on the Moon if you went out onto the surface completely exposed. The bad news is that the reason you wouldn't die from either of those is that you'd die from exposure to hard vacuum.

Lots of rocks

The surface of the Moon is covered in **regolith** – powdery dust and bits of rock 30 feet/9 metres deep. This is the residue of billions of years of meteorite and planetesimal collisions. Occasionally you find larger, concrete-like rocks that were formed when the heat and force of an impact fused regolith into chunks. These rocks are known as **breccias**.

The composition of the Moon's surface varies from place to place. In the bright highlands of the Moon, you'll find pale rocks comprised of aluminium and calcium called **anorthosites**. In the darker maria, you'll find **basalt**, dense rock whose mineral composition includes iron, magnesium and titanium. Apollo 11 and the majority of the Apollo missions landed in maria (so Neil Armstrong was walking over basalt), but Apollo 16 landed in the Moon's central highlands, near the Descartes Mountains, which means that astronauts John W. Young and Charles M. Duke walked through anorthosites.

The Moon is rock almost all of the way down to its core, with a **crust** of granite-like rock about 64 to 144km thick, followed by a thick **rocky mantle** that plunges 800km down. After this comes an **inner mantle** some 480km thick, which may be semisolid, and finally a **core** that is also possibly partially molten. If the Moon has an iron core – like the Earth does – it is very small; data from recent missions to the Moon suggest it has a radius no larger than 400km. Unlike the Earth, the Moon has no magnetic field, although some of the surface rocks brought back from the Moon showed slight remnant magnetism, suggesting the Moon had a magnetic field in its more violent, active youth.

In addition to being dusty, the Moon is also terribly dry. There is **no water** in Moon rocks, or anywhere on the Moon at all, the possible exception being small reserves of ice that may be hidden in shadowed craters at the Moon's poles, deposited there by impacts of icy comets. That water has played hide-and-seek with scientists: in July 1999, NASA took the **Lunar Prospector**, a spacecraft designed to explore the Moon's physical aspects, and rammed it into one of these polar areas in the hope of knocking up a dust cloud that might show the presence of water. No such cloud of dust was detected. However, in March 2001, the NASA scientists heading up the Lunar Prospector team anounced that they had indeed found between 10 and 300 million tons of ice tucked away in both the north and south polar regions, based on results from the mission's neutron spectrometer. In October of 2008, NASA's Lunar Reconnaissance Orbiter will be launched and will fly over the moon's polar regions, using four separate techniques to confirm the presence of ice on the moon.

The presence of water in the polar areas would be good news for future visitors: water can be used to create breathable oxygen, and its hydrogen could be used to refuel spacecraft. Water, like everything else, would be very expensive to ship to the Moon, so the discovery of water there would make a journey to the Moon that much less expensive.

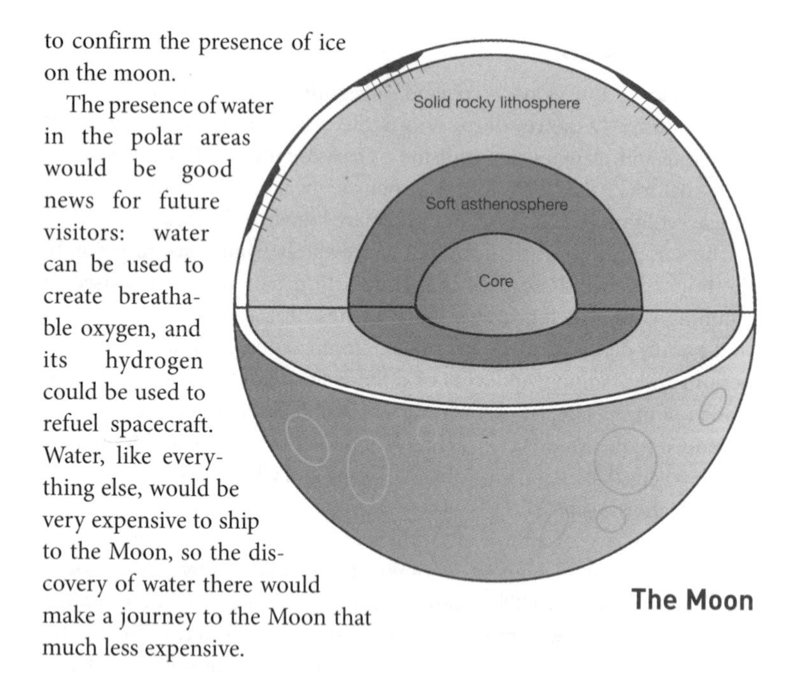

The Moon

Return journey?

When will humans return to the Moon? Since the final **Apollo mission** in December 1972, no one has returned. The US – the current leader in space exploration – has shown no interest in going back, concentrating instead on unmanned missions to other planets and Space Shuttle missions comparatively close to home. **China**, on the other hand, placed its first astronaut in space in 2003, with the eventual goal of landing a man on the Moon in 2020. International experts remain sceptical about their chances of success – but then again. in 1961 it seemed like a tall order for the US.

Viewing the Moon

The Moon is a singular astronomical pleasure, in that it is really the only astronomical object we can examine in some detail without the use of magnification. So before reaching for the telescope or the binoculars, it's worth taking time to explore the Moon with the naked eye. Here are some of the things to look out for.

The south polar region of the Moon, as photographed from Apollo 15. This view includes some of the Mare Australis (the Southern Sea), only a small fraction of which is visible from the Earth.

Phases of the Moon

The most obvious phenomena involving the Moon are its **phases**. Every 29.5 days, the Moon goes through a full cycle of phases, starting with the **new Moon** (in which the Moon is invisible to us) and culminating just over two weeks later with the **full Moon** (in which the face of the Moon is entirely visible). In between these two states, various amounts of the Moon's surface are visible, the proportion becoming larger (**waxing**) as the Moon moves towards full, and shrinking again (**waning**) as the Moon moves from full back to new.

The phases of the Moon reflect the position of the Moon in its orbit, relative to the position of the Sun. The Sun is always shining on one half of the Moon (except in the rare occurrence of a lunar eclipse), but that half of the Moon isn't always visible to us. In the case of the new Moon, the Moon is positioned between the Earth and the Sun, with the Sun shining on the far side of the Moon – the side we never see from Earth. The Moon is also invisible in the glare of the daytime sky, although it can be seen in the early morning and at twilight. When it's a full Moon, the Moon is behind the Earth in its orbit, and the Sun is shining onto the familiar face of the Moon, which then reflects the light into our night sky. In terms of brightness, the full Moon has an apparent magnitude of -12.7, making it easily the brightest object in the sky after the Sun. It's bright enough to be a hindrance to professional astronomers, with the glare from the Moon at times blotting out stars and deep-space objects in the night sky.

As you may have noticed, the period of the Moon's phases – 29.5 days – is longer than the time it takes the Moon to make a full orbit around the Earth – 27.3 days. This is because as the Moon is orbiting the Earth, the Earth is orbiting the Sun, and our planet's movement along its orbit means that it takes the Moon a little bit longer to get back into position relative to the Sun. Be that as it may, the period of the Moon's phase cycle is regular and obvious enough that ancient civilizations worldwide used it for their calendars. In fact the word calendar comes from the Latin meaning "onset of the month", while the word month derives from the Old English word for Moon.

As a practical matter, lunar calendars need to be heavily adapted to fit the solar year and the seasons, because 12 full phases of the Moon occur

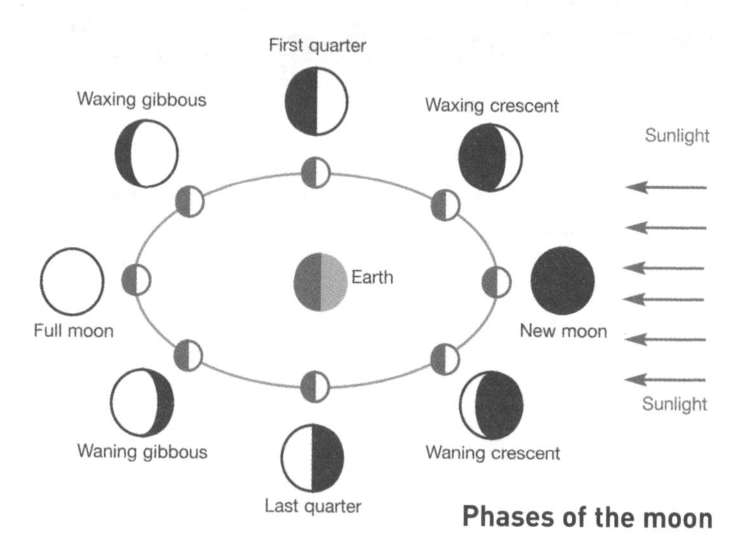

Phases of the moon

What is a blue Moon?

There are two explanations as to what a blue Moon is, the first being that it's the second full Moon in a calendar month. Given that the Moon's phase cycle happens every 29.5 days, these sorts of blue Moon happen fairly frequently, and can happen in any month but February (which is only 28 days long).

The second interpretation of blue Moon defines it as the fourth full Moon in a season (spring, summer, autumn and winter); normally a season will have three full Moons instead of four. These full Moons can only occur in the months of February, May, August and November.

Which one is correct? The second interpretation is older, but the first is more widely known.

in only 354 days, and there are 365.25 days in the solar year. However, lunar calendars are still employed culturally for the Jewish and Islamic calendars as well as the Chinese. The Gregorian calendar (the one we commonly use today) began as a Roman lunar calendar, and was then adapted over the centuries into a solar calendar.

The movement of the Moon

Just like every other celestial object in the Earth's sky, the Moon **rises** in the east and **sets** in the west, but unlike just about every celestial object in the sky, the Moon has a pronounced and noticeable eastward "drag" – that is it moves eastwards relative to the stars at a rate of about half a degree (about the width of the Moon itself) every hour. You can check this for yourself every moonrise by finding a bright star or planet just to the east of the Moon. After a few hours, take another look and you will find that the star's or planet's position is now further west of the Moon than it was before. Depending on the position of the star or planet (and when you return to look), it might even be behind the Moon, so if you come back and you can't find it, that's probably where it is.

The Moon appears to move eastwards because that is the direction in which it is orbiting around the Earth. The Moon is so close to the Earth that we can register the orbital movement over the course of a single night. Other objects in the night sky move in orbits, for instance the other planets in our solar system. However, they are too far away for us to notice their movement during a single night, especially when only using the naked eye.

The man in the Moon

Not everybody sees a "man in the Moon". A number of cultures look at the face of the Moon and see a **rabbit**, with the Mare Tranquillitatis as its head. A Chinese legend says that the rabbit was placed in the Moon by Buddha as a reward for a supreme sacrifice: Buddha had wandered into the woods disguised as an old man and asked the woodland animals for food, and the rabbit, having nothing to offer, promptly leapt into the fire and roasted himself for the benefit of the hungry traveller.

Whether you see a face or a rabbit in the Moon, what you're really looking at are the **maria**, the "seas" of basalt that flooded the Moon's surface billions of years ago. It was once thought that the maria of the Moon were merely a reflection of the Earth's surface, and that the Moon itself was perfect and unblemished; that thinking came to an end when **Galileo** fixed his telescope on the Moon's surface and saw shadows on the mountain ranges and craters.

One of the interesting facts about the Moon is that the maria are almost entirely on the side of the Moon that faces Earth. The reason for this is that the crust is thicker on the far side of the Moon, which made it more difficult for magma to seep out of the Moon's interior when the surface of the far side was hit by meteorites, comets and planetesimals. So the man in the Moon (or the rabbit) has no counterpart on the other side.

Lunar eclipses

Lunar eclipses occur when the Moon is directly in the path of the Earth's shadow, which blocks direct sunlight from hitting the Moon's surface. Why don't lunar eclipses happen every time there's a full Moon? After all, when the Moon is full, the Earth is between the Moon and the Sun – perfect conditions for an eclipse. The reason they don't is because the Moon's orbit is tilted about five degrees relative to the plane of the ecliptic, so most of the time during the full Moon the Moon is either above or below the ecliptic. We only get lunar eclipses when the sun and Moon line up on the ecliptic. This can happen up to three times a year and occasionally, as in 2009, four times – the final one of that year being on New Year's Eve.

As with solar eclipses, there are three different types of lunar eclipse: **total eclipses**, in which the Moon fully passes into the darkest part of the shadow of the Earth, known as the umbra; **partial eclipses**, when the Moon only partially enters the umbra; and **penumbral eclipses**, when the Moon only enters the Earth's penumbra (the less-dark portion of the Earth's shadow).

One useful difference between a lunar and a solar eclipse is that you can look directly at a lunar eclipse without damaging your eyes. Another difference is that a lunar eclipse is visible to anyone who can see the Moon, unlike a total solar eclipse which is only visible to people in a very small umbral path. The Earth's umbra is large enough to encompass the entire Moon (whereas the Moon's shadow on the Earth is relatively small).

A solar eclipse is over in a matter of minutes, whereas a lunar eclipse can take hours. It begins on the Moon's eastern edge, which darkens as the Moon enters the Earth's penumbra. If the Moon's path takes it into the Earth's umbra – the darkest part of the shadow – you have a total lunar eclipse on your hands. After the Moon passes through the umbra, the eastern edge will begin to lighten, and the Moon will slowly return to full brightness.

While the Sun is completely darkened during a total solar eclipse, the Moon's surface remains visible even when it is full in the shadow of the Earth, glowing a rather alarming mottled orange-red. This is because the Earth's atmosphere bends sunlight towards the Moon's surface. The Moon appears red because longer-wavelength light is scattered less by the atmosphere. If there's been a recent volcanic eruption that's thrown dust into the upper atmosphere, lunar eclipses will take on a particularly dark and sinister shade of red, and after very large eruptions, the Moon may become almost totally black (a *very* rare occurrence).

What would someone on the Moon see during a total lunar eclipse? A total solar eclipse, of course, with the Earth blocking the Sun. It would look different to a solar eclipse here on Earth, however, because though the Earth would be completely darkened, its atmosphere would be ablaze in the refracting light.

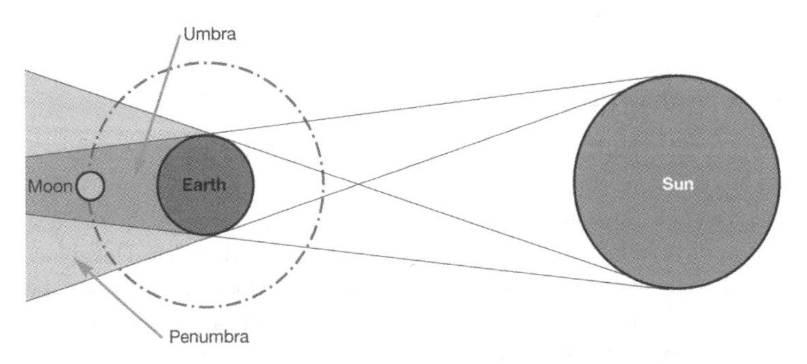

Lunar eclipse

Lunar eclipses

These eclipse timings were calculated by Fred Espenak of NASA/GSFC. The date is displayed first, then the time of greatest eclipse, as measured from Universal Time (also known as Greenwich Mean Time), and finally what sort of eclipse it will be.

2008

August 16 21.10 Partial

February 9 14.38 Total

2009

February 9 14.38 Penumbral

July 7 09.39 Penumbral

August 6 00.39 Penumbral

December 31 19.23 Partial

2010

June 26 11.38 Partial

December 21 08.17 Total

2011

June 15 20.13 Total

December 10 14.32 Partial

2012

June 4 11.03 Partial

November 28 14.33 Penumbral

2013

April 25 20.07 Partial

May 25 04.10 Penumbral

October 18 23.50 Penumbral

The Moon through the telescope

You don't have to wait for a full Moon in order to view the Moon through a telescope, and in fact a full Moon can be glaringly bright (although special lunar filters exist that help to cut down the glare). In fact, some of the most interesting views of the Moon happen while it is waxing or waning. So fire up your telescope at any time after the new Moon and catch some of the following exciting sights.

The terminator

Not Arnold Schwarzenegger, but the line that separates day from night on the Moon. Thanks to the lack of atmosphere, the Moon's **terminator** is rather sharp, with shadows in striking relief. This means that objects on the Moon at or near the terminator seem to pop right out in your field of view. This allows a better view of craters, maria and mountain ranges than you get when the Moon is full, when glare is at its height and shadows are at their minimum.

The terminator crosses the face of the Moon twice every month: first as the Moon grows from new to full, and then as it wanes from full back to new. During the waxing phase, the terminator represents sunrise; during the waning phase, sunset. This means there are two chances each month to get high-relief views of every part of the Moon's near side, each with its own sharply contrasting highlights.

Craters

The Moon's surface is covered in **craters**, ranging in size from microscopic pits all the way up to the Aitken Basin near the south pole, which is the Moon's largest crater at 2575km across. The sheer abundance of craters, and their variety in size and even age, make these objects a main draw for telescopic observation. Listed below are some of the most interesting craters you'll find in your lunar observations, most of which are named after prominent philosophers and scientists. All of them are observable with binoculars as well as telescopes.

- **Crater Tycho** What really catches the eye with Tycho is its immense and beautiful ray system – some of the rays stretch 2575 kilometres. The detail of its ray system shows that it's a youngish crater; older craters' ray systems have been eroded by later impacts and cosmic radiation. You'll find Tycho in the southern highlands (85km across and 4km deep), where during a full Moon it's impossible to miss.
- **Crater Copernicus** This is one of the Moon's most prominent craters, close to the equator, and off just a bit to the west. Copernicus has it all: a well-developed ray system; a crater wall that plunges 5180 metres to the crater floor and central mountains that rise over 792 metres from the crater floor. Just a short telescopic jog to the west of Copernicus, you'll find Kepler, a smaller crater whose ray system intermingles with that of Copernicus at full Moon.
- **Crater Plato** This northern crater is both big (108km across, 2440m deep) and dark; its darkness is due to the flood of basalt that forms its floor, and unlike many of the Moon's features, Plato is distinct even during a full Moon. On its floor are several small craters barely observable from Earth, and Plato is also where several instances of Lunar Transient Phenomena – momentary flashes of light – have been described. If you're looking for something new on the Moon, Plato is a good place to explore.
- **Crater Aristarchus** Named after the Ancient Greek astronomer (the first man to suggest that the Earth revolves around the Sun), Aristarchus is a mere 40km across. But it's the brightest crater on the Moon, its brightness enhanced by the fact that it's situated in the Oceanus Procellarum, one of the "darker" areas of the Moon. Just to the north of Aristarchus is Schröter's Valley, an interesting locale which will be discussed later in the book.
- **Crater Grimaldi** Aristarchus is the brightest crater on the Moon, and Grimaldi, on the far western edge near the equator, is the darkest. To the northwest of Grimaldi is another small, dark crater known as Riccioli. Both these craters were named after the Jesuit priests who mapped the Moon's surface and instituted its naming procedures in the seventeenth century. Because of their location, you won't be able to see Grimaldi or Riccioli until the Moon is full.

Maria

Thanks to their contrast with the rest of the Moon's surface, the **maria** are visible to the naked eye. In the telescope, however, the maria really come alive, becoming not just dark spots on the surface, but places of wonder, ringed and filled with fascinating objects to observe.

As with craters, the maria range in size from comparatively small to immense, and their official titles reflect their size. The largest maria are called **oceans** (like the aforementioned Oceanus Procellarum), while smaller maria are known as **bays** (sinus), **lakes** (lacus) and even **marshes** (palus). The names of the maria tend to be romantic and mystical, for instance Mare Tranquillitatis (the Sea of Tranquillity), Mare Nubium (the Sea of Clouds), Lacus Somniorum (the Lake of Sleep). Some have rather darker connotations: Oceanus Procellarum is the Ocean of Storms, Sinus Aestuum is the Bay of Seething, while for a disturbing time, check out the Palus Epidemiarium – the Marsh of Disease.

Of course, none of these oceans, bays or marshes have storms, clouds or disease. What they have is basalt, and lots of it. Here are three of the most prominent.

▶ **Mare Crisium** (Sea of Crises) One of the first large maria you'll see as the Moon waxes from new to full, **Mare Crisium** is also the only mare that is totally isolated from the other maria on the Moon. It's fairly circular to the naked eye (560km in diameter east–west and 435km in diameter north–south). Inside Crisium are two notable craters, Peirce and Picard, 18km and 24km in diameter respectively (and for you Trekkies: no, the latter is not named after Jean-Luc).

▶ **Mare Tranqillitatis** (Sea of Tranquillity) Best known to everyone as the landing spot of Apollo 11, which delivered the first humans to the lunar surface. **Mare Tranquillitatis** is also the head of the rabbit in the Moon (the ears being Mare Nectaris, to the west, and the aptly named Mare Fecunditatis to the east). A notable crater that borders Tranquillitatis (and Mare Serenitatis to its north) is **Plinius**, which is observable through binoculars. On the eastern "shore" of Tranquillitatis is **Palus Somnii** (Marsh of Sleep), made up of the basalt "overflow" from Tranquillitatis.

▶ **Mare Imbrium** (Sea of Rains) This magnificent mare in the northwest of the Moon is bordered by major lunar mountain ranges to the east, and features several striking craters, including **Archimedes**, **Aristillus** and **Autolycus**, all in the eastern portion of the mare. In the north are two of the Moon's rare freestanding mountains: **Piton**, which reaches 2440 metres and, to its northwest, **Pico**, also 2440 metres. Directly north of Pico is **Crater Plato** (see p.58); due west and slightly south from Plato is the semicircular **Sinus Iridum** (Bay of Rainbows), which features the **Moon Maiden.** This promontory, which juts out into the Mare Imbrium, when seen during Moon's sunrise and sunset reminded early astronomers of a young woman with flowing hair.

Anatomy of an impact crater

Craters are large holes in the ground created by the explosive force of a meteorite, comet or planetesimal slamming into the Moon's surface. Thanks to the Moon's airless, waterless and geologically inactive surface, craters formed on the Moon can last billions of years. NASA estimates there are over three trillion craters on the Moon larger than a yard across. Here are some of the specific features you can expect to see in your observations.

Crater Walls When an impact happens, a shockwave blasts through the immediate area and rams rock aside. Most of the rock is forced out, but some of that rock is pushed sideways and then upwards, forming a wall of material that lines the crater. Some of this material can fall back into the crater to form slumping terraces.

Rays Much of the material that used to be on the ground goes blasting up as ejecta, eventually falling back down to the surface of the Moon as rays – bright streaks of material that radiate outwards from the crater like spokes. Rays don't last forever but eventually fade or are broken up by other impacts. Craters with impressive ray structures are typically younger than craters without them. Large chunks of ejecta can also create secondary craters.

Central Peaks At the end of an impact, the bedrock struck by the impact will frequently "bounce back" at the centre of the crater to form central peaks. This only occurs in larger craters, those more than 20km in diameter. Very large craters (100 miles in diameter or more) have complex, ring-shaped uplifts; craters larger than 290km in diameter aren't called craters anymore, but impact basins.

Crater Copernicus was formed over a billion years ago when a mile-wide meteor slammed into its surface, gouging the nearly 100km crater into the Moon's surface. This view gives a good look at the crater walls.

NASA

Mountains and rilles

The Moon sports several impressive **mountain ranges**. Foremost among them is the **Apennine range** on the southeastern edge of Mare Imbrium, the peaks of which reach up to 6100 metres – far taller than the mountains in the Earth's Apennine range after which it is named. Directly to the north of the Apennines are the **Caucasus mountains**, while arching to the northwest of these is the **Alpine Valley**. Each of these ranges looks stunning when the terminator is on them or nearby. Moving eastwards to the border of Mare Tranquillitatis and Mare Serenitatis, are the **Haemus Mountains** to the west and the **Taurus Mountains** to the east. Moving west again, the Jura Mountains ring Sinus Iridum.

Mountains can also often be found in the centre of craters. These have been formed from the debris thrown up by the impact which formed the original crater. Copernicus has particularly notable central mountains, but you will discover many others in your explorations.

Rilles (Rimae in Latin) are great troughs on the Moon, some more than a mile wide and hundreds of miles long. These are not riverbeds, though they look very much like them, but are thought to be collapsed tubes of lava. One of the largest rilles can be found in **Schröter's Valley**, located just north of Crater Aristarchus. Another famous rille is the **Hadley Rille** (the landing place of the Apollo 15 mission), which can be found at the base of the Apennines. There are also rille systems near Crater Triesnecker on the north shore of Sinus Medii (the Bay of the Centre) with rilles **Hyginus** and **Ariadaeus** nearby. Rilles are best seen at or near the terminator; in the full Moon, they can only be seen as faint streaks.

Libration

The Moon always shows the same face to the Earth – more or less. But due to some eccentricities in the Moon's orbit, we are able to peek just a little beyond its usual face, in a process that is known as **libration**.

Because the Moon's orbit is tilted north–south relative to Earth, when the Moon is further north we can see a little more of its southern side and vice versa. As well as this, the Moon's orbit is slightly elliptical which means that when the Moon is closer to the Earth, it moves a little bit faster in its orbit. The Moon's rotation on its axis takes exactly the same time as its revolution around the Earth, but while the Moon speeds up and slows down in its orbit – depending on its position relative to Earth – its rotational speed is constant. So when the Moon speeds up as it gets closer

Moon myths

Here are some widely held, Moon-related beliefs and legends.

▶ **Werewolves** These nasty creatures – men and women who turn into wolves when the Moon is full – are a myth that has endured for centuries (inspiring some great B-movies along the way). There are a number of people who believe they turn into werewolves; they suffer from a legitimate psychiatric condition known as lycanthropy. But just because you believe you turn into a wolf, doesn't mean you really do.

▶ **Moon madness** The idea that the Moon somehow influenced people's mental states is so pervasive, that for many years the main word used to describe the mentally disturbed was lunatic (from *luna*, the Latin word for moon). Related to this, the belief that more "crazy" stuff happens during a full Moon has been shown to be statistically false. So don't use the full Moon as an excuse for doing something stupid.

▶ **The Moon's orbit influences menstruation** This longstanding notion is reflected in the fact that the word menstruation has the same root as the word for month which itself derives from moon. The average length of a woman's period, 28 days, is pretty close to the phase cycle of the moon, 29.5 days. This is almost certainly a coincidence, the most obvious reason being that menstrual periods vary from woman to woman, and are even more varied between different species of mammal – only five days for rats, for instance, but 37 days for chimps. The only mammal, other than humans, with a period of 28 days? The marsupial known as the opossum.

to Earth, we get to see a little more of the Moon by peeking over the side (when it slows down again as it gets further away, the process works again, but in reverse). Because of libration, we are able to observe 59 percent of the Moon's surface, although we can never see more than 50 percent of it at one time.

You can witness libration with the naked eye, but for a truly dramatic manifestation of the phenomenon, train your binoculars or telescope in the direction of Mare Crisium. Shortly after the new Moon (as the Moon's crescent begins to grow) you will notice a mare directly east of Crisium, just visible on the limb of the Moon. This is Mare Marginis (the Marginal Sea) which, as the Moon moves towards full, gradually disappears behind the eastern limb of the Moon while the Mare Crisium takes a noticeable path eastwards. When it comes to the waning half of the Moon's cycle, you will be able to observe craters Grimaldi and Riccioli move from the western limb of the Moon towards its centre.

Down to Earth

Most of us don't really consider the Earth as an astronomical object, if only for the simple reason that we're standing on it. But even as you read this, the Earth is moving through space and orbiting a star (the Sun) that is itself making a slow perambulation through our galaxy. You don't have to leave home to travel the universe: thanks to the Earth you're already doing so.

Our familiarity with the Earth keeps us from appreciating how unique our planet is. The fact is, that in all of our solar system, there is no planet like it – indeed, in all the known universe the Earth stands alone. And of all its singular characteristics, perhaps the most important is its inhabitants.

FACT FILE

Average distance from Sun 93 million miles (149.6 million km)

Diameter 7926 miles (12,755km)

Rotation 1 day

Orbital Period 365.24 days

Mean Temperature 59°F (15°C)

Axial Tilt 23.5 degrees

Gravity 32.1ft/s^2 (9.8m/s^2)

Moons 1

What makes the Earth unique?

Among other things, you do. Not because of who you are but because of *what* you are: a living thing. You, your family, your pets, the fleas that live on your pets, the bacteria that live in those fleas... every tree, shrub, floating bit of seaweed or mat of algae. **Life**, in all its myriad forms, is what makes our planet unique.

Despite speculation regarding Mars, the Jovian moon Europa and a few other places in our galaxy, the fact is that nowhere else has so much as a fossilized fragment of life been found, much less green-skinned, intelligent alien men and women. There is no Man in the Moon. The Martians aren't coming. Even if Klingons, Vulcans or Wookies do exist, they aren't sending signals our way.

There are several qualities that make the Earth amenable to having life forms on it. The first of these is its **location**, which in turn affects its **temperature**. The Earth's orbit averages a distance of about 93 million miles (150 million km) from the Sun, which places it in a zone where the planet is neither too hot (as Venus, Earth's near twin, is) nor too cold (as is Mars, our next closest planetary neighbour after Venus). Some people call this the **Goldilocks zone** – which is to say, it's not too hot, not too cold, but just right. What it's just right for is water.

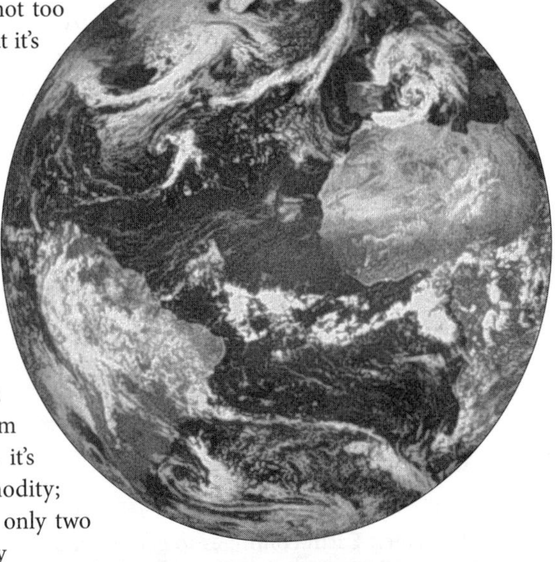

The planet Earth with the Americas to the left and Africa and Europe to the right. The view is actually a composite of over 70 pictures taken from the Clementine satellite.
Clementine/USGS

Water, water everywhere

That's the next thing the Earth has: **water** in liquid form, and lots of it. Other planets, moons and even comets also have water, but mainly in solid form as ice. Among planets, it's a pretty scarce commodity; **Mars**, for example, has only two small icecaps, and they are mostly frozen carbon dioxide. There is speculation that underneath Europa's frozen surface lies a vast dark ocean of liquid water, but with the cancellation of a NASA programme to explore the ocean there's no way of knowing. If you want liquid water in this solar system, Earth is where you need to be.

For life as we know it, liquid water is essential. Life was formed in water, billions of years ago. Humans carry our water-born legacy in the fact that our bodies are themselves more than seventy percent water. Water also aids life in more subtle ways as well. Earth's huge oceans act as a global heat sink, collecting the Sun's energy and retaining it to regulate the planet's **temperature**. Without our oceans, the earth's temperature would be much hotter during the day and far colder during the night – conditions that would not be very friendly to life. Of course, water is not the only regulator of global temperature. The Earth's atmosphere plays a part as well and interacts with Earth's liquid water in an ecological system that's very life-friendly – we know this system as **weather**.

A lively atmosphere

The earth's **atmosphere** is also unique: 21 percent of it is **oxygen**, an element that nearly all living things need; 78 percent is **nitrogen**; and the rest is trace gases such as argon, carbon dioxide and others. It's an atmosphere especially sensitive to life. The growth or reduction of plant life, for example, can cause a change in **carbon dioxide** levels in the air. Humans have also done their part to change the composition of the atmosphere – the rise in **greenhouse gases** and the reduction of the ozone layers in the polar regions can be at least partially traced back to human activity. Besides providing the air we breathe, the Earth's atmosphere aids life by insulating us from space debris – which typically burns up in the atmosphere – and from harmful radiation, such as ultraviolet rays, which are blocked by the now-thinning ozone layer.

Digging deep

Earth's unique qualities aren't merely skin deep: far below the surface of the planet, Earth continues to exhibit unusual qualities. On the surface and just below are Earth's **plate tectonics** – several interlocking plates that jostle each other at the edges, creating earthquakes and volcanic activity. These include the continents, but also extend below the ocean

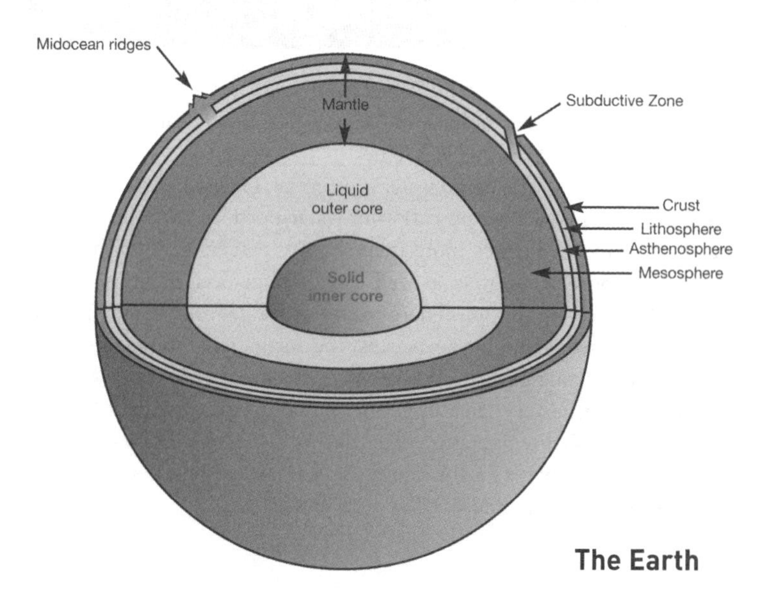

The Earth

to include the ocean floor, and their movement is often referred to as **continental drift**. Other planets and moons in the solar system have volcanoes, both active (Jupiter's moon Io) and inactive (Mars' immense Olympus Mons), but no other planet or moon has them as a result of plate tectonics, a phenomenon unknown anywhere else.

At the core

Further down, at the **Earth's core**, we find something that's also unique, at least among the small, rocky planets of the solar system – the Earth's outer core of **molten metal**. The other rocky planets in the solar system (Mercury, Mars, Venus) all feature solid cores, but the Earth's core is layered, with a solid core in the centre, and the liquid metal core around it. The core is as hot as the surface of the Sun (12,000°F/6700°C) and is made out of nickel and iron. The inner core is **solid** due in part to immense pressure; the core is, after all, bearing the weight of the whole world.

Magnetic field

As the molten iron steams around in the Earth's core, it generates the planet's **magnetic field** – a feature unique among the rocky planets. This magnetic field is why compasses work, but even more importantly, this

Why is the Earth's core so hot?

One of the many things that distinguish the Earth from the other, smaller, rocky planets of the solar system is that it has a core that is partially molten. And, since this core is made of iron and is under pressure, the temperatures involved are extraordinarily high – as hot or hotter than the surface of the Sun. Assuming the Earth has been around as long as the other planets, why is the Earth's core still molten, while the other rocky planets (Mercury, Venus, Mars) have solid cores?

One possible answer to what's going on down below may be over your head – namely, the Moon. It's thought that the Moon was formed as a result of a catastrophic collision between the Earth and a Mars-sized planetoid, whose impact tore off the rocky skin of our world, which eventually formed into our Moon. The incredible energies of such an impact would have also melted what remained of the Earth's surface; the heat at the core of our world could be a residual from that collision. This residual heat is augmented by frictional heat (generated by heavy elements slipping down into the core) and also by radioactive decay (parts of the Earth are radioactive). Eventually the Earth's internal heat will dissipate, since the solid portion of the core is growing steadily larger as heat slowly leaks out through plate tectonics and other means.

field helps to shield the Earth from electrically charged particles from space, which come in the form of **solar winds** and **cosmic rays**. This shielding helps to protect life on Earth, since these charged particles can be harmful to our DNA, and it also helps to generate the **aurora borealis**, the Northern and Southern lights (see p.36).

The Earth's magnetic field is persistent, but it's not what you would call stable; over the life of the planet the magnetic field has reversed polarity numerous times. Since the magnetic shifts are "frozen" into rock when it solidifies out of lava, geologists can see patterns of shifting magnetic fields over time. Among other things, this allows them to trace the shape of continental plates over millions of years. But your compass arrow is not about to point south – the last **polarity shift** occurred 780,000 years ago.

The Earth as an astronomical object

Even for a planet, Earth is not very big, just 7925 miles/12,755km in diameter. This makes it the largest of the rocky planets in our solar system, but a distant fifth overall behind the gas giants Jupiter, Saturn, Uranus and Neptune. However, the Earth distinguishes itself in other ways. Thanks to its large metal core, the Earth is the **densest** of all the planets: five and a half times as dense as water. It also boasts an exceptionally **large moon**, relative to its size – large enough for the Earth–Moon system to be often considered a "double planet".

Rotation and revolution

Everyone knows that the Earth **rotates** on its axis once every 24 hours. Or does it? Actually, it depends on what kind of "day" you're talking about. There's the **solar day**, which is the time it takes for the Earth to rotate once so that it is in the same position relative to the Sun. That's the 24-hour day we all know and love. But there's also the **sidereal day**, which is the time it takes for the Earth to rotate so that it is in the same position relative to the stars. That takes 23 hours, 56 minutes and 4 seconds. Why is the sidereal day shorter than the solar day? Because as the world turns, it also moves in its orbit, which (from the perspective of the Earth's surface) nudges the stars to the east in the sky.

Which brings us to **revolution**, not great struggles of class and society, but the path of the Earth around the Sun. The Earth makes a complete circuit around the Sun in 365.256 (sidereal) days, a figure that is not cleanly divisible by its rotation, which means that we're left with about six hours just hanging out there at the end of the year. That's why our calendar features leap years, adding a day to the calendar every four years. That pretty much takes care of the problem, although not entirely (see box below).

The seasons

Like every other planet in our solar system, the Earth does not have a perfectly circular orbit around the Sun. At some points in its orbit the Earth can be as far as 94.5 million miles/152 million km from the Sun, or as close as 91.4 million miles/147 million km. However, it would be wrong to assume that we have summer as we get closer to the Sun, and winter as we move further away. In fact, in the Northern Hemisphere – where most of the Earth's population lives – winter occurs when the

How fast are we going?

You're probably reading this sentence sitting down or possibly standing still, under the impression that you're not moving that much. Wrong. While the force of gravity has you pinned securely to the Earth's surface, you're moving in several directions at once, at speeds that are almost certainly greater than you think. Which directions and how fast?

▶ **West to East** The earth is rotating in an easterly direction, and it's taking you with it. The speed at which you are moving is related directly to where you are on the surface of the Earth. At the equator you would be zipping by at over 1000 miles per hour, while at the poles you'd experience very little motion at all. However, most people on the planet live in an area that is moving east at a speed of at least a few hundred miles per hour.

▶ **Around the Sun** As the Earth rotates, it also travels in its orbit around the Sun. The actual speed of the Earth varies slightly from day to day due to eccentricities in its orbit that make it go faster when it's closer to the Sun, and slower when it's further away. However, on average, the Earth orbits the Sun at a speed of 18.5 miles a second – 66,600 miles per hour.

▶ **Around the Galaxy** As the Earth moves around the Sun, the Sun moves around the Milky Way galaxy – a trip that takes over 225 million years per spin. That might seem like the Sun is moving slowly, but it's more that the Milky Way is a big galaxy, and in fact the Sun is blasting along at 135 miles per second, or 486,000 miles per hour.

▶ **Through the Universe** The Milky Way is on the move, too – heading in a southerly direction in our night sky (if you can see the southern constellation of Centaurus, that's the direction we're heading). Its speed? A mere million miles per hour.

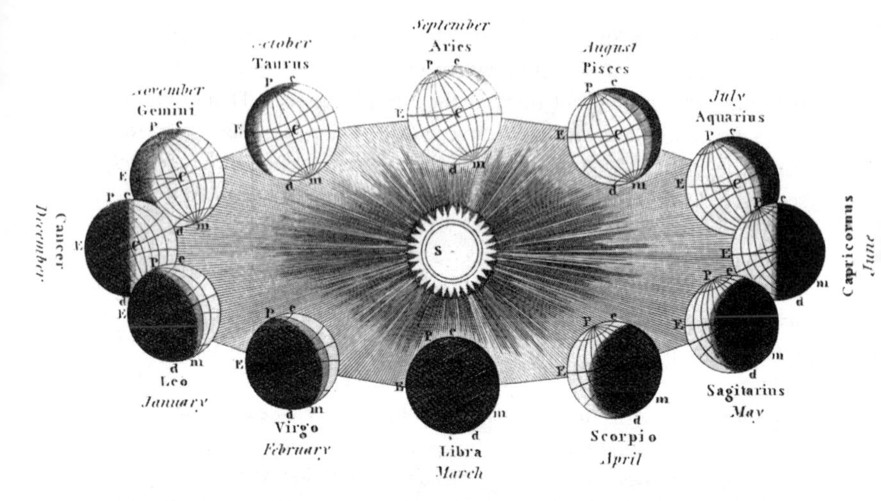

In this eighteenth-century engraving, the signs of the Zodiac correspond roughly to the months in which they are opposite to the Sun in the sky.

Earth is closest to the Sun. Our **seasons** come not from our location in orbit but from the tilt in the Earth's axis, which is tilted some 23.5 degrees with respect to the plane of the Earth's orbit around the Sun. When it's summer, the Earth's axis is tilted towards the Sun (which means longer days and more direct sunlight); when it's winter, the axis is tilted away (shorter days, less direct sunlight). This tilt also means that when it's summer in the Northern Hemisphere of the Earth, it's winter down south, and vice versa.

The ecliptic and precession

You can observe the tilt of the Earth's axis indirectly by tracking the Sun's path over the course of the year, as it moves through the sky relative to the stars (remember, the sidereal day is shorter than the solar day). This path is tilted 23.5 degrees relative to the earth's equator and is called the **ecliptic**. Besides the Sun, you'll also find the other planets of our solar system on or very near the ecliptic, and it's also where you'll find the constellations that make up the Zodiac.

The earth spins on its axis like a toy top; it also experiences another type of movement related to the axis – a slow, wobbly circular motion that's known as **precession**. In practical terms, this means that over the course of thousands of years, observers of the night sky will see the poles

Fun with the calendar

The **Gregorian Calendar** (the one now universally used in the Western world) was created in the late sixteenth century by Pope Gregory XIII when it was realized that the **Julian Calendar** (named after Julius Caesar) was inaccurate. Under the Julian system the seasons were slipping back, so that the vernal equinox (which starts off spring) was occurring on March 11, instead of on March 21 as it was supposed to.

The problem was that the Julian Calendar featured leap years to compensate for the fact that the Earth's year is slightly longer than 365 days (365.256 days to be precise). Adding an extra day every four years dealt with the extra .25 of a day – but it didn't deal with the extra .006 of a day that was still left over. And while .006 of a day isn't a whole lot (it's a little over eight and a half minutes), if you let an error like that sit for a couple of thousand years, your days get away from you, and that's precisely what happened.

The Gregorian Calendar (officially adopted by Britain and her colonies in 1752) corrects the Julian Calendar rather ingeniously, by eliminating the leap year in every year ending with double zeros (1700, 1800, 1900, etc), except those which are divisible by 400, which are allowed to remain leap years. This is why 2000 was a leap year, even though it was a double-zero year. BUT – if the year is divisible by 4000, then it's not a leap year after all. Something to look forward to, if you're planning to be around in a couple of millennia. Even then, the Gregorian Calendar is still off by a day every 20,000 years. But they can deal with that later.

point towards different stars as the axis goes through its "wobble". Our current pole star (for the North Pole, at least) is **Polaris**, located in the Ursa Minor constellation. In 2700 BC, the ancient Egyptians had **Thuban** (located in Draco) as their pole star; while 13,000 years from now, **Vega** (in Lyra) will point to true North. Polaris will resume its pride of place at the North Pole in 26,000 years, as the precession cycle comes full circle. Polaris is actually now still inching ever closer to the celestial North Pole – it gets as close as it's going to in 2017, after which point it starts moving ever so slowly away.

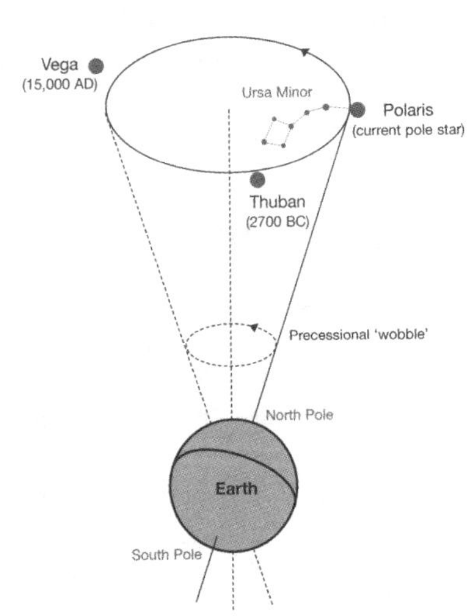

The precessional cycle

Mercury and Venus

frantic speed, fatal beauty

Earth is sometimes referred to as the "third rock from the Sun", because inside its orbit there are two other planets, Mercury and Venus, that spin around the Sun in tighter, faster orbits of their own. Their position relative to the Earth leads us to label this pair as "inferior planets", but both Mercury and Venus are fascinating objects in their own right, each with its own unique characteristics. Always near the Sun, and often obscured by its glare, Mercury features bizarre orbital dynamics that helped to verify Einstein's Theory of Relativity. Bright and beautiful, Venus's dazzling visage in the evening and morning skies masks the planet's true, hellish nature: in reality it's a place of blistering heat, acidic clouds and crushing pressure.

Mercury

In Roman mythology, **Mercury** was the messenger of the gods, zipping from place to place on winged sandals. The planet Mercury is also a fast mover, tearing along in its orbit at an average speed of nearly 108,000 miles per hour. The nearest of the planets to the Sun at 36 million miles average distance, Mercury's "year" is a brief 88 days long. It travels around the Sun four times for each time the Earth makes the same journey.

FACT FILE

Average Distance from Sun 36 million miles (58 million km)

Diameter 3032 miles (4880km)

Rotation 58.65 days

Orbital Period 88 days

Mean Temperature 333°F (167°C)

Axial Tilt .01 degrees

Gravity 12.1 ft/s^2 (3.7m/s^2)

Mercury, as seen from the Mariner 10 spacecraft. The planet's heavily pocked surface is reminiscent of that of our own Moon's, which Mercury is closely related to in terms of physical size.
NASA

Mercury was one of the five planets observable to the ancients, who noted its **rapid movement** relative to the other planets and named it correspondingly. It wasn't just the Romans who did this; Babylonian astronomers knew the planet as **Sihtu**, which translates as "jumping", while in China and Japan, Mercury was known as the "Water Star". The Greeks gave the planet two names – **Apollo** for when it showed up in the early morning sky, and **Hermes** (the Greek version of Mercury) for its evening appearance. Despite the two names, they knew it to be the same planet.

In addition to being fast, Mercury is also **small** – just over 3000 miles in diameter, which makes it the smallest planet in the system except for Pluto. Both Jupiter and Saturn have moons larger than Mercury (Ganymede and Titan, respectively), and our own Moon's diameter is only 900 miles smaller. But Mercury is far more massive than any of these moons, thanks to an immense **iron core** that astronomers believe accounts for nearly three-quarters of the planet's diameter. This core also makes Mercury the second **densest** planet in the system (5.43 times denser than water), trailing only our own Earth. It may also be what provides Mercury with its **magnetic field**, weak though that field may be (it's only one percent the strength of the Earth's field). Since Earth's magnetic field is created by movement of its molten core, Mercury's field suggests that its core may at least be partially molten as well – though no one knows for sure.

Given that Mercury is the planet closest to the Sun, it would be reasonable to assume it's also the hottest planet of the group, but in fact Venus has that honour for reasons we'll explain later. Mercury still gets pretty hot on the surface, however, and when the Sun's overhead at the equator (and three times larger in the sky than it would be on Earth), the planet's **temperature** can reach over 750°F.

Because Mercury has an extremely **thin atmosphere** (mostly of elements which are blasted off the planet's surface from relentless solar radiation, which then quickly leak into space), things cool down quickly on the night-time side of the planet to a frosty –290°F. This massive difference in daytime and night-time temperatures is unique in the solar system. One major contributing factor for this range of temperature is the fact that one "day" on Mercury (that is, the time from one sunrise to the next) is exactly two of Mercury's "years" (176 of our days). That comes out to 88 days of sunlight and 88 days of darkness.

Weird orbital fun

If you think a two-year-long Mercurian day sounds weird, then consider this: from sunrise to sunrise on Mercury takes 176 days, but the planet **spins on its axis** once every 58.65 days – exactly one third the length of the Mercurian "day". To put this in perspective, the Earth spins on its axis once every 24 hours – exactly the length of our day. So how does Mercury pull off having only one "day" when the planet itself has rotated three times? How can the Mercurian "day" be so wholly unlike the Earth day?

The answer lies in the fact that while Mercury is slowly turning on its axis, it is also orbiting the Sun at high speed. Mercury's **orbital speed** counteracts some of the action of Mercury's **rotation**, making the Sun appear to move much more slowly in Mercury's sky than other celestial objects like stars, whose speed in the sky is wholly determined by Mercury's rotation.

As well as the weirdness of a slow-moving Sun, Mercury's orbit around the Sun is also highly **elliptical** – at its closest approach to the Sun (**perihelion**) it's only 28.6 million miles away, fifty percent closer to the Sun than when it's at its furthest point in orbit (**aphelion**) of 43.4 million miles. (In comparison, Earth at perihelion is only three percent closer to the Sun than when it's at aphelion.) The laws of planetary motion dictate that planets move faster the closer they get to the Sun, and Mercury certainly does that

Mercury and Einstein

No planet's orbit is perfectly circular; all are elliptical, with the Sun located at one of the foci. A critical point of any planet's orbit is the perihelion – the point at which the planet is closest to the Sun. One interesting thing about planetary perihelions is that they move slightly over time, largely due to the gravitational pull of other planets. Mercury's perihelion moves like the rest of them, but it moves more than twice as much as can be accounted for by the gravitational interaction of other planets, according to the Newtonian laws of physics.

How can this be explained? Before the early twentieth century it couldn't: the best physicists could come up with was to posit the existence of another previously unobserved nearby planet (tentatively named Vulcan) which was performing the additional tugs. Then along came Albert Einstein, who solved the problem by creating a new understanding of gravity as part of his general theory of relativity (published in 1916). Basically, Einstein noted that gravity causes space to "curve" rather like a funnel, and that the closer you get to an object, the more curved space becomes around it. Mercury's orbit causes the planet to skate along different parts of the Sun's gravity curve, which helps to create the additional perturbations of Mercury's orbit.

Einstein's theory explained Mercury's perihelion puzzle quite handily, and was an early confirmation of the theory of relativity; more confirmation came in 1919 during a total eclipse of the Sun which showed the Sun's gravitational field bending the light of distant stars.

Albert Einstein outside his Berlin laboratory in 1920. His theory of relativity helped explain some idiosyncracies in the orbit of Mercury.

– moving so fast, in fact, that at perihelion its orbital speed is faster than its rotational speed. So if you were on the surface of Mercury, what you'd see would be the Sun moving west, stopping, and then moving east for a time before stopping again and starting its journey westward again.

Mercury rotates only three times for every two revolutions around the Sun. When rotation and revolution are locked in like this scientists call it **spin–orbit coupling**. It occurs mostly with moons, whose days are as long as their orbital time, making for a spin–orbit coupling ratio of 1:1. Mercury is the only body we know of with a 3:2 spin–orbit coupling ratio.

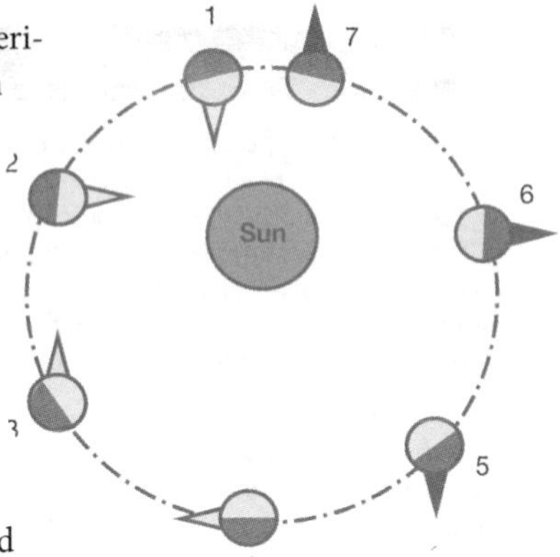

Spin-orbit coupling

Mapping Mercury

Mercury is relatively close to Earth, but its astronomical position close to the Sun made it very difficult for astronomers to find out much about its surface; in order to get a proper look we had to send the **Mariner 10** probe, which in three fly-bys, in 1974 and 1975, mapped about half of Mercury's surface.

The pictures sent back look as though they could be of our own Moon which, like Mercury, is small and airless and pocked with surface **craters**. Mercury's craters are named after famous artists, with Tolstoy, Goethe, Mozart, Twain and Matisse among the larger ones. The largest, at nearly 380 miles across, is Beethoven. Mercury also has numerous **intercrater plains**, which may be formed either from

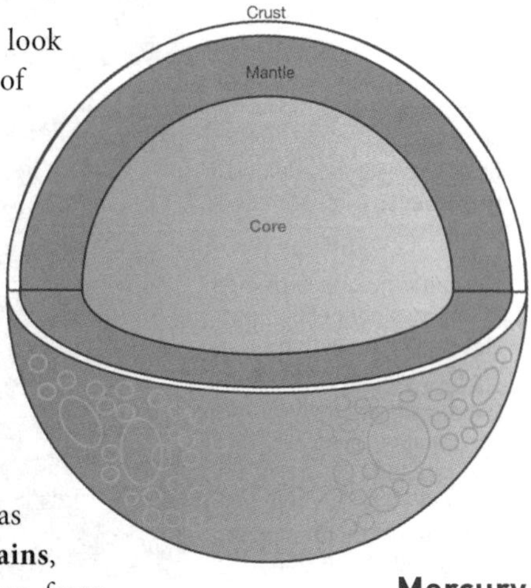

Mercury

The Caloris Basin is the single largest physical feature on Mercury.

ejecta (the dust from the aftermath of a meteorite hit) or, more controversially, from volcanic activity. No one's entirely sure.

The largest single feature on Mercury's surface is **Caloris Basin**, a massive basin some 800 miles across (a quarter of Mercury's diameter), which was gouged out by some immense collision in the depths of time. The collision did more than just blast the Caloris Basin into existence. Some astronomers suspect that shockwaves from the Caloris hit converged on the opposite side of Mercury's globe, rearranging the surface into a blocky, jumbled landscape (referred to as "weird terrain").

Another strange feature on Mercury is its **scarps** – winding cliffs that can stretch for hundreds of miles and reach two miles high into the sky. The suggested explanation for these scarps is that they're "wrinkles" caused by compression from when the planet shrunk slightly during its core formation.

It's close to three decades since Mercury got a visit from a spacecraft, but it won't be much longer until the next. In August 2004, NASA launched **MESSENGER** (Mercury Surface, Space Environment, Geochemistry and Ranging mission), and will evenutally end up in a stable orbit of the planet in late 2009. One of its stated missions is to discover if **water**, in the form of ice, exists at the poles of Mercury (in craters which don't receive direct sunlight). If ice did exist, it would be just one more strange feature of this strange little planet.

Big core, little planet

The most interesting thing about Mercury is something we can't even see – its metal **core**. This is huge (relative to the size of the planet), taking up three-quarters of its diameter. One writer has described Mercury as "a huge metal ball dipped in mud", which is as apt a description of the planet as any. So how did its core get so big?

Perhaps it's not the core that got big so much as the planet that got small. Most astronomers believe that a collision between Earth and a stray Mars-sized object shaved off a portion of the Earth's crust, which eventually reformed into the Moon. By the same token, a **collision** between Mercury and another large object could have performed a similar shearing of crust into space, reducing the planet's size to its current, meagre dimensions and making its core a much larger proportion of its overall diameter (no moon this time, however). This sort of massive collision would also help to explain Mercury's highly **eccentric orbit**.

Venus

Venus is the planetary equivalent of the girl your mother warned you about. It's stunningly beautiful on the outside: the brightest object in the sky outside the Sun and the Moon, a glimmering jewel which has captivated astronomers for centuries. And as a virtual twin of our own planet in size and composition, it was the planet most realistically suited for human life.

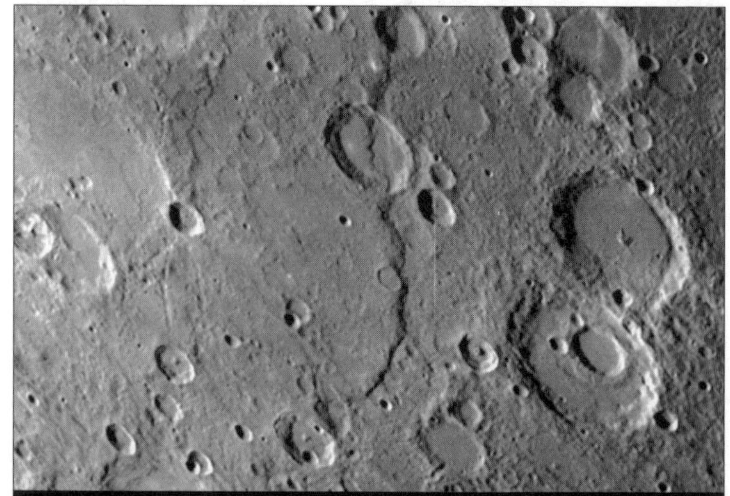

The Discovery Scarp, which stretches for nearly 220 miles and reaches almost two miles in height. Scarps are unusual "wrinkles" in the surface of Mercury, thought to be the result of planetary cooling.

FACT FILE

Average Distance from Sun 67.2 million miles (108.2 million km)

Diameter 7521 miles (12,103km)

Rotation 243 days

Orbital Period 224.7 days

Mean Temperature 867°F (464°C)

Axial Tilt 177.4 degrees

Gravity 29.1 ft/s^2 (8.9m/s^2)

Get past the glorious façade, however, and Venus shows its true face: violent, brutal and furious. Its carbon-dioxide atmosphere would suffocate you, its sulphuric-acid clouds would pick your bones clean and its intense heat would scorch whatever remained into dust. This planet will suck you dry and spit you out. If Venus is the Earth's twin, it's an evil twin, the twisted sister of the solar family.

Venusian hard data

Venus is the second planet from the Sun, 67 million miles out on average, with an equatorial diameter of ,521 miles (a mere 404 miles short of Earth's own equatorial diameter). It has eighty percent of our planet's mass and ninety percent of its gravity. It's not enough to solve any serious weight problems, but more than enough to put an extra spring into your step, if you really wanted to visit.

Each planet in our system has its own weird little tics, but Venus has more than most. Most notably, Venus is the only planet in the solar system with a **retrograde rotation**. Every other planet in the neighbourhood has the Sun rising in the east and setting in the west, but Venus has the Sun rising in the west and setting in the east. No one knows exactly why Venus exhibits this wild streak of individuality, but the most plausible explanation is that it's a remnant of a violent encounter with another planetary body long ago. Venus's backwards rotation is the planetary equivalent of a scar from a bar brawl.

That bar brawl may also be the reason why Venus revolves so slowly. Venus rotates on its axis just once every 243 days, which is longer than the Venusian "year" (224.7 days) by more than eighteen days. This makes it the only planet whose **rotation** takes longer than its **revolution** (remember that Mercury's "day" is actually longer than its rotational period). In a strange coincidence, Venus rotates three times on its axis for every two times the Earth orbits the Sun. There's some speculation that the Earth and Venus are locked into some sort of tidal resonance, but there's no real scientific evidence for it. Venus also has the distinction of having an almost perfectly **circular orbit**. There's a bare million miles between perihelion and aphelion, which creates an orbital eccentricity of less than one hundredth of one percent. Earth's orbital eccentricity is twice that of Venus, while Mercury's is nearly thirty times as eccentric.

Finally, Venus has almost no **magnetic field** to speak of, which indicates that its core, probably of iron and nickel like Earth's, is no longer molten.

A cultural history of Venus

Human civilization has been well acquainted with Venus for thousands of years. The planet's brightness and closeness to the Sun made it known as the **Morning Star** and the **Evening Star**, and a number of civilizations gave different names to the planet's dual incarnations. The Romans called it **Venus** or **Vulcan**, depending on whether it was up in the morning or in the evening, and the Greeks and Egyptians also gave the planet a double identity.

Venus's morning phase was at one time known as "Lucifer", which means "light bringer", not inappropriate for an object that precedes the Sun in the morning sky. It was also known by the less troubling name Phosphorus, which also means "bringer of light". Currently, Venus is the only major planet other than Earth (in its alter ego of Gaia) that is named after a goddess rather than a god.

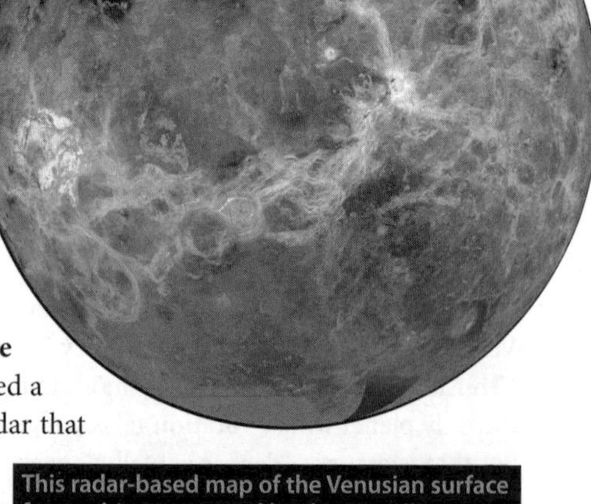

This radar-based map of the Venusian surface from orbit was created by the Magellan spacecraft. Venus shows evidence of volcanic activity in the past but has few impact craters thanks to a thick atmosphere that burns up all but the largest of meteors.
NASA

The planet's passage across the sky is remarkably regular, so much so that it makes an extremely **reliable timekeeper**. The Mayans used a 584-day, Venus-based calendar that was far more accurate than the one used in Europe at the same time.

Ancient observers did not have the benefit of telescopes, so they didn't know

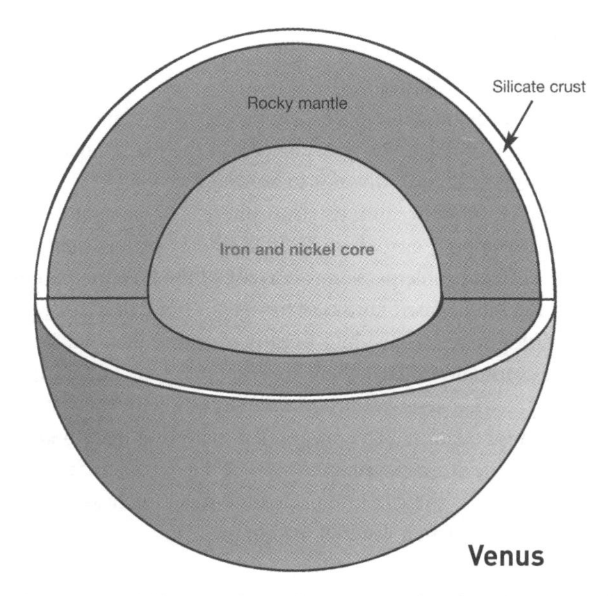

how similar Venus was in size to Earth. But once this fact was established, astronomers and others began speculating that Venus might have **life** much like our own. The popular vision of Venus was one in which the planet, thickly covered in clouds and warmed by its proximity to the Sun, supported a pole-to-pole swamp jungle, swarming with creatures that we might have found in our planet's own, swampy past.

There were even hints of **intelligent life** – or so it seemed. While Venus is in crescent phase, the shadowed part sometimes appears to be faintly illuminated. This phenomenon is called "ashen light" and appears similar to our planet's aurorae. In the early nineteenth century Franz von Paula Gruithuisen, head of the Munich Observatory, suggested that this ashen light might be the result of fires or even a massive firework display – evidence of an advanced (if somewhat pyromaniacal) Venusian civilization.

Because Venus's thick swaddling of **clouds** obscured investigation from Earth, it wasn't until man developed spacecraft that we got our first close-up look at our bright sister. What they found destroyed forever any idea of it as a jungle planet.

Hot stuff

Venus is the **hottest planet** in the system. Hot enough, in fact, to make lead melt and drip away like water. Venus's surface temperature hovers at around 900°F/480°C degrees, everywhere on the planet, nearly all of the time. Occasionally it cools down – to around 835°F/445°C.

The planet is so hot because its atmosphere is 97 percent **carbon dioxide**, the heat-trapping gas which is at the heart of our planet's current **greenhouse effect** problems. Venus is a case of the greenhouse effect gone out of control, but it wasn't always as hot as it is now. One theory proposes that Venus might once have had oceans of water, but that its closeness to the Sun eventually caused them to evaporate into water vapour. This started the greenhouse-effect process, beginning with the water vapour's own greenhouse qualities. The oceans also served to help to keep carbon dioxide out of the atmosphere. Our own oceans serve to dissolve carbon dioxide and deposit the gas into carbonate rocks such as limestone, but this doesn't happen when water is in vapour form.

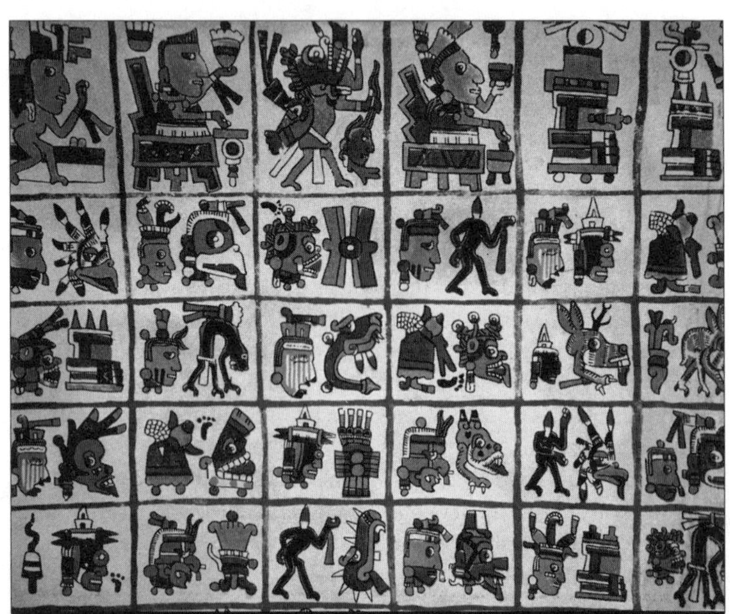

Part of a Mayan calendar (Codex Cospi). The journeys of the planet Venus, which rises as a morning star approximately every 584 days, were integral to the calendar's timekeeping system.

© Archivo Iconographico, S.A./CORBIS

Life on Venus

With a suffocating carbon-dioxide atmosphere and a surface hot enough to melt metal, Venus is the last place in which anyone could reasonably suggest the existence of life. And yet, in September 2002, two scientists at the University of Texas proposed just that.

According to Dirk Schulze-Makuch and Louis Irwin, certain levels of the Venusian atmosphere are "relatively hospitable" and could be home to airborne, bacteria-like creatures. Furthermore, these bacteria could be the cause of some interesting aspects of the atmosphere, such as the presence of hydrogen sulphide and sulphur dioxide and the absence of carbon monoxide. These bacteria-like creatures may have arisen during an earlier, less torrid period of the Venusian life cycle, and then evolved to fit into the few niches available after the greenhouse process got out of control.

Needless to say, at this point, the general scientific community is pretty sceptical. But if life could actually survive on Venus, it's possible it could survive just about anywhere – and that has exciting implications for life in the universe. Could Venus ever have been cool enough to harbour life as we know it? It all depends on when Venus's runaway greenhouse effect kicked in. The standard theory has this taking place within a few hundred million years of the birth of the solar system, which all points against life. But some recent theorietical models, notably advanced by David Grinspoon of the Denver Museum of Nature and Science, posit Venus possibly retaining ocens of water on its surface for up to two billion years – enough time for the idea of life to possibly catch a foothold.

Without the liquid water to scrub the atmosphere, the percentage of carbon dioxide continued to rise, and the temperature of the atmosphere with it. Eventually, the water vapour dissociated and the hydrogen atoms leaked into space, leaving Venus with a carbon-dioxide atmosphere and a severe heat problem.

Dense atmosphere

The **atmosphere** on Venus is incredibly **dense**. At the planet's surface, the pressure is 96 times higher than Earth's atmosphere at sea level, making it closer to the sort of pressure that exists half a mile down in our oceans. When the Soviet Union sent probes to the planet's surface in the late 1960s and 70s, the pressure was sufficiently great for the probes to be almost instantly crushed into scrap metal (the heat didn't help, either). The photos the probes managed to send back show one interesting effect of this intense atmospheric pressure: the ground seems to curve towards the horizon, giving the effect of the probe sitting at the bottom of a fishbowl.

Could what happened to Venus happen to us?

It's been suggested that at one point Venus might have been quite a pleasant place to live, but that over the course of time **greenhouse gases** collected and helped to turn the planet into the searing planetary rotisserie it is today. Here on Earth, where we've been having our own debate about greenhouse gases, inquiring minds wish to know whether our own planet could finish up in the same state.

Let's leave aside the question of whether humans could trigger a chain reaction of greenhouses gases, and look towards another culprit – the **Sun**. As we know, the Sun is an immense nuclear reactor, burning hydrogen in its core. As the amount of hydrogen shrinks, the Sun contracts; as the Sun contracts, it burns slightly hotter, and the amount of energy the Earth receives from the Sun grows. Polar icecaps will melt; land will be flooded. If it gets hot enough, water will start to evaporate, increasing the amount of water vapour in the air, forcing some of that vapour into the stratosphere, where sunlight is intense enough to break the bonds that connect hydrogen to oxygen in the water molecule. The hydrogen slips out of the atmosphere. Repeat this until the oceans run dry and all water has gone, and runaway global warming begins. The Earth fries in its own gases.

The theoretical timetable for this is about a billion years or so. Because, although the Sun is burning hydrogen, it's got a lot of hydrogen to burn. So we've got a little time before we have to leave the planet – presuming, of course, we don't manage to speed things up ourselves.

Venus's atmosphere has distinct and clearly **defined levels** to it, separated by temperature and density. One way to think of the atmosphere is that it is like one of those exotic drinks, in which one kind of liqueur sits on top of another, heavier type. At the bottom level, near the surface, the atmosphere is thick and stagnant; wind speeds at this level are 3 to 11 miles per hour at best, so any hope for a wind-chill factor goes right out the window. Further up are three levels, at 30, 33, and 37 miles respectively, where sulphuric acid condenses into **clouds**, forms droplets and rains down (the rain, because of the intense heat, never makes it to the surface). These clouds are what gives Venus its brightness and, because of the sulphur content, its distinctive yellowish tinge.

The upper level of the Venusian atmosphere has **winds** that course over the planet at incredible speeds of up to 225 miles per hour/360kph, which can cross the globe in about four days. These fierce winds are probably caused by a vast temperature difference in the upper atmosphere between the day side of the planet and the night side (100°F/37°C on the day side, and -270°F/-167°C on the night). Imagine a worldwide tornado, and you have a good idea of what these winds are all about. More recent discoveries about Venus' atmosphere have come from the **Venus Express** spacecraft,

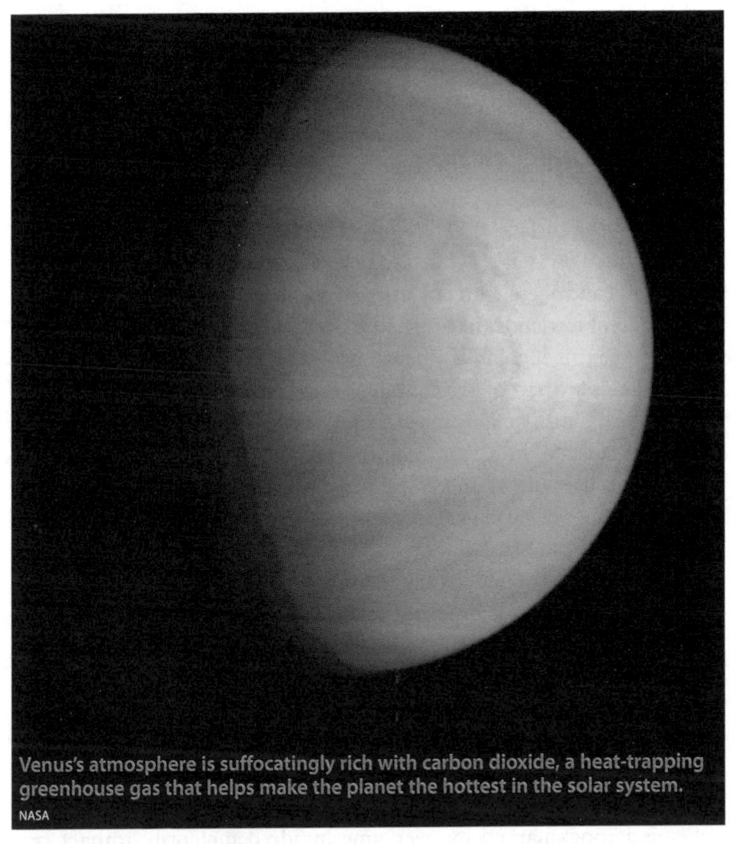

Venus's atmosphere is suffocatingly rich with carbon dioxide, a heat-trapping greenhouse gas that helps make the planet the hottest in the solar system.
NASA

launched by the European Space Administration in November 2005, which has been in orbit around the planet since April 2006. The spacecraft's mission was to explore the planet's atmosphere, and it's been doing just that. Among its discoveries are florescent clouds of oxygen, contributing to a phenomenon called "airglow" which was first noticed on Venus via observations from Earth. It has also revealed a huge vortex of clouds at Venus' south pole, complementing a similar vortex at the north pole, discovered during the Pioneer Venus mission in 1978. The Venus Express has also found an exotic form of carbon dioxide that can absorb even more energy than the more common form – possibly contributing to the runaway greenhouse effect that the planet suffers under today. The spacecraft's mission is currently slated to run until May 2009.

Down on the surface

Although clouds obscure the surface of Venus from casual view, US and Soviet spacecraft have nevertheless charted the **surface** of the planet using radar. The most notable of these was the **Magellan**, which generated a detailed map of the planet during its two-year mission in the early 1990s.

Basically, the surface of Venus looks like the surface of the Earth would if you drained away the oceans. sixty-five percent of it is a rolling plain, which you can equate with our own ocean floor. A quarter of the surface is made up of lowlands (like our deep ocean pockets), and the remaining fifteen percent is highlands (the equivalent of Earth's dry land). Most of the highlands are part of two continent-sized masses of land: **Ishtar Terra**, the northern mass (roughly the size of Australia), and **Aphrodite Terra** in the south (close to the size of Africa).

Venus has its share of impressive geological formations, including **canyons**, **lava flows** of varying ages, **mountains** on Ishtar Terra (which dwarf the Himalayas), and fault lines which have their parallels in Africa's Great Rift Valley. These formations suggest that Venus has some **tectonic activity**, though the planet is probably not as active as the Earth.

There's also strong evidence of recent **volcanic activity**. Spacecraft have recorded variable amounts of **sulphur** in the atmosphere, which is indicative of recent eruptions. They have also recorded large, highly reflective areas that could be **pyrite** (better known as "fool's gold"). This mineral decomposes quite rapidly, so its presence in large quantities suggests it is of fairly recent origin.

Venus is pockmarked by over nine hundred meteorite **impact craters**, most of which are quite large (the smallest recorded example is three miles across). All are named after notable women, from Sappho to Florence Nightingale, with the largest (168 miles in diameter) named after anthropologist Margaret Mead. Their size reflects the fact that small meteorites would burn up in Venus's thick atmosphere. These impacts are also relatively young, which is further evidence of fairly recent volcanic activity. One interesting aspect of the Venusian surface is that most of it is young (geologically speaking): about 700 million years old. One theory to explain this uniform age of the surface is that, once the Venusian oceans boiled away from the rising heat, plate tectonics stopped working. The internal heat of the planet then built up in the crust, melting the surface rock.

Looking at Mercury and Venus

Because Mercury and Venus are "inferior" planets (closer to the Sun than the Earth), observing them through telescopes and binoculars is challenging but rewarding. One of the primary challenges is that these two planets never stray far from the Sun – Mercury is never more than 28 degrees from the Sun and Venus never more than about 47 degrees. This means that both planets are visible in the sky only in the **early morning** before the Sun rises, or in the **early evening** just after the Sun sets. Mercury presents the additional problem of always being viewed low in the sky where the atmosphere is thicker, thus leading to viewing distortion.

On the other hand, one of the pluses for new observers is that Mercury and Venus display **phases**, just like our Moon does. This is due to the fact that the orbits of these planets are within ours, which allows us to see both the day and night sides of the planets from our perspective.

One interesting fact is that both of these planets are at their brightest from Earth not when they are near full but when they are crescent-shaped. When the planets are full, they are directly at a point in their orbit called the **superior conjunction**, which puts them at the opposite side of the Sun from Earth. So not only are they hidden in the glare of the Sun, but what light does reflect off them is having to travel a significant distance to reach us. The **inferior conjunction**, when the planet is between us and the Sun, means that the dark side of the planet is towards us, making the planet difficult to see (not to mention that it's again hidden in the glare of the Sun).

For the amateur astronomer, the best times for viewing both Mercury and Venus are when they are at points in their orbits we call their **greatest elongation**. This is a twofold phenomenon: there's a greatest eastern elongation and a greatest western elongation. At these points, both planets are as far from the Sun as they get in our sky. This often places the planets higher in the sky, which makes for easier and better observing. Both Mercury and Venus appear half full when they are at their greatest elongations, but even so, Venus will outshine all stars in the sky (Mercury outshines most of them as well), so in terms of pure brilliance, you won't be missing out.

Finding Mercury

Amateur astronomers find Mercury quite difficult to locate, for a variety of reasons. As noted, it's very close to the Sun in the sky – no more than

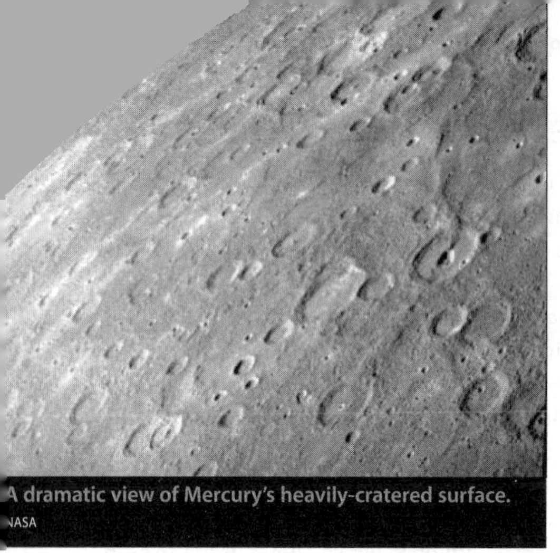

A dramatic view of Mercury's heavily-cratered surface.
NASA

28 degrees from the Sun at its most distant – and because Mercury's orbit is so eccentric, sometimes the separation can be as little as eighteen degrees even at greatest elongations. The **greatest western elongation** happens before sunrise, so to see it you will have to haul yourself out of bed at an ungodly hour (devoted sky-watchers do this a lot).

Its proximity to the Sun means it's most often viewed at or near the horizon (which means more air to look through), while its tight orbit around the Sun means that most of the time the planet is lost in the Sun's glare. When it is visible in the sky, it is visible only for a brief period. In short, you'll need to make an appointment to see this planet.

Mercury's **separation** from the Sun is more or less along the **ecliptic** ("more or less" because Mercury's orbit can take it up to seven degrees above or below that celestial line), so your ability to view the planet, even at a greatest elongation, will depend on your latitude on Earth, and the time of year. For example, if you were in the equatorial city of Nairobi on May 26, 2010, Mercury (then at greatest eastern elongation) would be almost directly straight up from the Sun in the sky, well above the horizon. Whereas someone in the northern city of Oslo would see Mercury off to the side of the Sun and substantially lower on the horizon, making the planet more difficult to view. So, in general, the more northerly or southerly you are, the lower Mercury will appear in the sky. This is true of all planets, since they all follow the ecliptic, but you'll notice this more with Mercury because it's so close to the Sun.

When you do find Mercury, you'll realize that it's very small; about ten arcseconds in diameter at most (in comparison, the full Moon takes up 1800 arcseconds). Its size combined with its generally poor viewing-angle means that even when you train your telescope on Mercury, there's not all that much to see. On the other hand, there are professional astronomers who have never viewed Mercury through a telescope at all, so it is definitely worth aiming at.

Best viewing times

If you want to put Mercury on your observing list, catching the planet at its **greatest elongation** is likely to be your best bet. At each of these dates, depending on which greatest elongation you are viewing, Mercury will rise or set within two hours of the Sun. Consult the chart opposite for dates of greatest elongation over the next few years so you'll know when to look.

If you're looking for Mercury during an **eastern elongation**, note where the Sun sets, and the path of the Sun prior to setting. About a half-hour after the Sun sets, Mercury should become visible along the path of the Sun. If you're looking for Mercury during a **western elongation**, you must get up early to observe the rising Sun and to plot its path into the sky. Since Mercury follows the same path as the Sun but precedes it, this will give you the approximate route of Mercury (and the Sun) for the following morning. (Of course, because Mercury precedes the Sun, you will have to get up even earlier.)

If you're looking in the right place, you'll easily be able to view Mercury with the naked eye during its greatest elongation. Mercury's apparent magnitude is about 0, which means it outshines all but the brightest five stars in the sky, but you won't see much in the way of detail. **Binoculars** are useful for helping you locate the planet at dusk and dawn, and a **telescope** with a 75mm or larger aperture should enable you to catch all of the planet's phases. With a larger home telescope more of the planet's disc becomes visible, but the claims of some amateur astronomers that they are able to make out surface detail should be viewed with scepticism.

Using filters

Finding Mercury can be made easier with **filters** that polarize light (eliminating glare) and with orange or red filters. One popular Wratten filter colour for viewing Mercury is the #25 red, which helps with observations at twilight.

Advanced sky-watchers keen on observing Mercury will search for it during the day, using a combination of filters to block out the Sun's rays. They're also on the lookout for "transits", during which Mercury crosses directly across the disc of the Sun and can be viewed as a silhouette on the blazing surface. Beginners should probably stick with viewing the planet in the evenings or early mornings.

Elongations for Mercury

When's the best time to view Mercury? Here's a table with the greatest eastern and western elongations up to 2012. Remember that eastern elongations are in the evening, and western elongations in the morning.

Elongation	Date	Separation	Magnitude
2008			
Eastern	May 14	21.8˚E	+0.6
Western	July 1	21.8˚W	+0.6
Eastern	September 11	26.9˚E	+0.4
Western	October 22	20.209˚W	-0.3
2009			
Eastern	January 4	19.3˚E	-0.3
Western	February 13	26.1˚W	+0.2
Eastern	April 26	20.4˚E	+0.4
Western	June 13	23.5˚W	+0.7
Eastern	August 24	27.4˚W	+0.5
Western	October 6	17.9˚W	-0.3
Eastern	December 18	20.3˚E	-0.3
2010			
Western	January 27	24.8˚E	+0.1
Eastern	April 8	19.3˚E	+0.3
Western	May 26	25.1˚W	+0.7
Eastern	August 7	27.4˚W	+0.6
Western	September 19	17.9˚W	-0.1
Eastern	December 1	21.5˚E	-0.2
2011			
Western	January 9	23.3˚W	-0.0
Eastern	March 23	18.6˚E	+0.1
Western	May 7	26.6˚E	+0.7
Eastern	July 20	18.1˚W	+0.7
Western	September 3	24.248˚E	+0.0
Eastern	November 14	22.7˚E	-0.0
Western	December 23	21.8˚W	-0.2
2012			
Eastern	March 5	18.2˚E	-0.1
Western	April 18	27.5˚W	+0.6
Eastern	July 1	25.7˚E	+0.7
Western	August 16	18.7˚W	+0.2
Eastern	October 26	24.1˚E	+0.1
Western	December 4	20.6˚W	-0.3

Regarding Venus

Venus shares some viewing similarities with Mercury because it is relatively close to the Sun, limiting its time in the sky to evenings and early mornings. But unlike Mercury, Venus is never difficult to spot because, for one thing, it is visible in the sky over much longer periods and the amount of time it spends hidden in the glaring skirts of the Sun is relatively small. With the exception of the Sun and the Moon, there is no brighter astronomical object in the sky, and with all the sunlight bouncing off its highly reflective clouds, Venus's apparent magnitude ranges from -3.7 to -4.5. That's bright enough for it to be visible, on occasions, in broad daylight, and to be able to cast shadows on clear, dark, moonless nights. UFO reports inevitably go up when Venus is at maximum brilliance.

This brilliance makes Venus easy to spot with the naked eye, even when it is not at a greatest elongation. The planet is actually brightest shortly before its inferior conjunction when it's in a crescent phase. Viewing with the naked eye will not reveal phases, but a good pair of binoculars should show them easily enough. Binoculars will also let you find Venus in the daytime. An easy way to do this is to track Venus's path in the early morning sky, before the Sun has risen, and then plot the path forwards. Later in the day, aim your binoculars to the area you believe Venus will be in, always remembering not to look anywhere near the Sun.

While Venus is a rewarding site both with the naked eye and binoculars, it can be tricky with **telescopes**, if for no other reason than that its sheer brightness can cause your telescope to scatter its light in odd ways, making for loss of detail. Many sky-watchers combat this through the use of **filters**: a Wratten #38A blue filter can help to bring out low-contrast features on Venus' disk, while a Wratten #25 red can help you to spot irregularities on Venus' terminator (the dividing line between night and day on the planet). A polarizing filter can also reduce glare.

Venusian transits

As with Mercury, Venus is occasionally spotted making a **transit** across the globe of the Sun. These are very rare indeed – there was a transit of Venus on June 8, 2004, but the last one before that occurred in 1882. In fact, it is currently an exceptional period for Venusian transits: there will be another on June 6, 2012. This will be best viewed from eastern Asia or Australia. If you choose to view it, of course remember to set up your telescope with the appropriate filters, since the action takes place inside the disc of the Sun. If you miss it, then that's it – until the year 2117.

Clouds and phases

What will you see on Venus? In terms of surface features, you won't see anything at all, because Venus's planetary surface is completely obscured by **clouds**. There's never a break in the clouds, so when you're looking at Venus, you're looking at cloud tops – all the time.

Be that as it may, there are still things to look out for. One interesting exercise is to watch the **phases of Venus** as it progresses in orbit, and to observe how dramatically it changes in size. Near superior conjunction, when Venus is furthest away from Earth, the planet is only ten arcseconds across (as small as Mercury at its largest) and viewable in "gibbous" phase – that is, almost full. Near inferior conjunction, the planet grows nearly five times larger and can be seen as a thin crescent, a particularly impressive sight.

Terminator shadows and "ashen light"

If you are using a filter to bring out contrasts, you may well be able to observe variations in Venus's cloud cover; and if you view the terminator, you may get to see **terminator shadows** – dark spots which get lighter the closer you get to the planetary edge. These are best seen when the planet is at half phase (at or near greatest elongation). When Venus is at greatest

Elongations for Venus

Venus is at its greatest elongation far less frequently than Mercury, and for the novice, the need to catch this brightly visible planet at this point is less urgent. Still, viewing Venus at or near greatest elongation offers a good combination of observational factors. Here are Venus's greatest elongations up to 2016. Remember that eastern elongations are in the evening, and western elongations in the morning.

Elongation	Date	Separation
Eastern	January 14, 2009	47.471°E
Western	June 5, 2009	45.458°W
Eastern	August 20, 2010	46.460°E
Western	January 8, 2011	47.0°W
Eastern	March 27, 2012	46.0°E
Western	August 15, 2012	45.8°W
Eastern	November 1, 2013	47.1°E
Western	March 22, 2014	46.6°W
Eastern	June 6, 2015	45.4°E
Western	October 26, 2016	46.4°W

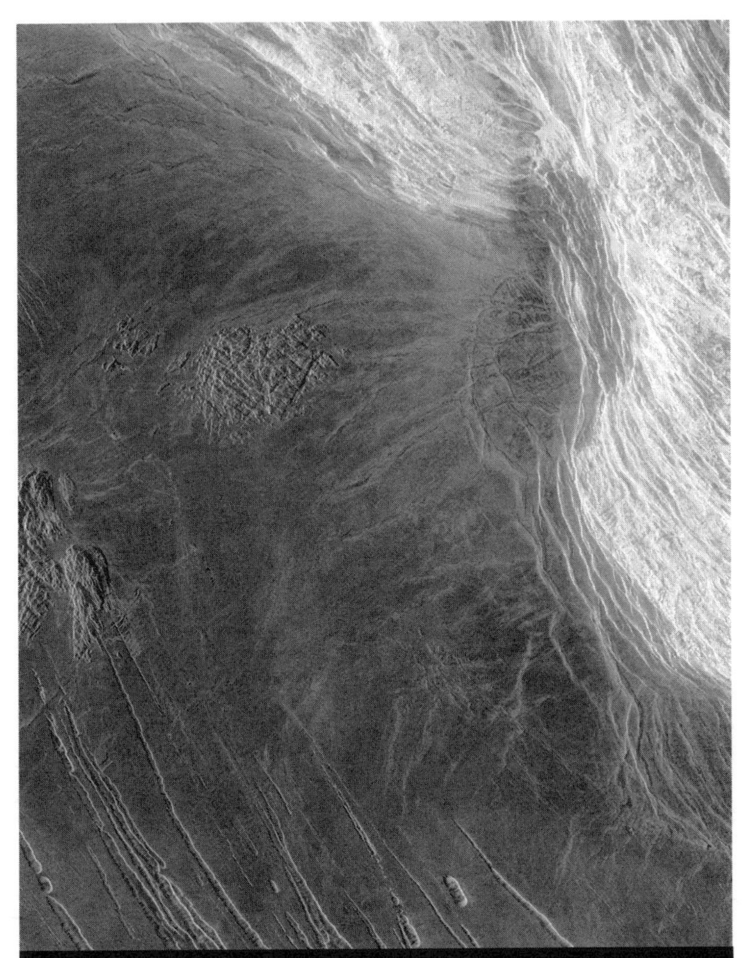

Maxwell Montes, the highest mountain on Venus, rising above the adjacent plains in Lakshmi Planum.
NASA

elongation and moving toward inferior conjunction (or vice versa), it's worth looking out for **ashen light**, that mysterious phenomenon in which reflected light gives Venus's unilluminated face a faint glow. If you do spot it, ask yourself where that reflected light is coming from. After all, Venus has no moon to reflect light back onto it. The amazing answer is that it's sunlight reflected back in Venus's direction from our own planet, Earth.

Mars

the red planet

Mars looms large in our imaginations. A near neighbour, the "red planet" is one onto which men have imposed their own imaginings, whether it's Percival Lowell's canals, Robert Heinlein's deep-thinking Martians, Edgar Rice Burroughs' jungles, H.G. Wells' flu-susceptible invaders, or the alleged "face" supposedly carved into the planet's rock.

As tantalizing as the fantasies about Mars have always been, the science is just as wild – and in many ways even more intriguing. Recent discoveries on Mars have made it more likely than ever that humans will one day cross the distance between the two planets. Before that happens, it's worth taking a look at how it is today.

FACT FILE

Average Distance from Sun 141.6 million miles (228 million km)

Diameter 4222 miles (6794km)

Rotation 24.6 hours

Orbital Period 687 days

Average Temperature -85°F (-65°C)

Axial Tilt 25.2 degrees

Gravity 12.1ft/s^2 (3.7m/s^2)

Moons 2

The real Mars

Most of us assume that Mars is very much like Earth, a little redder perhaps, but overall a place we might easily colonize and build on. In fact, Mars is quite a lot different from our own blue planet – beyond just the colour scheme.

There are certain things our planets do have in common. The **rotational period** of Mars is nearly the same as Earth's – just 37 minutes longer than our own 24-hour day (a Martian day is called a "sol"), while its **axial tilt** of just over 25 degrees is also the closest of any planet to the Earth's own 23.5-degree tilt. Mars also exhibits distinct **seasons**, and can be defined as a "terrestrial" planet, i.e. one with a rocky crust and mantle and a metallic core. It even has polar ice caps.

The planet Mars. The dark area near the centre is Syrtis Major, while the snowy region in the south is the Hellas Basin. Mars' north pole is just visible at the top of the picture.
NASA

How Mars differs from Earth

The first and most obvious difference between the two planets is that Mars is substantially smaller than the Earth, with a **diameter** of only 4222 miles/ 6794km, which is just over half that of Earth's. If ever humans step onto Martian soil, they will feel quite a bit lighter: the **gravity** of Mars is less than forty percent that of Earth's, so a 180lb astronaut would feel as if he was carrying less than 70lb. An astronaut would also need to bring their own air supply, because while Mars does have an **atmosphere**, it's very thin: on the planet's surface the atmosphere has less than one percent of the pressure of the Earth's atmosphere. It's also more than 95 percent **carbon dioxide**,

Why Mars is Red

Mars is famously the "red" planet, but what exactly makes it red? The answer is simple: the entire surface of Mars is rich in iron ores, which have bonded with oxygen to create **iron oxide** or, in everyday language, rust. And it's not just the ground that's rusted: the sky of Mars often turns to a rusty pink because of the dust kicked up by massive dust storms.

with oxygen and other trace gases taking up less than 0.2 percent of total atmosphere. And it's pretty cold, with an average temperature of -85°F.

Mars also differs from Earth in that its **orbit** is highly eccentric – more than five and a half times as much as Earth's. At perihelion – its closest approach to the Sun – Mars is 128 million miles/206 million km away; at aphelion, its furthest point, it's 154 million miles/248 million km.

Any human colonists who make the journey across space to Mars will find that it truly is an alien world, one which won't welcome their presence or make it easy for them to stay and thrive. The only advantage Mars has for humans is that every other planet in the system is even less appropriate for human settlement. As difficult as Mars may be for those who wish to

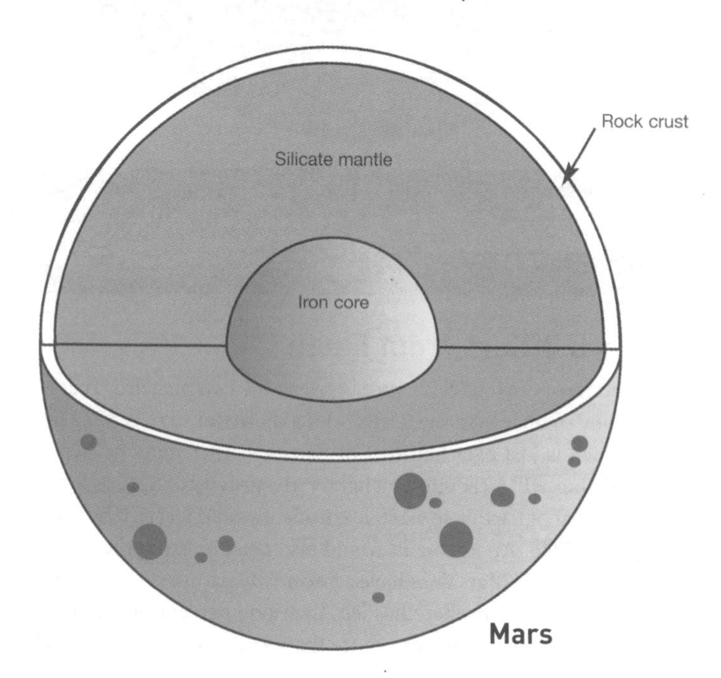

Mars

Water on Mars

For many years, one of the great debates about Mars was whether the planet – which in its warmer past may have held a great abundance of liquid water – had any water left on it at all, either as ice or in liquid form. Thanks to the **Mars Global Surveyor** spacecraft, this is one debate that has been settled. Mars does indeed have water, in the form of ice. So much of it, in fact, that NASA announced in March 2007 that if all the water ice in Mars' southern polar cap were to melt at the same time, the entire planet would be flooded to the depth of eleven metres. In addition to this ice, Mars sports a layer of permafrost which extends far beyond the polar regions. The existence of water ice once again raises the possibility that life might have once existed on Mars – and it also makes manned exploration to Mars more likely. If there is water on Mars that is both accessible and in large amounts, it makes it possible that humans who visit can refuel their spacecraft with hydrogen (and maybe have a drink too).

Water on Mars may also have had a significant effect on the topography of the planet. In June 2002, researchers at the US National Air and Space Museum suggested that liquid water could have created a Grand Canyon–sized **valley** on Mars in just a few short, violent months. The researchers said that a **lake**, with more than five times the volume of all the Great Lakes, overflowed into an impact crater, the wall of which collapsed, sending the water rushing out across a plain, carving a channel 6900ft/2100m deep and over 500 miles/800km long. The valley is still there, known as **Ma'adim Vallis**. The surface water, alas, is all gone.

stay, and as little as it really has in common with our own planet, it's still as Earth-like as we get in our solar system.

It wasn't always like this. There's a great deal of evidence to suggest that Mars had a warmer, wetter past. Billions of years ago, the still molten interior of Mars helped give the planet a thicker atmosphere through the gases spewed out from erupting volcanoes, while liquid water lay in oceans on its surface. When the interior cooled, its atmosphere (only tenuously held by the planet's weak gravity) leached away into space, leaving the thin residue that exists there today.

Mars is roughly divided into two hemispheres, each with its own characteristic topography. The north is dominated by vast, smooth **lava plains**, while the south is made up of rugged, **cratered highlands**.

The canals of Mars

The surface of Mars is more famous for what isn't there, than what is – the so-called **canals**. The idea of canals on Mars grew out of a simple mistranslation. In 1877, Italian astronomer **Giovanni Schiaparelli** gazed at the planet through his telescope and saw what he thought were straight

One of Schiaparelli's maps of the famous (and fictional) Martian canals. It was a case of poor optics; no canals of intelligent design exist on Mars.

lines on the Martian surface. He called these channels (Italian *canali*), which was misunderstood by English-speakers to mean "canals"; that is, waterways created and maintained by some intelligent beings.

The prospect of canals on Mars so intrigued the American astronomer **Percival Lowell** that he devoted a great deal of time to viewing and charting them. He identified over five hundred, crisscrossing the entire surface of the planet. Lowell speculated that the canals moved water from the polar ice caps to farms and other agricultural sites located around the planet – proof, for him, of intelligent and industrious life on Mars.

The problem is that the canals don't exist, either as the intentional work of sentient beings or as the natural end result of some very precise Martian planetary tectonics. In 1965 the **Mariner 4** spacecraft settled the issue when the pictures it sent back showed a planet without them. What Schiaparelli and Lowell had seen were illusions, resulting from the limitations of the human eye and of Earth-based telescopes, and also, perhaps, from the psychological need to see something amazing on the Martian surface.

Mega-volcanoes and valleys

For such a small planet, Mars likes to do things in a big way, and it boasts huge **mountains** and **rifts** that dwarf anything to be seen on Earth. Perhaps the most prominent feature of its surface is **Olympus Mons**, an

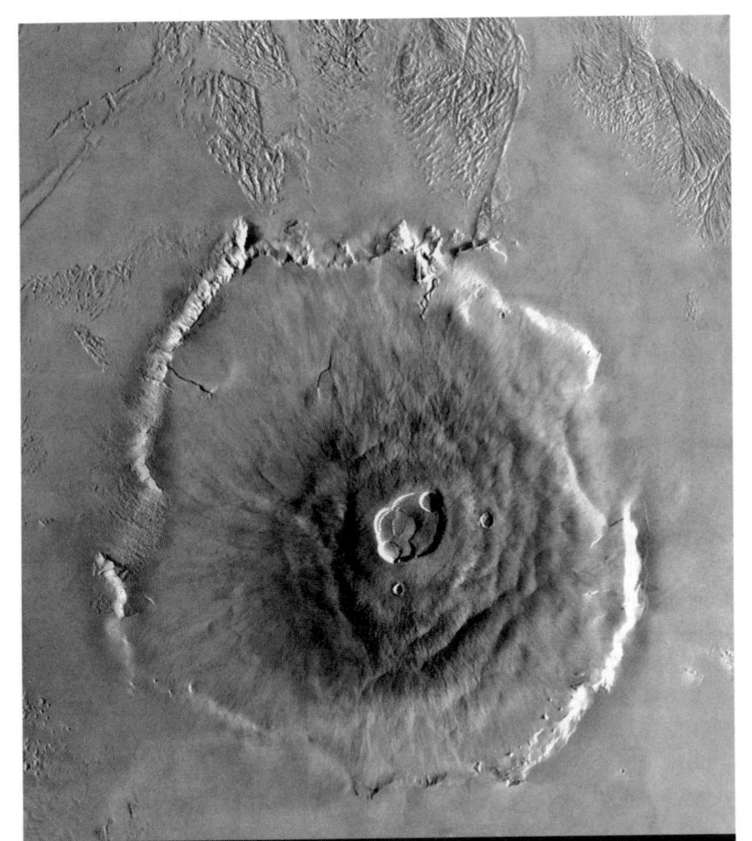

The inactive volcano Olympus Mons is the largest mountain of any planet. With a diameter of over 335 miles/540km and a summit of 16 miles/25km, it towers over the Tharsis plain.
NASA

immense shield volcano 335miles/540km across, whose peak rises over 88,000ft/26,800m from its base. In comparison, the largest volcano on Earth is Hawaii's Mauna Loa, which is 74 miles across and about 32,000 feet high (much of Mauna Loa is located underneath the Pacific Ocean). Olympus Mons is also surrounded by a scarp (basically, a cliff), which is itself up to 32,000ft/9750m high, as tall as Mauna Loa, and some 3000ft/915m taller than Mount Everest.

Olympus Mons rests on the **Tharsis plain**, a vast northern plateau 3000 miles/4000km across, situated over three and a half miles/five and a half kilometres above Mars' mean radius. The Tharsis plain is such a notable bulge on the Martian surface that astronomers suspect that it could have forced a change in Mars' axis, which in turn had an effect on the climate of the entire planet.

There are three other volcanoes on the Tharsis plain, **Arsia Mons**, **Pavonis Mons** and **Ascraeus Mons**, each of which is larger than any volcano on Earth. None of these volcanoes is currently active, although there is evidence of volcanic activity on Olympus Mons as little as ten million years ago. It's possible (though unlikely) one of these Tharsis volcanoes could blow again.

Off to the side and south of the Tharsis plain is another dramatic feature of the Martian surface, the **Valles Marineris** (Valley of the Mariners). This valley is actually a canyon of enormous proportions. Overlay the Valles Marineris on a map of the United States and it would literally stretch from sea to sea. It's some 2800 miles/4500km long and up to 400 miles/640km wide, with an average depth of just under 5 miles/8km. The closest thing we have to it here on Earth is the

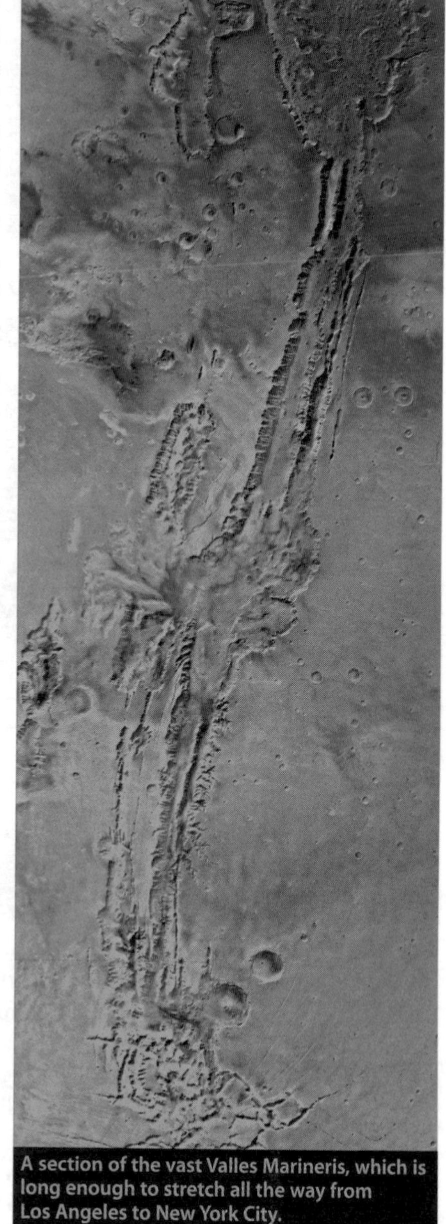

A section of the vast Valles Marineris, which is long enough to stretch all the way from Los Angeles to New York City.
NASA

East African Rift Valley, which is longer than the Valles Marineris (it stretches 4000 miles/640km from Jordan to Mozambique) but is not anywhere near as wide, deep or dramatic.

New types of crater

Craters can be found in the southern hemisphere of Mars, which is more rugged, and older in geological terms than the volcanic plain-filled northern hemisphere. The largest crater, the **Hellas Basin**, is roughly 1300 miles/2100km across and 6 miles/10km deep, and was probably gouged out of the planet's surface by an **asteroid**. Much of the southern hemisphere's height is due to material ejected out of the Hellas Basin by the asteroid impact.

Mars also sports two unusual types of crater – rampart and pedestal. **Rampart craters** are impact craters with unusual ejecta blankets, an ejecta blanket being the area around the crater where the material gouged out of it falls back to the ground. These blankets show evidence that the ejecta was flowing, because they extend further out from the crater than can be explained by pure ballistics. The presence of ground ice, which turns the

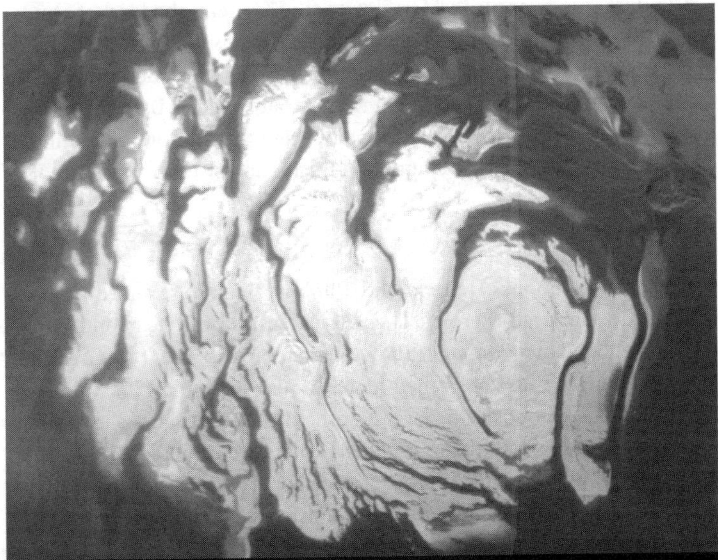

The south pole of Mars, under which was discovered a "whopping large" hydrogen signal, implying the presence of a large store of water, possibly in ice in the Martian soil.
NASA

ejecta into mud, is one explanation for this crater type. **Pedestal craters** are craters that rise up from the surrounding terrain. One explanation for these is that the debris field around the crater protects it from wind erosion. As far as we know, both these crater types are unique to Mars.

Ice caps and atmosphere

Mars has ice caps at each of its poles, the ice consisting of frozen carbon dioxide, as well as water ice, which in recent years has been discovered to be more common than previously assumed (see p.97). The size of these caps varies dramatically with the seasons; the amount of carbon dioxide released as the ice caps sublimate in warmer temperatures has a noticeable impact on the atmospheric pressure, which can increase by up to 25 percent. The **atmosphere** on Mars is thin, but it has enough power in it to create winds that can reach hundreds of miles per hour. Coupled with the low surface gravity, this can create some amazing storms, ranging from twister-like "dust devils" that can tower several miles high, to dust storms which can engulf the entire planet. The largest **wind storms** are created when Mars is closest to the Sun, though smaller can kick up at any time.

The dust clouds occur in Mars' lower **troposphere**, the level of atmosphere closest to the surface. Higher up in the troposphere, isolated clouds of either carbon dioxide or water vapour have been identified, and carbon dioxide clouds can also be found in Mars' stratosphere. **Mists** are also visible around the polar ice caps, especially when the caps are rebuilding in cooler temperatures. These clouds and mists can be seen from Earth.

The moons of Mars

Mars has two moons, **Deimos** and **Phobos**, which are more properly thought of as captured asteroids. Their Greek names mean "fear" and "panic" respectively, and in mythology they were attendants to **Ares** (the Greek god of war) whom the Romans called Mars. At just 16.5 miles/26.5km across its longest axis, Phobos is the largest of the two, while Deimos is a mere 9 miles/14.5km across. Both are irregularly shaped – rather like dark, dusty potatoes – being too small to compact into a sphere under their own gravity.

Each of the moons has an **albedo** (reflectivity) of just six percent, half that of our own moon. Deimos isn't particularly interesting, but Phobos has a six-mile-wide crater, **Stickney**, named after the wife of astronomer Asaph Hall, who discovered both moons in 1877. Stickney is so large that the impact that caused the crater almost blew the moon apart; this is

About face

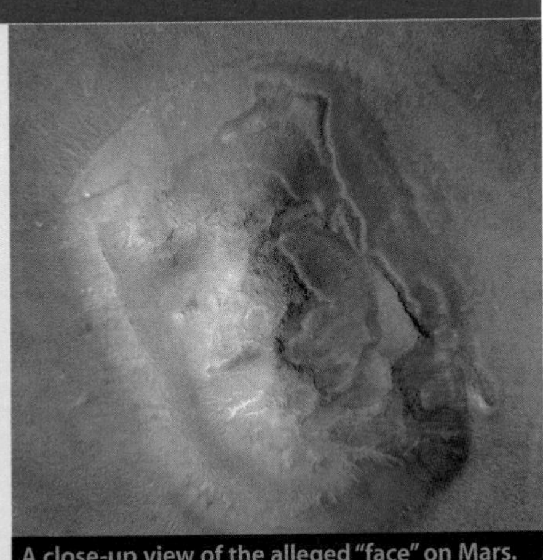
A close-up view of the alleged "face" on Mars.
NASA

Contrary to popular belief, there is no "face" on Mars. The "face" in question – a two-mile-long plateau as photographed by the Viking spacecraft in 1976 – is merely a trick of the light, exaggerated by the less than perfect resolving power of the spacecraft's cameras. When the Mars Surveyor spacecraft took new pictures of the area in 2001, it looked like what it really is – a big pile of rocks. But if you are really keen to see a "face" on Mars, check out the crater called **Galle**, whose natural features look rather disturbingly like a "happy face".

The southern hemisphere also sports another interesting geological feature: long, meandering valleys that resemble river-cut valleys. These may be leftovers from the warmer, wetter period of Mars' past, when running water did exist on the surface. An alternative explanation is that a "sapping" process, caused by emerging groundwater (rather than raging rivers) weakened the terrain. These, and not Schiaparelli's and Lowell's fantasies (see p.99), are the real "channels" of Mars.

shown by the network of grooves, hundreds of feet wide and dozens deep, that are a prominent feature on the rest of the moon's surface.

It's likely that Deimos and Phobos were both **wayward asteroids** that came too close to Mars and were captured by the planet's gravity. Deimos orbits Mars from a distance of 14,500 miles/23,330km, and makes an orbit once every thirty hours. Phobos, on the other hand, is snuggled in quite a bit closer – just 3700 miles/5950km from the surface of Mars – and races along its orbit, making one complete round in just over seven and a half hours. Phobos is so close to the surface of Mars that it can't actually be seen from Mars' polar latitudes (it's always below the horizon there), and moves so quickly that it seems to race backwards in the Martian sky, rising in the west and setting in the east. Phobos' orbit will ultimately lead to its destruction – it's so close that it's being pulled in by Mars at the rate of about two yards per century. Eventually Mars' gravity will pull Phobos in completely or tear it apart. This means that in about

Mars' inner satellite, Phobos, whose name translates as "fear". Crater Stickney, Phobos' major geographic feature, is visible at top left. Phobos is slowly being pulled in by Mars' gravity and will eventually break up or impact into the planet.
NASA

Is there life on Mars?

As much as we like to imagine there's life on Mars, to date there is no firm evidence that it exists there or even ever existed billions of years ago. The closest we've come to confirming evidence is a **meteorite**, located in **Antarctica** in 1984, which came from Mars and carried tantalizing suggestions of tiny, microbe-sized life. The meteor, **ALH 84001**, featured squiggles that looked similar to fossilized Earth **bacteria**, as well as carbon and organic molecules. In 1996, NASA suggested that this evidence added up to a possibility of life once existing on Mars. Since then, sceptics have countered by suggesting the "squiggles" were too small to have been bacteria, the "fossils" could have been created chemically rather than organically, and that the organic materials in the rock could have come from Antarctica, where the meteorite had lain for 13,000 years or so. In short, the Mars microbes might really be microbes; then again, they might not... More recently, spacecraft orbiting Mars have found traces of methane and formaldehyde on the planet, compounds which break down quickly in the atmosphere – which means they are relatively fresh and possibly a by-product of life processes. The problem is, these compounds can also be produced by nonorganic means as well. So we're back where we started.

One additional wrinkle to the "life on Mars" debate came in January 2007, when scientists Dirk Schulze-Makuch of Washington State University and Joop Houtkooper of Justus-Liebig University argued that data from the Viking missions in 1970 showed we saw life on Mars – we just didn't recognize it because its life processes rely on hydrogen peroxide, outside the scope of what the Viking missions were searching for. This hypothesis, however, is yet to be proven.

fifty million years Mars will have just the one moon, Deimos, but it also means the planet is likely to sport a ring made of the debris from Phobos.

Observing Mars

The good news about **observing Mars** is that when the planet is close to Earth, it can be a marvellous sight in your telescope. The bad news is that Mars isn't often all that close to Earth.

The reason for this lies in the planet's **orbit**, which, as mentioned before, is rather eccentric compared to the orbit of the Earth or most other planets. At **perihelion**, Mars is quite a bit closer in than at **aphelion**. How close Mars gets to Earth depends on where each is in their respective orbits when they make their closest approach (or **opposition**). If we catch Mars at perihelion, which we do every 15 to 17 years, a mere 35 million miles separates our planets; if we reach opposition when Mars is at aphelion, which happens every 9 to 12 years, Mars is more than 60 million miles/96.5 million km distant. Other oppositions take place at distances between those two.

The difference for the observer between oppositions at perihelion and those at aphelion is striking. At a **perihelion opposition**, the apparent diameter of Mars swells up to 25 arcseconds, which makes it (and its surface features) more visible in larger telescopes. At an **aphelion opposition**, Mars can be just thirteen arcseconds across, a much smaller target for planet-watchers. The average opposition has Mars showing up as seventeen arcseconds across. In 2003, Mars had one of its closest approaches ever – 55.75 million kilometres. The next opposition, in January 2010, will find Mars sitting back some 62 million km from us. The next comparably close opposition will be in July of 2018, when Mars is 57.6 million km away. (Other, more distant oppositions will take place in March 2012, April 2014 and May 2016.)

There are some disadvantages to the perihelion opposition – at least in the northern hemisphere. Because Mars is relatively low in the sky during these oppositions, your observations will be more affected by the state of our atmosphere. Aphelion oppositions find Mars higher in the sky in the North, mitigating some of the problems the additional distance between our planets imposes.

When Mars is not at opposition it can look very small indeed; as small as 3.5 arcseconds across at its most distant from Earth, with its brightness many times dimmer than when it comes close to us. It's almost as if Mars was a planet with a divided personality: sometimes bright and open for

Mars opposition dates

When do the oppositions of Mars occur? Where will you find Mars on those dates? How large will Mars be in the sky? Here are the answers up to 2025. Oppositions in bold are those at perihelion. For a reminder of how right ascension and declination works, see p.17.

Date	Right Ascension	Declination	Diameter
2010 Jan 29, 19:42 UT	08h 54m	+22° 09'	14.0"
2012 Mar 3, 20:10 UT	11h 52m	+10° 16'	14.0"
2014 Apr 8, 21:03 UT	13h 14m	-05° 08'	15.1"
2016 May 22, 11:16 UT	15h 58m	-21° 39'	18.4"
2018 Jul 27, 05:12 UT	20h 33m	-25° 29'	24.1"
2020 Oct 13, 23:25 UT	01h 22m	+05° 26'	22.3"
2022 Dec 8, 05:41 UT	04h 59m	+24° 59'	16.9"

inspection, sometimes dim and closed off to the casual observer.

Depending on where it and we are in our orbits, Mars can be seen by the naked eye as a brightly shining orange-red "star", ranging in magnitude from -2.7 (as bright as Jupiter) to 2 (still bright but not a standout in the sky). If you are watching Mars approach opposition over a series of weeks, you'll definitely notice the planet getting brighter, and then just as rapidly losing brightness as we move away from opposition.

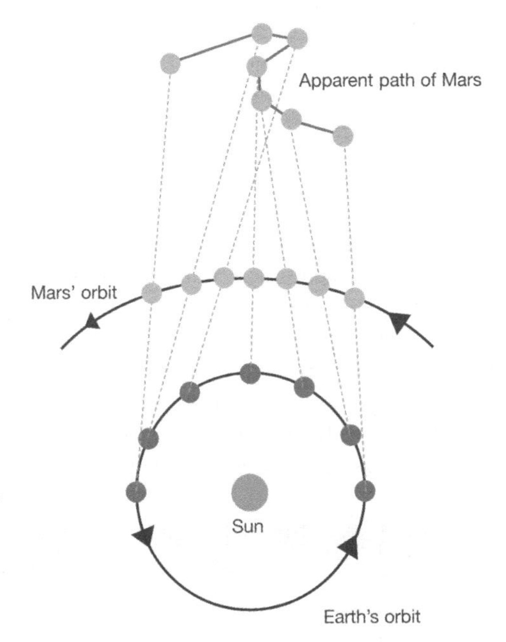

Mars' retrograde motion

Retrograde motion

A time of opposition is also the time to get a glimpse of Mars exhibiting **retrograde motion**. Because Earth's orbit is inside that of Mars, as we catch up with and then pass Mars' position, the planet will appear to move backwards against the backdrop of stars, describing a smart little loop. This retrograde motion begins in the weeks leading up to opposition and continues for a few weeks immediately after. If you track Mars during this time, you'll see it first appear to stop in its eastward-heading direction, before moving back west, only to eventually stop again and head eastwards once more.

Telescope viewing

Mars appears as a disc when seen through binoculars (particularly near opposition), but does not reveal much in the way of detail. To see some features of the Martian surface, you will definitely need a **telescope**.

A small telescope will allow you to make out the polar caps of Mars, but for other features, a larger scope (with a 6- or 8- inch aperture) is necessary. You will also need to view during a clear night with Mars high in the sky.

Polar caps

Among the most noticeable features are the planet's **polar caps**, which are prominent at opposition. Which polar cap you see depends on what type of opposition it is. At oppositions that occur at perihelion, you'll spot the south polar cap, while aphelion oppositions bring the north polar cap into view. Both caps grow and shrink over the course of the Martian year, and this activity is responsible for weather on Mars. This takes the form of **clouds** and **frosts**, some of which amateur sky-watchers can observe at closer oppositions, especially with the use of **coloured filters** to highlight

The top four Martian sights

Hellas Basin This is the big crater gouged out of the southern hemisphere of Mars by an asteroid hit in the distant past. Hellas is big and bright, and it's also a place where you'll see a lot of weather activity. In the winter, Hellas is often completely covered in frost; in the summer, it's a place where dust storms often get started.

Syrtis Major This will be visible as a dark triangular patch a little to the north of the Martian equator, directly north of the Hellas Basin. It's one of the most prominent dark features on Mars, and was also one of the earliest to be observed (the Dutch astronomer Christiaan Huygens drew it in 1659). Syrtis borders another important basin on Mars, **Isidis Planitia**, to the northeast. It's been noted that the features of Syrtis Major change from year to year.

Solis Lacus It means "Lake of the Sun", but it's also known as the "eye of Mars" because its shape reminded astronomers of a pupil. Located in southern mid-latitudes, Solis Lacus has changed considerably over the time it's been observed; it's much smaller now than it was in the 1970s, for example. One of the reasons for this is that the **Solis Planum**, on which Solis Lacus is located, is a common spot for the formation of dust storms. North of Solis Lacus is **Valles Marineris**, which you may see in your telescope as a dark line. You won't be able to discern the actual canyon, however, large though it is.

Tharsis The great bulge of Mars, located north and west of Solis Lacus. While the Tharsis plain is quite prominent, in your telescope it'll seem featureless. Even the great Olympus Mons will not resolve in most amateur instruments. However, you may be able to spot the clouds that form around Olympus Mons, clouds that have been given their own name, **Nix Olympica** – the snows of Olympus.

Mars up to the minute

For an up-to-date guide to what's happening on Mars, turn to the NASA website of choice for all self-respecting Mars-watchers, entitled "Mars Today", at **www-mgcm .arc.nasa.gov**. This creates a poster each day that provides you with six panels of information, including the positions of Mars and Earth in their orbits; how large Mars appears from Earth in terms of arcseconds (and also how large Earth looks from Mars); a snapshot of the Martian surface at noon UT; and even a weather map, featuring temperature and wind speed. There are also links to other Mars sites.

If you want to know how Mars looks at this very minute, check out the NASA Solar System Simulator at **space.jpl.nasa.gov**. Enter "Mars" as the planet you want to see and "Earth" as the place from which you are viewing, and then enter the time (it's keyed to UT, so adjust your time zone accordingly). What then appears is an image of how Mars will look at that time. This is very useful for assisting your own observations, though bear in mind that the simulator is not a live picture, so some surface details may be different and the image in your telescope will be rather less detailed.

cloud features. Violet filters (such as the Wratten #47) will help you identify high-level clouds, while a blue filter (Wratten #80) and a blue-green filter (Wratten #64) will help you to identify clouds at lower levels. For frosts on the ground, try a green filter (Wratten #58).

Dust storms

Another atmospheric phenomenon you may come across is one of the famous Martian **dust storms**. These aren't just local events on Mars; large dust storms can occasionally completely engulf the planet in a haze of particles, completely obscuring the view of the surface. Early warnings of dust storms include yellow clouds, which can be best observed with the use of a yellow filter (Wratten #8). Orange filters like the Wratten #21 will highlight cloud boundaries, so you can see how big a Martian dust storm can get.

There's some irony in the fact that dust storms are more common in the time frame close to an opposition at perihelion. Mars is closer, which affords better observation, but dust in the Martian atmosphere can cause observers on Earth to lose details or see features appearing to be "washed out". Oppositions at aphelion often feature much clearer Martian skies – although cloud-watchers will be happy to know that clouds are more common in the northern hemisphere of Mars during those oppositions.

Blotches

Moving away from ice caps and atmospheric phenomena, observers with larger telescopes will notice that Mars appears to sport **grey-green blotches** on its surface. These are the **highlands** of Mars, which are not actually green but merely look that way in contrast to the more orangey tint of the planet's plains and deserts. Interestingly, as the polar caps melt during the Martian seasons, the highland areas on Mars appear to grow and darken. Astronomers in the past saw this as evidence of plant life on Mars, growing in the Martian springs and summers. The truth is that it's not that the highlands are darkening, but that the deserts and plains surrounding them are brightening as frosts melt away and dust in the atmosphere filters back down to the ground.

The Mars mission that could

Mars has a somewhat notorious reputation for "eating" the spacecraft humans send to it – to the extent that the 2007 science fiction movie *Transformers* "explained" the failure of the European Space Agency's **Beagle 2** lander in 2004: the craft landed safely but then was squashed by a giant alien robot. So it's always a pleasant surprise to have a Mars mission not only fulfil its intended purpose but to go above and beyond it. Such is the case with NASA's **Mars Exploration Rover** mission.

The missions began in 2003, when the rovers *Spirit* and *Opportunity* were independently launched, both arriving at Mars in January 2004. The missions of the small, wheeled rovers, designed to explore the Martian surface and investigate the planet's geography, were supposed to last just ninety sols each – not a huge amount of time considering the nearly $900 million price-tag of the missions. However, once those ninety sols were done, the rovers were still largely functional, so NASA extended their mission... and extended it... and extended it again. The rovers mission has been extended a fifth time and, should they last long enough, could extend into 2009.

This is not to say the rovers are in perfect condition: they have been plagued with various mechanical issues that have caused the mission handlers to improvise, and being solar powered, they are susceptible to dust storms and other energy-sapping events. The greatest crisis of this kind happened in July 2007, when a massive global dust storm threatened to drop the rovers' energy level below the point of no return. Fortunately, the dust eventually cleared and the rovers went back to work.

Ironically, one of the most famous achievements of the Mars Exploration Rover mission took place not on the ground but in the sky. On the 63rd sol of the mission, the *Spirit* rover lifted its camera to look into the night and snap a picture of the Earth, hovering above the Martian horizon. It was the first picture of Earth taken from the surface of another planet. And it caught you blinking.

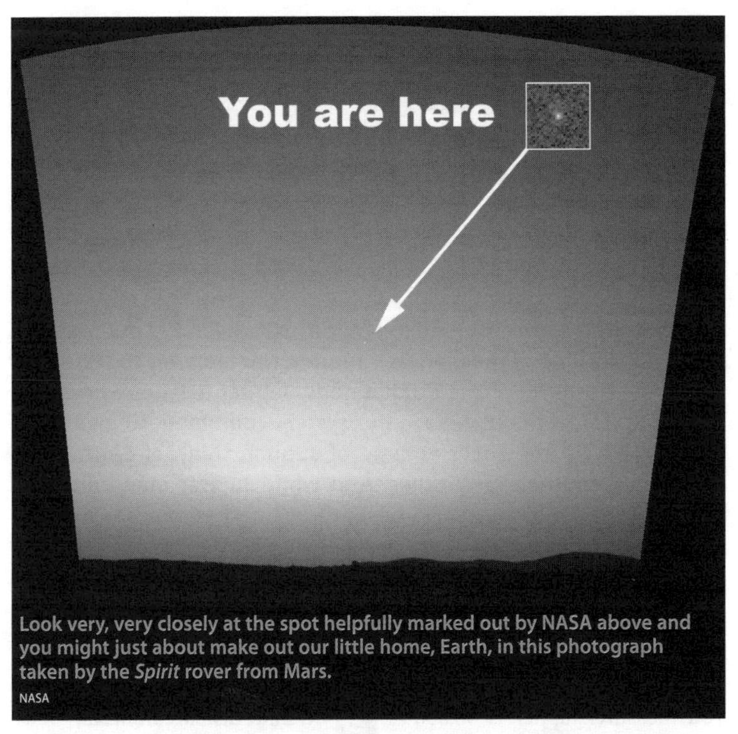

Look very, very closely at the spot helpfully marked out by NASA above and you might just about make out our little home, Earth, in this photograph taken by the *Spirit* rover from Mars.
NASA

Spotting specific features on the face of Mars is something that requires patience, and a willingness to stick to observing the planet for a reasonable period of time in order to become familiar with its features and seasons. Many committed amateur Mars-watchers find it useful to sketch what they see in their telescopes in order to help them remember what they've seen and to help them locate fine details they may have missed in other observations. Whatever your level of interest, casual or committed, Mars is a highly rewarding planet to view.

Jupiter

king of the planets

Four hundred eighty four million miles/780 million km out from the Sun lives the king of the planets – Jupiter. Named after the principal and most powerful of the Roman gods, Jupiter is twice as massive as all the rest of the planets in our system combined. To give some perspective on the sheer volume of Jupiter, think of Earth. Then imagine it another 1300 times. And while Jupiter may only be a quarter as dense as Earth, overall it's three hundred times as massive as our own planet. If you were to set eleven Earths side by side, they still wouldn't equal Jupiter's equatorial radius. Even Jupiter's storms can be larger than some planets, most famously the Great Red Spot which is over twice the size of Earth.

FACT FILE

Average distance from Sun 483.8 million miles (778.6 million km)

Diameter 88,846 miles (142,984km)

Rotation 9.9 hours

Orbital Period 11.85 years

Average Temperature -166°F (-110°C)

Axial Tilt 3.1 degrees

Gravity 75.9ft/s^2 (23.1m/s^2)

Moons 63

Size, however, depends on where you measure. Jupiter is 88,850 miles/143,000km in diameter at the **equator**, but only 84,200 miles/135,500km in diameter at the **poles**. That's a difference of 4650 miles/7500km, greater than the diameter of Jupiter's nearest planetary neighbour, Mars. Why is Jupiter so flattened at the poles? Because of its **rotation**. Immense as Jupiter is, the entire planet rotates in just under ten hours – over twice as fast as the Earth. At the planet's equator, the **rotational speed** is a blisteringly fast 28,000mph/45,000kph, a speed that causes the planet's equator to bulge. If Jupiter appears a little oval, do not adjust your telescope. It's just that way naturally.

A composite of four photographs of Jupiter taken by the Cassini spacecraft on December 7, 2000. The shadow of the moon Europa is the black dot in the lower left.
NASA

Jovian atmosphere

Jupiter's rapid rotation has also created distinct and long-running weather patterns in the planet's **outer atmosphere**, giving Jupiter its distinctive **banded appearance**. These bands are made up of **belts** and **zones.** Belts are regions where air is descending, and are dark in appearance, while zones feature rising cloud masses and are visibly lighter in appearance. **Wind speeds** on Jupiter are fast, and can be as high as 250mph/400kph at the equator. The air on Jupiter is not like that of Earth: almost ninety percent is **hydrogen**, with the rest mostly **helium** along with traces of methane, ammonia, sulphur and water. These trace elements contribute to the appearance of the planet's **clouds**, giving the atmosphere its opaque, mottled and constantly shifting appearance.

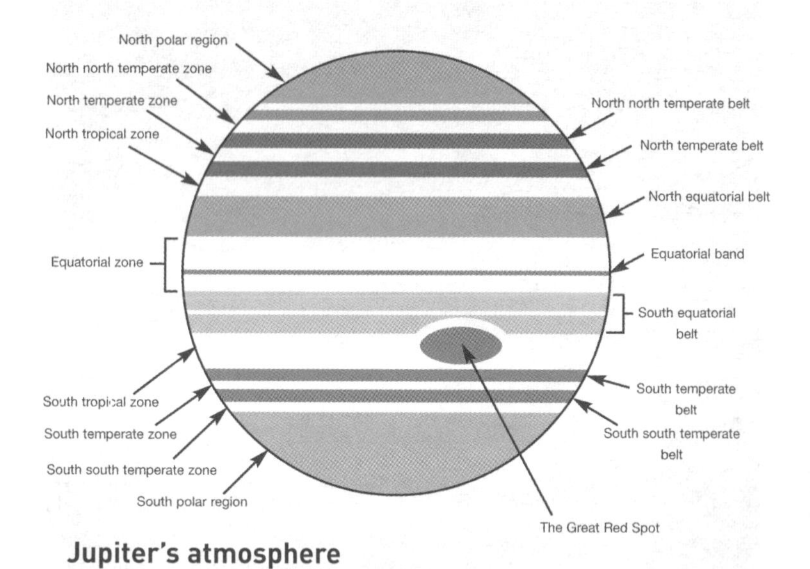

Jupiter's atmosphere

Although we know what **Jupiter's atmosphere** looks like from the outside, what would it look like if we were standing on the planet's surface looking up? It's a trick question, because the fact is that Jupiter doesn't have a surface, at least in the sense that Earth or other terrestrial planets have a surface of rock or ice which can be stood on. What it does have is the previously mentioned hydrogen (and other gases), slowly compressed by greater and greater pressure the further you get towards Jupiter's heart.

Inside Jupiter

Our current understanding is that Jupiter's atmosphere is made up of three decks of clouds. The highest of these consists of wispy, cirrus-like clouds of **ammonia ice**. Below this is a second deck of **ammonium hydrosulphide** clouds, the result of water vapour combining with sulphur. It is these orangish clouds that give Jupiter most of its colour. Jupiter's final deck of clouds is made up of **water**, in the form of ice or droplets.

Pressure builds the deeper you get into Jupiter until you reach liquid in the form of a vast global ocean, made up of tremendously compressed, hot hydrogen and helium. Further down, as the pressure continues to

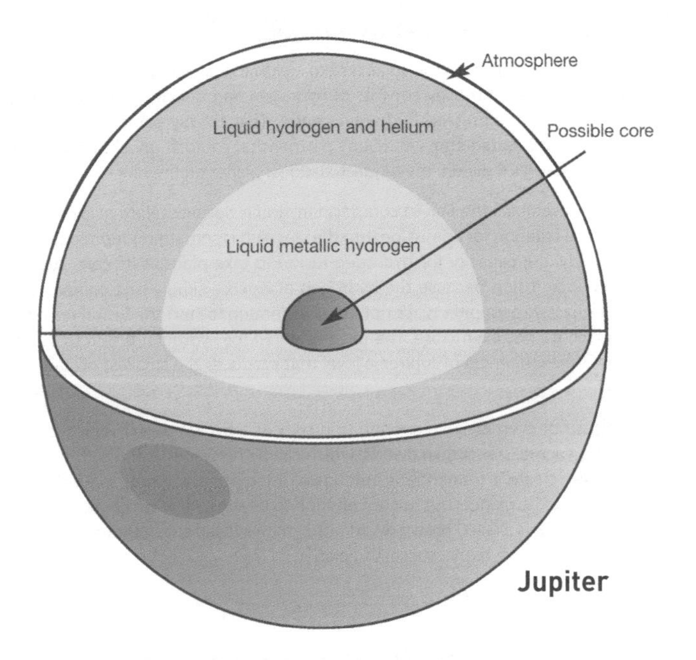

build, the already immensely compressed hydrogen compresses still further, changing its state to become **liquid metallic hydrogen**. The pressure here is staggering: one hundred million times the atmospheric pressure at Earth's sea level. At the very **core** of Jupiter there may finally be something solid. Scientists speculate about a core of heavier metals roughly ten times as massive as the Earth but – because of the pressure – compressed to a diameter only twice the size of that of our planet. It's only speculation, however, since clearly there's no way of going down to check.

The heat generated from all this crushing pressure is immense: at Jupiter's **core**, the temperature is a toasty 50,000°F/30,000°C – substantially hotter than the surface of the Sun. This heat moves outwards through convection currents, fueling the immense storms – such as the **Great Red Spot** – which roil across Jupiter's atmosphere. It's so much heat that Jupiter actually radiates out seventy percent more energy than it receives from the Sun. Interestingly, the energy Jupiter is radiating away is causing it to shrink over time – but as the shrinkage is on the order of two centimetres a year, it's not something the casual observer would notice (unless he or she had lots of time).

Failed star?

In relation to other objects in the solar system – with the exception of the Sun itself – Jupiter is huge. But it mostly consists of hydrogen and helium, generating heat down to a highly compressed hydrogen centre. That is why some astronomers dismiss Jupiter as a **failed star** – the Sun's partner that couldn't go all the way. But Jupiter isn't a star, it's a planet, the biggest, baddest planet on the block.

True, Jupiter resembles the Sun in composition, and it does generate its own heat. But the critical criterion for admission into the star club is something Jupiter doesn't have – namely, the capacity for hydrogen fusion to take place at its core. Jupiter doesn't produce fusion because the hydrogen at its core simply isn't compressed enough; as massive as Jupiter is, it's not massive enough to start the fusion reaction going. In fact, it's not even close. The mass required for hydrogen fusion to begin is about 84 times the mass of Jupiter, a level that produces the smallest of stars – a **red dwarf**.

But Jupiter is not even massive enough to qualify as a **brown dwarf**, which is one step below an actual star. Brown dwarfs, which can be between 10 and 83 times the size of Jupiter, create a tremendous amount of internal energy but no fusion. So, they're too hot to be planets but not hot enough to be stars. All of which means that, if anything, Jupiter is a failed brown dwarf rather than a failed star. Although by most people's accounts, it's a highly successful planet.

Jupiter's magnetic field and other wonders

Jupiter's liquid metallic hydrogen also generates a **magnetic field**, just as the Earth's molten, metallic core creates ours. This being Jupiter, however, everything is bigger and better. Jupiter's magnetic field is huge – fifty times larger than Earth's – and is swept by the Sun's solar wind past the orbit of Saturn, hundreds of millions of miles/kilometres into space. Enough **radiation** is trapped in Jupiter's magnetic field to kill any unshielded human who might have the unlikely misfortune of wandering into its inner portions.

Other Jovian wonders include **lightning**, which is a hundred times brighter than that of Earth's (it was first observed on Jupiter's night side by the Voyager spacecraft and more recently during the day by the Galileo spacecraft). Then there is the planet's **gravity**, which is two and a half times stronger than Earth's, and its **aurorae**, which are a thousand times more powerful. Jupiter's aurorae are caused by ionized particles from one of its volcanically active moons, **Io**, being funnelled to Jupiter's poles by the planet's magnetic field.

The Galilean moons

The first person to witness the moons of Jupiter was the renowned Italian astronomer **Galileo Galilei**, who in January 1610 trained his telescope on the planet and was astonished to find four bright star-like objects orbiting the planet. Actually, someone else – a certain **Simon Marius** – claimed to have viewed the moons in 1609, but (irrespective of the validity of Marius's claim) Galileo published first and so got all the credit. Galileo called the moons the Medicean planets as a way of currying favour with the enormously influential Medici family; and it worked, since after the discovery, Galileo got a job as court mathematician for the grand duke of Tuscany, Cosimo II de' Medici. Nowadays, in honour of the great astronomer, the four largest moons of Jupiter are known as the **Galilean moons**. Although, ironically their individual names – **Io**, **Europa**, **Ganymede** and **Callisto** – were provided by none other than Simon Marius.

In the four hundred years since Galileo made his discovery, observers, scientists and visiting spacecraft have identified many more Jovian moons, and the total now stands at a whopping 63, most of which have been discovered since the turn of the century. Jupiter's moons range in size from very large (Ganymede is larger than the planet Mercury) to very small (S/2003 J12 being a measly 1km in diameter). The larger ones have names, but most of the recently discovered small moons are provided with designations only, simply noting the year and order in which they were discovered: S/2003 J1 through to J23, for example.

For astronomers, the Galilean moons are still tremendously exciting: two in particular, Io and Europa, are the focus of considerable interest. **Io** for its hyperactive volcanic activity, and **Europa** for the tantalizing possibility that it may have the conditions necessary for life.

Io

Io, the closest of the Galilean moons to Jupiter, is roughly the same size as our own moon (about 2256 miles/3630 km in diameter), but that's about all the two have in common. When **Voyager 1** passed by Io in March 1979, NASA scientists were surprised to discover the moon doing what no other moon had been known to do – erupting. Not just from one volcano, either, but from nine separate volcanoes. When **Voyager 2** sped by several months later, a new volcano was erupting. So Io is not just **volcanic**, it's the most volcanic object we know of in the solar system.

SIDEREVS
NVNCIVS
MAGNA, LONGEQVE ADMIRABILIA
Spectacula pandens, suspiciendaque proponens
vnicuique, præsertim verò
PHILOSOPHIS, atq; ASTRONOMIS, quæ à
GALILEO GALILEO
PATRITIO FLORENTINO
Patauini Gymnasij Publico Mathematico
PERSPICILLI
Nuper à se reperti beneficio sunt obseruata in LVNÆ FACIE, FIXIS INNVMERIS, LACTEO CIRCVLO, STELLIS NEBVLOSIS,
Apprime verò in
QVATVOR PLANETIS
Circa IOVIS Stellam disparibus interuallis, atque periodis, celeritate mirabili circumuolutis; quos, nemini in hanc vsque diem cognitos, nouissimè Author depræhendit primus; atque
MEDICEA SIDERA
NVNCVPANDOS DECREVIT.

R. 2. 27

VENETIIS, Apud Thomam Baglionum. M DC X.
Superiorum Permissu, & Priuilegio.

The title page of Sidereus Nuncius ("The Starry Messenger"), which reported Galileo's discoveries in the night sky, including the first four moons of Jupiter.

Master and Fellows of Trinity College Cambridge

Io's volcanic activity causes it's entire surface to be resurfaced every few thousand years; material ejected from Io into orbit funnels towards Jupiter and contributes to that planet's gigantic auroral displays.
NASA

Io erupts because it has a **molten interior**, and it has a molten interior because it is being tugged and pulled by Jupiter on one side and by Europa and Ganymede on the other. This causes immense **tidal forces** that can stretch Io by up to 300ft/90m. Jupiter's tidal pull is self-explanatory, but the reason Europa and Ganymede exert such force on Io is that the three moons are locked in a 4:2:1 orbital resonance. What this means in effect is that for every four times that Io orbits Jupiter, Europa orbits it twice and Ganymede orbits it just once. And when all three moons line up, there's a tug. (Io orbits Jupiter once every 1.76 days.)

The moon's intense **volcanic activity** shows on its surface, which looks rather like a pizza dropped from a great height. Scientists attribute these great swaths of red, orange and yellow coating to various **sulphur**

Naming the moons

Simon Marius thought hard about the naming of the first four Jovian moons, finally adopting an idea, suggested to him by fellow astronomer Johannes Kepler, of identifying them with mythological characters who had some relationship (usually sexual) with Jupiter, the king of the gods. Here's a rundown on some of them.

▶ **Io** A priestess of Hera (Jupiter's wife) who was seduced by Jupiter in the form of a cloud. Hera then turned her into a cow and set the hundred-eyed Argus to watch over her. She was eventually returned to human form and bore Jupiter a child, Epaphus.

▶ **Europa** Another of Jupiter's girlfriends; this time it was Jupiter who took on bovine form, transforming himself into a bull and carrying her across the sea to Crete, where she bore him three sons.

▶ **Ganymede** A beautiful boy carried off by Jupiter, in the form of an eagle, to be his cupbearer. It's Ganymede who is represented in the constellation **Aquarius**.

▶ **Callisto** A companion to Diana, goddess of the hunt; Jupiter seduced Callisto and she bore him a son, Arcas. In revenge, Hera turned her into a bear. When the adult Arcas almost killed her, Jupiter transformed her into the constellation **Ursa Major**, and Arcas into the constellation **Boötes**.

▶ **Metis** Jupiter's lover and the goddess of prudence. Jupiter swallowed her during her pregnancy, afraid their child would be more powerful than he; a month later, Metis gave birth to Minerva, who sprang forth from Jupiter's forehead.

▶ **Leda** The wife of King Tyndarus of Sparta; Jupiter had sex with her by turning himself into a swan. Her children included Castor and Pollux, the twins of the constellation **Gemini**. Pollux was fathered by Jupiter, Castor by Tyndarus.

▶ **Amalthea** The goat that suckled the baby Jupiter after he had been placed in a cave to protect him from his cannabalistic father Cronos. He turned one of her horns into the cornucopia.

▶ **Thebe** Daughter of the river god Asopos; Zeus carried her off to the town of Thebes which was subsequently named after her.

compounds undergoing phase changes. There's so much volcanic activity that Io is entirely resurfaced every few thousand years. Craters are less part of the landscape than volcanoes, lakes of molten sulphur, and long viscous lava flows. It also has a very thin atmosphere, mostly of sulphur dioxide. Surface change is so rapid that there are distinct differences in surface features from between the time the Galileo probe photographed the moon in 1999 and when the New Horizons spacecraft, en route to Pluto, photographed it in 2007.

Io's orbit places it within Jupiter's magnetic field, creating a large enough **current** (an estimated one trillion watts) for particles to be stripped from

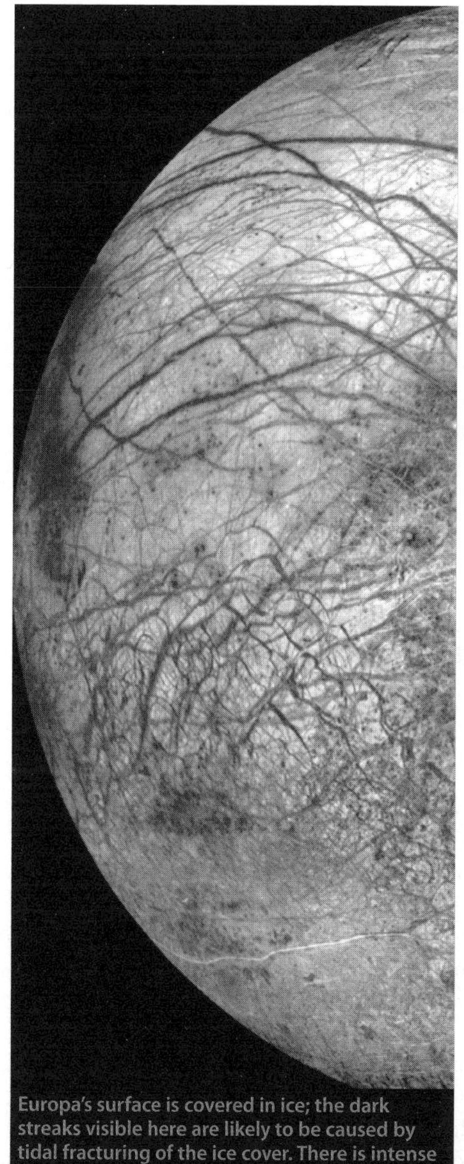

Europa's surface is covered in ice; the dark streaks visible here are likely to be caused by tidal fracturing of the ice cover. There is intense speculation that a sea of liquid water lies underneath Europa's ice pack, and with it the possibility of life in some form.
NASA

it. These ionized particles cause **aurorae** on Jupiter and form a torus of intense **radiation** around the planet. All of this, added to the volcanoes and molten sulphur, make Io a highly unwelcoming place (if it were ever possible to visit). Think Hell, orbiting round a ball of gas.

Europa

Europa is slightly smaller than Io at 1950 miles/3140km in diameter, but it is similar in that it doesn't offer much of the typical lunar landscape of craters and plains – only three craters larger than three miles wide have been identified. By planetary standards, Europa is utterly **flat**. There's not a single feature on the planet surface more than about a thousand feet/300m high. If God were to play billiards, Europa would be his cue ball.

What Europa does have is **ice**, and plenty of it. Ice entirely covers the surface of the moon and provides it with what topography it does have: vast, smooth plains; knobby, fractured areas called **chaos**; wedge-shaped bands; and high (for Europa) ridges, formed through tidal stretching and heating. Also visible on the Europan surface are dark streaks, some several miles wide and hundreds of miles long. Some astronomers link these with geysers and eruptions, or with fractures in Europa's ice cover.

Europa's icy cover is interesting, but what's more interesting to astronomers (and everyone else) is the possibility that underneath the ice is **water** – liquid water, protected from Europa's -260°F/-162°C surface temperature and vacuum by the ice (Europa has only the most tenuous of atmospheres). Since water is where life on Earth got its start, it would seem logical that if life exists elsewhere in the solar system, its best chance would be somewhere watery.

What's the evidence for liquid water? Well, the composition of the **chaos areas** on Europa's surface suggests the area could have thawed and refrozen at some point in time. Basically, it looks like sea ice here on Earth. There are also the recent measurements of Europa's magnetic fields, which exhibit erratic characteristics consistent with sea water. But there's nothing conclusive in terms of liquid water – much less of life – just yet. That said, NASA has taken steps to protect what microscopic life might possibly exist there. One of the reasons the Galileo spacecraft was made to burn up in Jupiter's atmosphere at the end of its mission in 2003 was to avoid the possibility of contaminating Europa with any Earth bacteria that might have hitched a ride.

Ganymede

At a diameter of 3260 miles/5250km, **Ganymede** is the **largest satellite** in the solar system, larger than the planet Mercury, although not as dense. Where Mercury is made up of iron and rock, Ganymede is iron, rock and ice mostly in distinct layers. Despite the profusion of ice, there's no evidence of liquid water. Ganymede, like Io and Europa, also has the thinnest of atmospheres and its own magnetic field is likely to be generated from a molten metal core.

The moon's appearance is of blotchy dark areas and streaky lighter ones. The darker areas are cratered and very old; the lighter ones are not so old, and feature **craters** but also **grooves** and **ridges** probably of tectonic origin. Ganymede's craters are flatter than the craters found on Mercury or on our Moon, probably because Ganymede's icy crust has flowed over millions of years. The oldest craters aren't even really craters any more; rather, they're flattened areas where craters used to be, called **palimpsests**.

Callisto

The final Galilean moon, **Callisto**, is the second largest after Ganymede at just less than 3000 miles/4830km in diameter. Like Ganymede, Callisto is comprised of rock, iron and ice, but unlike Ganymede, these components are jumbled up instead of in distinct layers, which makes it compositionally less complicated. The puzzle for astronomers is how these two otherwise similar moons could be so radically different in this respect.

A further difference from the other Galilean moons is that Callisto is heavily and dramatically **cratered**. Its largest crater, **Valhalla**, is over 1800 miles/2900km across and features a multi-ringed basin. Another dramatic feature is **Gipul Catena**, a series of impact craters strung in a link like pearls in a necklace. This was probably created by an object torn apart by Jupiter's gravity, which then slammed into Callisto's surface.

Not forgetting the ring!

It's not much when you compare it to Saturn's, but Jupiter has a **ring system** that was discovered by the **Voyager 1** spacecraft as it zoomed toward the planet in 1979. It includes a **main ring** that is roughly 80,000

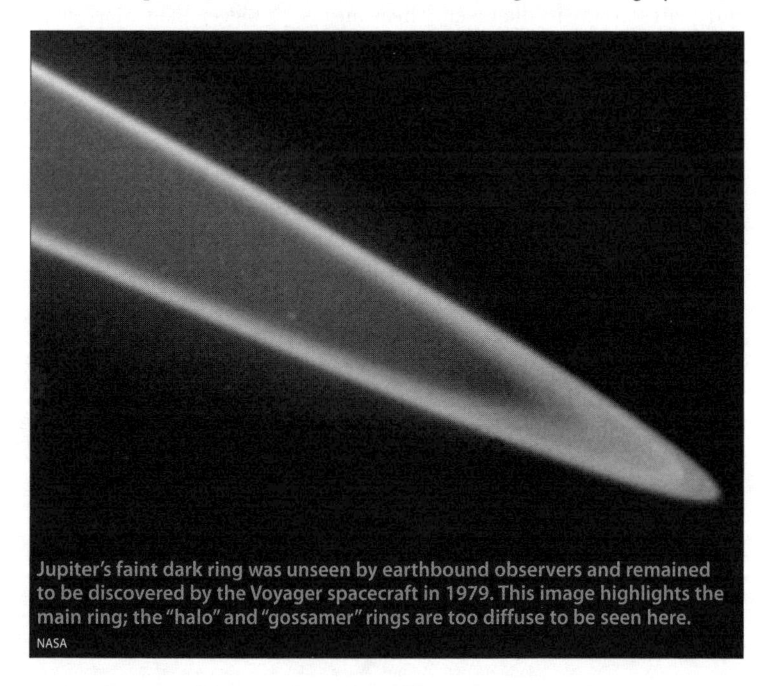

Jupiter's faint dark ring was unseen by earthbound observers and remained to be discovered by the Voyager spacecraft in 1979. This image highlights the main ring; the "halo" and "gossamer" rings are too diffuse to be seen here.
NASA

miles/128,750km from Jupiter and about 4300 miles/6920km wide. Inside of this are two of Jupiter's smaller satellites, **Metis** and **Adrastea**. When these tiny, irregular moons are struck by equally tiny meteorites, the material dislodged may be the origin of the ring itself. These two moons are doomed, because their close orbits mean that Jupiter's gravity will eventually suck them down into the planet. Inside the orbit of the main ring is a faint torus-shaped **halo ring** that extends halfway down to the planet; outside the orbit of the main ring are two very faint **gossamer rings** formed by debris from two other small inner moons, **Amalthea** and **Thebe**.

Viewing Jupiter

Once Venus sets in the evening sky (or alternately, before it rises in the early morning), Jupiter is usually the brightest object in the night sky. Its **magnitude** ranges from -1.3 to -2.7, with an average magnitude of -2.3 at opposition. Since the planet's orbit is outside our own, it's visible all night long most months of the year and is therefore easy to spot in the sky.

Jupiter is also a gratifying object to view through a **telescope**, primarily because it is so large relative to other planets; even at its smallest, it is thirty arcseconds in diameter, which means it's bigger than Mars at its largest. When Jupiter is at its largest, it's fifty arcseconds, bigger even than Venus when it's at its closest to Earth. Jupiter moves through the night sky at a stately pace: one orbit takes 11.86 years, which means (roughly) that the planet moves through one Zodiac sign in a year.

If you want to see Jupiter in **detail**, you'll want at least a 4-inch (10cm) **refractor** or a 6-inch (15cm) **reflector** telescope. With a telescope of this size or larger, you'll start to be able to make out Jupiter's features. Remember, of course, that what you'll be seeing is not Jupiter's "surface" but its **atmosphere**. The good news is that its atmosphere is easily one of the most interesting things you'll view in your telescope: constantly on the move and changing, and with no two views of it ever the same.

Since Jupiter **rotates** in less than **ten hours**, you could have the opportunity to view quite a bit of it in just one night. Many amateur astronomers like to make sketches of the planet they're observing; if you're making a sketch of Jupiter's face, you'll have to work fast, since even ten minutes is more than enough time for features to change and shift position.

Belts, zones and what's inside them

The first things you should be able to resolve are Jupiter's north and south **equatorial belts**, dark bands that girdle the planet's equatorial region, and the lighter **zones** (see p.114). As your eye settles in, other belts to the north and south will begin to show themselves, and you may begin to see details within the belts and zones themselves. These details are given names according to whether they are light or dark phenomena. Some dark features you might see are **festoons** (a dark filament across a zone), **projections** (an area that extends out from a belt), **columns** (a vertical projection jutting out from a belt into a zone) and **disturbances** (large, well-defined dark areas). Lighter features include **ovals** (well-defined bright areas), smaller versions of the same (called **nodules**), **bays** and **notches** (semicircular indentations on belt edges) and **rifts** (bright streaks within a belt).

Viewing these features may take some patience, but keep at it. Using **filters** on your telescope can help to increase contrast on the planet, which will make it easier to spot details. Try a **light-blue Wratten #80A filter** to increase the contrast between belts and zones; a **yellow filter** can help you bring out festoons and columns.

The Great Red Spot

Jupiter's biggest feature is the **Great Red Spot**, which can be found at the southern edge of the **south equatorial belt**. It's the oldest feature seen on Jupiter, first identified in 1664. Despite the name, it's important to note that the Great Red Spot's colour fluctuates: it can be noticeably red, but it can also become paler and harder to see. Remember also that it's not always visible from Earth – sometimes it's on the side of the planet away from us. There are other long-lived storms visible in the belt south of the Great Red Spot, including one – cheekily described as "The Little Red Spot" – created in the merger of three previous storm systems. Little Red and Big Red swoop close by each other every couple of years, but to date the two haven't merged or collided. It would be something for observers to look forward to if they did.

How does the Red Spot last?

In the final analysis, Jupiter's **Great Red Spot** is nothing more than a storm; a vast storm (in fact, a huge, high-pressure anti-cyclone), but a storm nonetheless. Down here on Earth, even the mightiest storms eventually come to an end, so how is it that the Great Red Spot has managed to exist for at least three and a half centuries?

For a start, Jupiter lacks one feature that disrupts storms on Earth – land. Storms break up over land because they're unable to draw additional energy; the Great Red Spot doesn't have that deterrent, since below Jupiter's atmosphere there's a vast sphere of liquid hydrogen. Then there's the fact that the Great Red Spot is tucked in between the south equatorial belt's southern edge and the southern tropical zone's northern edge. Because the winds in these areas flow in different directions – westerly in the north, easterly in the south – this adds energy to the Great Red Spot (which rotates in an anti-clockwise direction every six days); it also isolates the system, keeping it going for a long period of time.

The Great Red Spot may one day finally dissipate – but don't hold your breath waiting for it to happen.

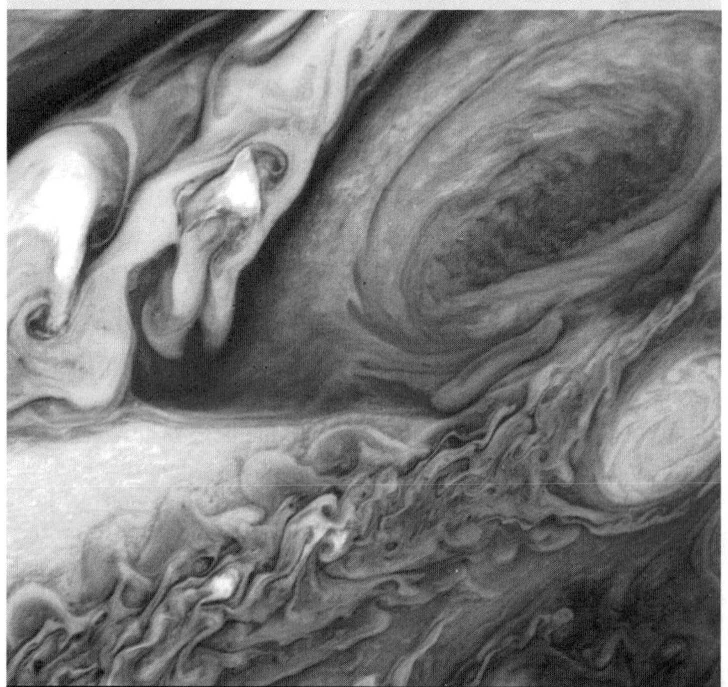

Jupiter's Great Red Spot, a massive, long-standing storm located in the upper right of this picture and accompanied by several other, relatively smaller storms.

Viewing the Galilean moons

While Jupiter is easily visible to the naked eye, its **Galilean moons** need a little more resolving power to be seen clearly. A small telescope (or even a good pair of binoculars) held very steady can help you see them, strung along Jupiter's waistline.

Transits, eclipses and occultations

An exercise worth trying with a small telescope is to map the movements of **Io**, **Europa**, **Ganymede** and **Callisto** over the course of a month. What you'll see is Io zooming around in its orbit (it makes a complete trip around Jupiter in under two days), and the other moons proceeding at their own steady paces.

The moons' dance around Jupiter presents some highly dramatic viewing events. **Transits** occur when a moon crosses in front of Jupiter. Catching such a moment can be difficult, since the moon can often appear to blend in to Jupiter's **atmosphere**. You'll probably have more luck viewing the moon's shadow as it crosses Jupiter's cloud tops – the shadow will be sharp and black and fairly unmistakable.

An **eclipse** occurs when a Jovian moon enters Jupiter's shadow. Observation is possible when the Earth is off to the side of Jupiter, relative to Jupiter's position to the Sun. For the observer, this means that the moon seems to disappear from view, well before it is obscured by the bulk of Jupiter itself. Alternatively, depending on where the Earth is relative to Jupiter, the moon will reappear well after it has moved out from behind the planet. This is fascinating to watch – the eclipsed moon slowly fading out (or fading in) like a ghost. Such eclipses can be partial or total.

Occultation happens when a Jovian moon moves behind Jupiter and then reappears some time later. Catching it as it reappears can also be tricky, this time because of the **glare** of the planet.

With both eclipses and occultation, some times are better for viewing than others. Thanks to the positions of the Earth and Jupiter, much of 2003 was ideal, and 2009 will be as good. You don't have to aim your telescope at Jupiter and hope to randomly catch an eclipse or occultation: a list of the two phenomena can be found on the US Naval Observatory website, **aa.usno.navy.mil**

Saturn

the ringed planet

Whatever else Saturn has going for it, the thing people notice first and foremost are its amazing and awe-inspiring rings. As it happens, there's considerably more to this beautiful planet than its most conspicuous feature. For a start it is nine times larger than Earth yet light enough to float in water; then there are its sixty moons (including one, Titan, with an actual atmosphere), not to mention some of the fastest winds of any planet, including Jupiter. But it's the rings people remember, so that seems like a good place to begin.

The rings of Saturn

FACT FILE

Average Distance from Sun 890.8 million miles (1.43 billion km)

Diameter 74,897 miles (120,535km)

Rotation 10.7 hours

Orbital Period 29.42 years

Average Temperature -220°F (-140°C)

Axial Tilt 26.7 degrees

Gravity 29.4ft/s^2 (9.0m/s^2)

Moons 60

While all the gas giant planets (Jupiter, Saturn, Uranus, Neptune) have **rings** of some kind or another, none of them are displayed as gloriously as Saturn's. Indeed, the discovery of other planets' rings had to wait until late in the twentieth century, and the advent of high-powered telescopes and visiting spacecraft.

Saturn's rings, however, have been known about for centuries. **Galileo** was among the first to see them in the early seventeenth century, though at first he didn't know what he was looking at. He initially thought Saturn was actually three planets close together – a large planet with a smaller one on either side of it. A few years later, when Saturn's position in orbit put the rings' edge within the view of Earth and effectively made them disappear, Galileo wondered if the larger planet hadn't somehow eaten

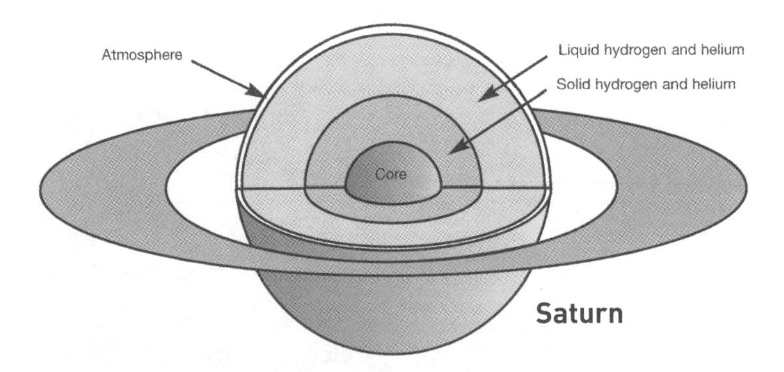

the other two. In 1655, Dutch astronomer **Christiaan Huygens** aimed his (then) technologically advanced telescope towards the planet and deduced that Saturn had a ring.

In fact the planet has several rings. Beginning 4000 miles/6400km above the cloud tops of the planet proper, the rings extend out to 46,000 miles/74,000km for a total diameter of nearly 170,000 miles/273,600km. And yet, what is really interesting is that the entire ring system is less than a thousand feet thick. Clumped all together into one satellite, the material in the rings would amount to very little – perhaps the equivalent of

A glorious view of Saturn and its rings.
NASA

While Saturn's rings are classified into several "primary" rings, close observation reveals that there are in fact thousands of them. Also visible as bright patches are the mysterious "spokes" of the ring system.
NASA

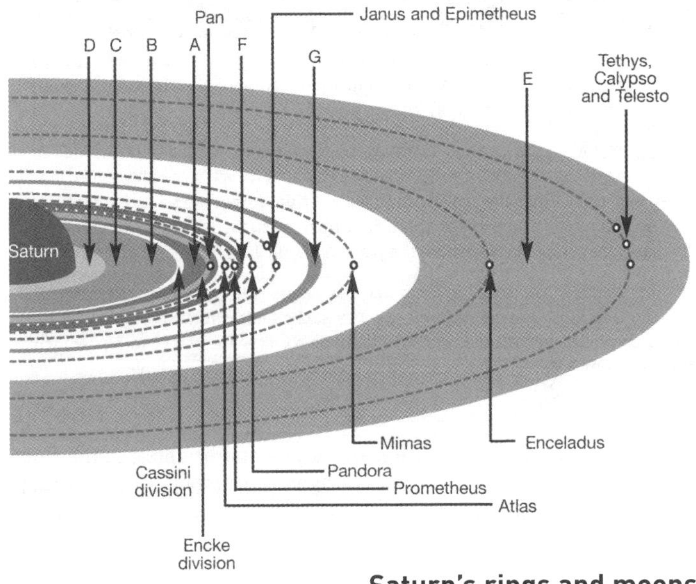

Saturn's rings and moons

Saturn's moon **Mimas**, which is just under 250 miles/400km in diameter. How can the rings be so thin? Well, for one thing, the particles of **rock** and **ice** (mostly ice) they consist of are pretty small, ranging in size from microscopic particles to chunks several yards across. Though there are large objects in the rings (including some interesting satellites), these are few in number compared to the smaller pieces.

From the Earth, the rings look like entire structures, and early observers suggested they might actually be solid. Kepler's laws of planetary motion argued this wasn't the case because the inner portions of the rings would have to rotate faster than the outer ones. It wasn't until 1895 that proof of the rings' particulate nature was provided, when **James Keeler**'s spectral analysis showed variable orbital speeds in different areas of the rings.

Saturn has **five primary rings**, each of which is identified by a letter, although, somewhat confusingly, these are not arranged in alphabetical order. The major rings are – in order from closest to the planet to furthest out – D, C, B, A and F. Two additional rings, G and E, lie even further out but both are faint and tenuous. Rings A, B and C are observable from Earth, although the C ring, sometimes called the "crepe" ring, is darker

Taking it to the (Roche) limit

Saturn's rings are certainly pretty to look at, but why are they there at all rather than a moon or moons? The reason is that any large moon (or moons) in the same position as Saturn's rings are, would simply be torn apart.

The rings reside within an area known as the **Roche limit**, named after the French astronomer who first noted it in 1848. Inside this area (which is roughly equivalent to two and a half times the radius of the planet) **tidal forces** are so strong that any large moon would literally break apart. Saturn's rings, in fact, may be what remains of a moon that originally lay within the planet's Roche limit.

But what about those "**shepherd moons**" that reside within the rings themselves? How can they exist if every moon inside the Roche limit is torn apart? Well, these moons are irregularly shaped and small enough for tidal forces not to have much effect on them. It's only larger moons that have to mind the Roche limit.

and harder to spot than A or B. B is the largest and brightest of the rings and is separated from the A ring by the 2800-mile/4500km **Cassini division**. It is called a division, because to observers on Earth (including Gian Domenico Cassini who discovered it in 1675) it seemed as if there was nothing but empty space separating rings A and B. Other divisions also exist, including the **Encke division**, which bisects the A ring in two. While these divisions look empty from Earth, close-up observation (by the **Voyager spacecraft**) revealed that within these "gaps" are thousands of ringlets.

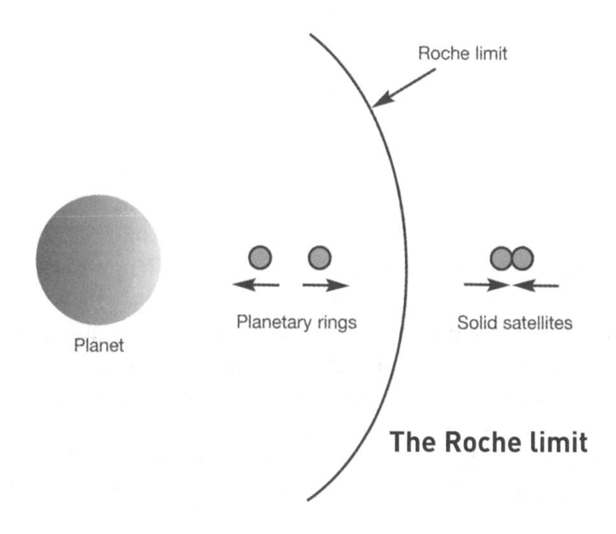

The Roche limit

That's not the only thing Voyager found. Inside the rings are some amazing structures, like the "spokes" that stretch across the entire length of a ring. Given the differences in rotational speeds in varying parts of the rings, these shouldn't exist. The fact that they do is probably the result of electrostatically charged dust above the actual ring structure. Voyager also found that the gravitational influence of two tiny "**shepherd moons**" orbiting on either side of ring F (the outermost and thinnest of the major rings), pushes and pulls at the ring material causing it to "zip" and "unzip" whenever these moons go by. The result is a peculiar "braiding" effect. The recent visit to Saturn by the **Cassini** spacecraft revealed more secrets of the rings, including the Keeler Gap, a small (42 km) division of the A ring, which has its own small moon, Daphnis, orbiting within it to keep it clear. Cassini also found additional ring structures in existing rings, such as the one found within the E ring in September 2006. The mission also discovered two faint additional rings, currently labelled R/2004 S1 and R/2004 S2. Doubtless these rings will acquire more official names (presumably "H" and "I") at some point...

The planet itself

Saturn is the **second largest planet** in terms of both mass and diameter: it's 95 times the mass of Earth, and nearly 75,000 miles/120,700km at the equator (although, like Jupiter, it's measurably flatter at the poles thanks to a less-than-11-hour rotational period). As massive as it is, however, Saturn is not very dense at all, having only an eighth of the density of Earth, and only seventy percent of the density of water. If there were an ocean large enough to set the planets into, Saturn would float while the rest of the planets sank beneath the waves. Its **gravitational pull** is less than Earth's, which means if you could stand on it, you'd weigh about ten percent less than you do on Earth.

As a planet, Saturn shares many qualities with Jupiter. Its composition is largely the same – primarily **hydrogen** and **helium**, with trace amounts of other elements. The planet was also constructed much as Jupiter was, with a **gaseous atmosphere**, followed by a vast globe of **liquid hydrogen** and **helium**, then by a core of **liquid metallic hydrogen**, and finally a (relatively) small **rocky core**. Like Jupiter, Saturn also generates more **energy** than it receives from the Sun; this energy is most likely due to a small but continual contraction of the planet, and by a settling of helium into the inner part of the planet, releasing energy as it sinks. Also like Jupiter, Saturn has a strong **magnetic field**.

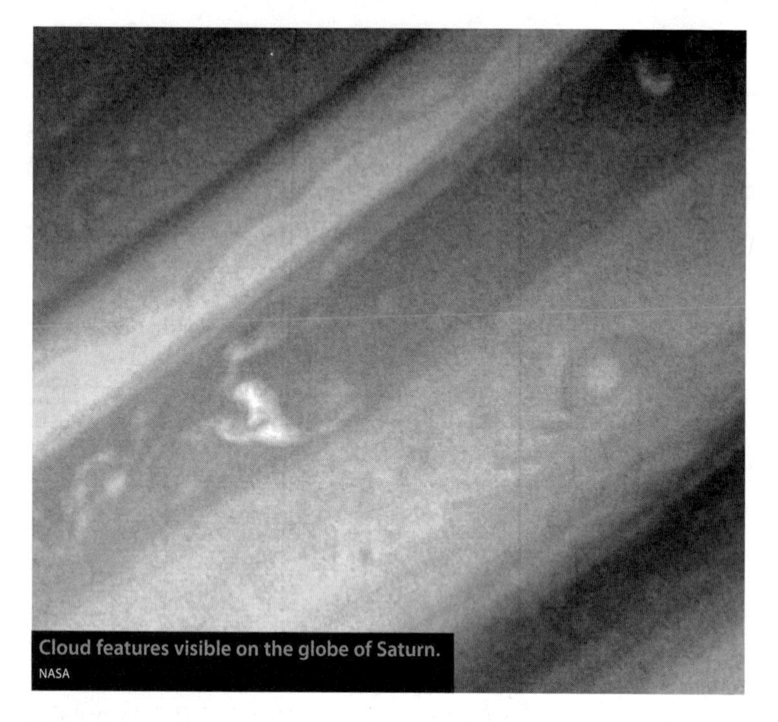

Cloud features visible on the globe of Saturn.
NASA

Winds and storms

What Saturn offers, however, are variations on the theme. Unlike Jupiter, whose prevailing winds alternate up and down its globe in tightly defined zones and belts, most of **Saturn's winds** travel in an easterly direction, except near the poles. Saturn's cooler temperatures (it's about 100°F/55°C cooler than Jupiter) give its clouds a less dramatic aspect than its bigger brother, but what Saturn lacks in cloud complexity, it makes up for in **wind speed**: at the equator, winds can reach 1000 miles/1600km per hour.

Saturn doesn't offer up any permanent, massive storms like Jupiter's Great Red Spot. However, some long-lasting smaller cyclones have been noted, including a spectacular, massive hurricane discovered by the Cassini spacecraft hovering at the planet's south pole. At 8000km across, it's both the largest hurricane humans have ever seen, and the first hurricane-like storm seen outside of our own planet. There's also a notably strange hexagon-like storm at the north pole, first seen by Voyager 1 and spotted again by Cassini. No one knows what forces are at work to cre-

Could Titan sustain life?

Titan has the right atmosphere for life and it also has the organic compounds that are created when ultraviolet radiation hits that atmosphere – so, is it possible that when probes are sent to Titan, we'll find life waiting there?

Almost certainly not, because at -280˚F/-170˚C, Titan is just too cold. It's so cold that any water that exists on it is in the form of ice. Both these factors significantly lessen the chances of life taking hold.

On the other hand, perhaps time will sort out the cold. Billions of years from now, the Sun will swell up as a red giant, making it cooler but, at the same time, larger, thereby increasing the amount of energy Titan receives from it.

Astronomers Christopher McKay of the NASA Ames Research Centre California and Jonathan Lunine and Ralph Lorenz at the University of Arizona have modelled how the Sun's expansion would affect Titan. They suggest that the moon would increase in temperature to about -148˚F/-100˚C, which, while still extremely chilly, would be warm enough to allow for the possibility of life there.

This would not be life as we know it on Earth, since it would need nitrogen, and not carbon, as its fundamental building block. The catch is that Earth's inhabitants are unlikely to be around to see it happen.

ate it, although there's no doubt it's a natural process and not the work of geometry-minded aliens.

From time to time, Saturn puts on a real show with **massive white storms** the size of two Earths, the most recent of which was in 1990. These storms, collectively known as the **Great White Spot**, seem to appear on a rough cycle of once every thirty years, which is in line with the length of Saturn's orbit (which takes 29.5 years). The Great White Spot lasts for about three weeks before it begins to dissipate; start planning now for 2020.

The moons of Saturn

Saturn has **sixty moons**, more than any other planet except Jupiter. But it's not simply a question of quantity: the moons themselves are some of the most unusual and intriguing in the solar system. These include moons that share orbits, a moon with an atmosphere, and one moon that looks disturbingly like the Death Star from *Star Wars*.

Titan

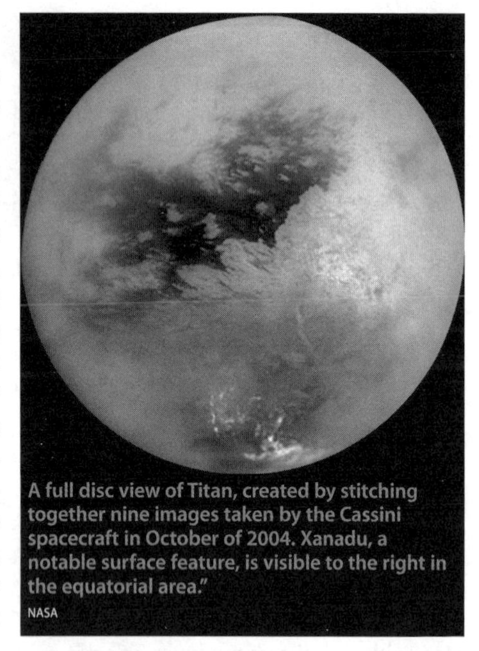
A full disc view of Titan, created by stitching together nine images taken by the Cassini spacecraft in October of 2004. Xanadu, a notable surface feature, is visible to the right in the equatorial area."
NASA

First among Saturn's moons is **Titan**, the second largest moon in the solar system with a diameter of 3200 miles/5150km. Only Jupiter's Ganymede is larger, and only barely so. But Titan has something that Ganymede doesn't (indeed, that no other moon in the system has): a thick, genuine atmosphere, composed primarily of **nitrogen** just as the Earth's is, but with a **surface pressure** sixty percent greater than that of Earth.

But don't imagine that Titan is much like our own planet. For one thing, at -280°F/-170°C, it's very much **colder**. For another, aside from nitrogen, Titan's atmosphere doesn't have much in common with our own. The most common gases – after nitrogen – are **methane**, followed by **hydrogen**, **carbon monoxide** and **carbon dioxide**. This combination of gases, and other trace gases, reacts with ultraviolet radiation from the Sun to produce "**smog**" – organic particles which have turned Titan's atmosphere into a reddish and opaque veil. It was only recently, in fact, that humans penetrated that veil, thanks to the Cassini spacecraft and the landing craft it carried. Named **Huygens**, after Titan's seventeenth-century discoverer, Christiaan Huygens, the lander penetrated the atmosphere of Titan and landed on its surface on January 14, sending scientists ninety minutes' worth of data. Among the things we've learned from the spacecraft is that lakes of liquid hydrocarbon exist at the moon's polar regions – the first time liquid (albeit much colder liquid than our bodies could stand) has been found on the surface of a solar-system world. One of these lakes appears to be about the size of the Caspian Sea.

That said, it does not appear as if there are vast hydrocarbon oceans on the moon, as some scientists had hypothesized. The geology of the moon, however, is interesting, and described by scientists as being "geologically young" – evidence that forces have worked upon them, including, possibly, the liquid on the surface of the moon as well as **cryovolcanoes**, which erupt not with lava but with a mixture of ice and ammonia. There are recognizable surface features, most notably **Xanadu**, an Australia-sized area first discovered in the 1990s by the Hubble Space Telescope and examined in more detail by Cassini.

Other interesting features of Titan include elongated "dunes" similar to ones we find in desert areas here on Earth, and likely formed by more or less the same atmospheric processes; these dunes can reach to up 330 meters in height. And Titan has at least one mountain range, with peaks almost 1.6km high. These mountains are icy and indeed may be entirely made of ice. The speculation is that they may have been formed in the aftermath of a massive impact, which unleashed dormant tectonic forces.

Tethys

Tethys is essentially a huge **ball of ice**, some 650 miles/1046km in diameter. Its most distinctive feature, a huge crack in it's surface called the **Ithaca Chasma**, runs some 1200 miles/1930km – nearly three-quarters of the moon's circumference – with parts of it more than two miles deep and sixty miles wide. As well as this there's a huge impact crater named **Odysseus**.

Tethys shares its orbit with two other moons, **Telesto** and **Calypso**, both of which are less than 20 miles/32km in diameter. These two moons are parked sixty degrees in front and behind Tethys in its orbit.

Tethys, one of the larger of Saturn's moons, is almost entirely made up of ice.
NASA

Dione

Slightly larger than Tethys at just under 700 miles/1125km in diameter, **Dione** exhibits unusual bright streaks on one of its hemispheres. This seems to suggest some sort of **volcanic activity** that has resurfaced that portion of the moon, "volcanic" being a qualified term because – like numerous other Saturn moons – Dione is largely composed of **ice**. Its other hemisphere, on the other hand, is heavily cratered. Dione shares its orbit with another tiny moon, **Helene**, which is about 20 miles/32km in diameter.

Enceladus

This is another ball of **ice**, but an usually shiny one. **Enceladus** has the **highest albedo** – the extent to which an object diffusely reflects light – of any object in the solar system, reflecting back more than ninety percent of the light that shines on it. By comparison, Earth reflects 39 percent of light, and our moon, a mere seven percent. Enceladus is located within the orbit of Dione and shares a 2:1 orbital resonance with it.

Enceladus became rather more interesting to scientists in the wake of the Cassini mission to the Saturn system – partly because, during a flyby in 2005, the spacecraft caught site of a huge lifting plume of water at the moon's south pole. This plume, along with other evidence, suggests that Enceladus is currently geologically active. Enceladus' eruptive personality also strongly suggests that it is largely responsible for Saturn's diffuse "E" ring: material flung off the moon ends up orbiting the planet.

Mimas

Mimas is the one that looks like the Death Star in *Star Wars*. Its appearance is due to an impact crater, named **Herschel**, the largest in the solar system relative to the size of the body it impacted on. Herschel is fully one quarter the diameter of Mimas, which is itself about 250 miles/400km in diameter.

Mimas has a 2:1 orbital resonance with Tethys, and also with the ring particles near the Cassini Division in Saturn's rings, which means it is largely responsible for keeping that gap clear. It is also in orbital resonance with the gap between the C and B rings.

Iapetus

This large moon (about 900 miles/1450km in diameter) is one of the most unusual-looking objects in the entire solar system: one half is bright, reflecting fifty percent of the light that hits it, while the other half reflects as little as three percent, making it darker than soot. A possible explanation is that **Iapetus**'s dark half is coated with material thrown off by another moon, **Phoebe**, which is similarly dark. Another explanation may be that **ultraviolet radiation** interacts with **methane ice** on the moon's surface to create darker organic compounds. However it happens, it makes for quite a sight.

Observing Saturn

Saturn is a breathtaking sight in a telescope, especially when the rings are tilted substantially towards Earth. At between 15 and 21 arcseconds in diameter in the night sky, the planet is also a decent size. Getting a good view of the rings and of features in Saturn's atmosphere does take a little work, however.

Finding Saturn with the **naked eye** is not usually a problem. The planet is visible ten months out of the year and shows up against the night sky, bright and yellow, with a magnitude that varies between 1.5 and 0. Looking at Saturn through a pair of **binoculars** doesn't do much to resolve the rings, but a good pair of binoculars might enable you to find **Titan** (the largest of Saturn's moons) with a magnitude of about 8.5.

Ring viewing

Saturn's rings become visible with a **small telescope**, but larger scopes will, of course, bring out more detail. For example, the **Cassini division** (the mostly empty area that separates rings A and B) is visible through most telescopes that can resolve the rings, but larger scopes (six inches and greater) can also pull out details, such as the irregular appearance of the inside portion of the A ring. **Larger home telescopes** can also show the **Encke division** within the A ring and bring out colour variations in the B ring, the widest of Saturn's rings. Generally speaking, the A and B rings are the onesyou are most likely to see; the C ring can also be observed, but it's faint, and best seen against the background of the planet's disc.

The appearance of Saturn's rings changes as the planet makes its way through its orbit; the rings dip and rise according to the inclination of Saturn in its orbit, relative to our own orbit and the axial tilt of the planet itself. At maximum tilt, Saturn dips 27 degrees relative to the Earth, allowing for truly spectacular views of the rings. When the rings are viewed straight on, as they will be for a brief time in 2009, they actually disappear from view (remember that relative to their width, the rings are razor-thin) before appearing again. Currently, we're viewing the **southern face** of the rings but after 2009, we'll be seeing the **northern face**.

Cloud watching

Saturn's clouds are a little bit harder to see. Although its atmosphere is banded like Jupiter's, neither Saturn's bands nor its cloud colour are as distinctive as those on its bigger neighbour, and if you're interested in seeing the details in Saturn's atmosphere, a filter is in order. A Wratten #80A **light-blue filter** will help up the contrast on the belts and zones, and it may also help you to spot **large storms**, visible as white ovals. Unlike Jupiter's longer-lasting storms, these storms usually only last for a few weeks, so enjoy them while you see them.

Moon spotting

Small telescopes can discern five of the largest of Saturn's sixty moons. These are (in order of distance from Saturn): Tethys, Dione, Rhea, Titan and Iapetus. Reddish **Titan**, with a magnitude of 8.5, is the easiest to spot; **Tethys**, **Dione** and **Rhea** each clock in at around magnitude 10. **Iapetus**, with its unusual half-bright, half-dark surface, ranges in magnitude from 10 to 12. If you watch Saturn over a period of several weeks, you may be able to notice Iapetus fade and lighten.

Uranus, Neptune

and the dwarf planets

For most of the recorded history of civilization, people only knew about six of the planets – Earth, Mercury, Venus, Mars, Jupiter and Saturn. But out there in the dark more planets lurked: the gas giants Uranus and Neptune, and then strange, icy Pluto, a planet so unlike the others that astronomers eventually "demoted" it from planet to a new classification: Dwarf Planet. Their discovery – when it eventually happened – proved to be relatively easy; learning about them was rather more difficult. Even today, these far-off neighbours continue to provoke controversy, particularly Pluto – now the most famous of its new class of astronomical entity.

Discoveries, accidental and otherwise

The accidental discovery of **Uranus** occurred in 1781. English astronomer **William Herschel** was looking for stars (he was trying to record all stars down to the eighth magnitude), when he came across what he described as "a curious either nebulous star or perhaps a comet" exhibiting a disc shape. Eventually he realized it was neither a star nor a comet, but an entire new planet orbiting the Sun. Herschel, who like Galileo before him realized the value of flattering the powerful, proposed calling his find the Georgian Planet after Britain's King George III. The French wanted to call it Herschel, after its discoverer. Eventually the new planet was named – in the tradition of the other planets – with reference to Roman mythology. **Uranus** is the father of Saturn, who in turn is the father of Jupiter.

The 40ft telescope of William Herschel (1738–1822) which he built in the grounds of his home near Slough and with which he discovered Uranus. Only the foundations still survive.
Private collection

After Uranus had been observed for several decades, it became apparent that something else – apart from the Sun and the known planets – was influencing its orbit. There had to be another planet out there. In the 1840s, scientists in France and England brought out their pencils and tried to determine, through physics and mathematics, where this new planet might be located. In 1843, English mathematician **John Couch Adams** began efforts to calculate the planet's position but failed to get the support of the Astronomer Royal, Sir George Airy, and so never verified the planet's existence. Meanwhile, in 1846, **Urbain-Jean-Joseph Le**

Verrier of France made his own attempt to work out the location of the planet, unaware that Adams had already done so.

Neither Adams nor Le Verrier actually spotted **Neptune**, but their data was used by others, and it was finally observed in 1846 by Berlin astronomer **Johann Gottfried Galle** and his assistant **Heinrich Louis d'Arrest**, using Le Verrier's data. Adams' findings, championed by John Herschel (the son of Uranus's discoverer) were handed over to Cambridge astronomer James Challis. Since both Adams and Le Verrier accurately and inde-

John Couch Adams (1819–92), the English astronomer who is credited, along with Le Verrier, with the discovery of the planet Neptune.
Hulton-Deutsch/CORBIS

Could there be other planets?

Since 1930, the planetary retinue of the solar system has appeared to be complete: Mercury, Venus, Earth, Mars, Jupiter, Saturn, Uranus, Neptune and, for several decades at least, Pluto. What, then, are the chances that another planet might still exist to be discovered outside the orbit of far distant Pluto?

If you're thinking about another object like Pluto, the answer is that objects of its size have already been found – and indeed caused the contretemps that resulted in Pluto's "demotion." We'll be discussing this later in the chapter.

As for the chances of an even larger planet being discovered? Well, a few astronomers suspect that a massive planet, the size of Jupiter or possibly larger, is lurking, trillions of miles away from the Sun. The evidence for this large, dark stranger lies – so it is claimed – in the orbits of several comets which suggest that they have been shunted towards the Sun by the same large object.

So there might well be other large planets out there. Just don't expect them to leap out of your telescope.

pendently plotted where the new planet would be, they are both credited with Neptune's discovery.

Even after the discovery of Neptune, astronomers still thought that there was another planet out there, because of what they perceived to be irregularities in the orbits of both Uranus and Neptune. Chief among them was American astronomer **Percival Lowell** – the man who claimed there were canals on Mars (see p.99) – who made more than one attempt to find the ninth planet, but died in 1916 before it was discovered.

However, when **Pluto** was finally discovered in 1930, it happened at the **Lowell Observatory** in Flagstaff, Arizona, on a telescope especially designed for the purpose. The man who found it was **Clyde Tombaugh**, who was an amateur astronomer at the time (proof that you don't need a PhD in order to make astronomical history). The (then) planet was named Pluto, and a craze for the new member of the solar system hit the US, to the extent that Mickey Mouse's dog was named Pluto to mark the occasion.

As it turns out, the supposed irregularities in the orbits of Uranus and Neptune were just that – supposition. As an astronomical object, Pluto is far too small to cause the predicted perturbation. So while astronomers were looking for the planet they thought would be out there, they ended up, quite by chance, finding another object entirely.

Uranus and Neptune – turbulent twins

FACT FILE

URANUS

Average Distance from Sun 1.784 billion miles (2.871 billion km)

Diameter 31,763 miles (51,117km)

Rotation 17.2 hours

Orbital Period 83.75 years

Average Temperature -320°F (-195°C)

Axial Tilt 97.8 degrees

Gravity 28.5ft/s^2 (8.7m/s^2)

Moons 27

FACT FILE

NEPTUNE

Average Distance from Sun 2.793 billion miles (4.494 billion km)

Diameter 30,775 miles (49,495km)

Rotation 16.1 hours

Orbital Period 163.72 years

Average Temperature -330°F (-200°C)

Axial Tilt 28.3 degrees

Gravity 36.0ft/s^2 (11.0m/s^2)

Moons 13

Venus and Earth are often regarded as near twins, but **Uranus** and **Neptune** are the real twin planets of the system. Uranus is the larger of the two with a diameter of about 31,800 miles/51,180km, but Neptune's **diameter** is only 1000 miles/1600km smaller – closer in percentage terms to Uranus than Venus is to Earth. Both have a diameter roughly four times that of Earth's, and half that of Saturn's.

Uranus and Neptune also both have several **moons** and **ring systems**, and their planetary and atmospheric compositions have far more in common with each other than do those of Earth and Venus. Both planets are bone-numbingly **cold** – about -320°F for Uranus and -330°F for Neptune. Despite their similarities, however, it would be a mistake to think of Uranus and Neptune as identical twins, since both possess several highly distinctive features of their own.

First among the unique features of Uranus is the fact that, relative to most of the rest of the solar system, the planet is **perpendicular**, with its axis having a tilt of over 97 degrees. A possible reason for this is that early in the history of the solar system, the planet was hit by a **massive object** (or objects) that knocked it "on its side". On Uranus, the Sun doesn't rise in the east and set in the west (or vice versa, since Uranus's axial rotation is in the opposite direction to Earth's), it slowly **spirals** down the sky, from north to south, as the planet makes its nearly 84-year-long orbit around the Sun. So while **Uranus's rotation** only takes a little over seventeen hours, the time between sunrise and sunset on Uranus can literally be measured in years. This means each pole on Uranus experiences nearly 42 years of sunlight, followed by 42 years of darkness.

If you were lost on Uranus, there would be no point in pulling out the compass to point you in the direction of the pole, because the planet's magnetic poles are 59 degrees apart from its rotational poles, and the centre of the **magnetic field** is out from the physical centre of the planet by about 10,000 miles/16,000km. Uranus's **magnetosphere** is also unusual in that it's generated not at Uranus's core (as with Earth) but from highly pressurized layers between the core and the planet's nominal surface. This factor, along with Uranus's axial tilt, gives the magnetosphere a structure that is literally twisted, a magnetic "tail" that extends behind the planet like a corkscrew.

Uranus's featureless atmosphere suggests the planet is a relatively bland world; but from its nearly horizontal axial orientation to its unusual interior composition, there's a lot going on under the surface.
NASA

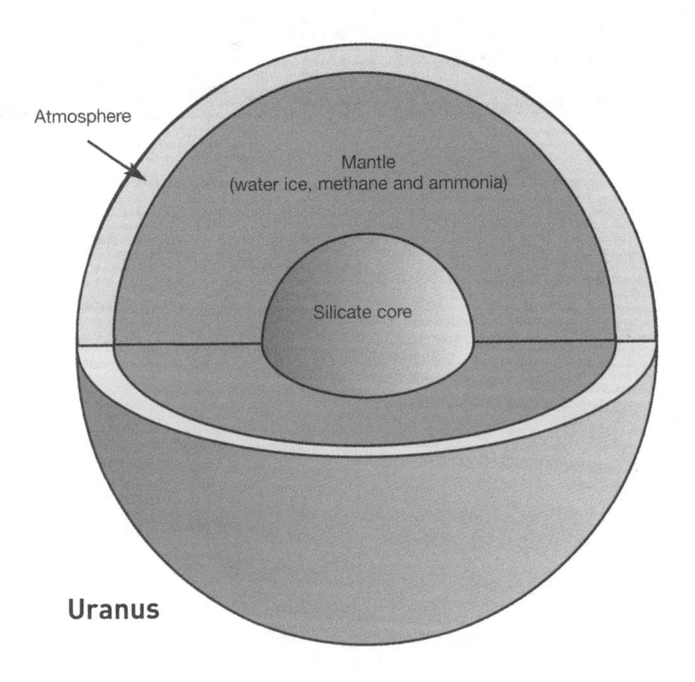

The **atmospheres** of both Uranus and Neptune are a mixture of **hydrogen**, **helium** and **methane**, and their interiors are a slushy mixture of **water**, **methane** and **ammonia**. Uranus's atmosphere is unusually calm for a gas giant – it has the fewest storms of the four planets in this classification – although **wind speeds** can still reach 360 miles/580km per hour. The planet's upper atmosphere dramatically reveals the presence of methane through its sea-green tinge.

Uranus also differs from Neptune, and the other gas giants, in that it produces relatively little heat from its **interior**. All the other gas giants radiate far more energy from their interiors than they get from the Sun; Uranus, on the other hand, puts out only a fraction more **energy** than it takes in. No one quite knows why it is such an underachiever in this respect, although it could be that bombardment during the early years of the solar system's formation caused Uranus's planetary development to happen differently from the other gas giants.

While Uranus is a little twisted (literally), Neptune comes across as positively normal. Its axis is tilted 29 degrees, making it in line (so to speak) with most of the other planets, while its aquatic name is confirmed by its

This photo of Neptune, taken by the Voyager 2 spacecraft, features the Great Dark Spot, an Earth-sized storm which, unlike Jupiter's Great Red Spot, does not appear to be a permanent feature.
NASA

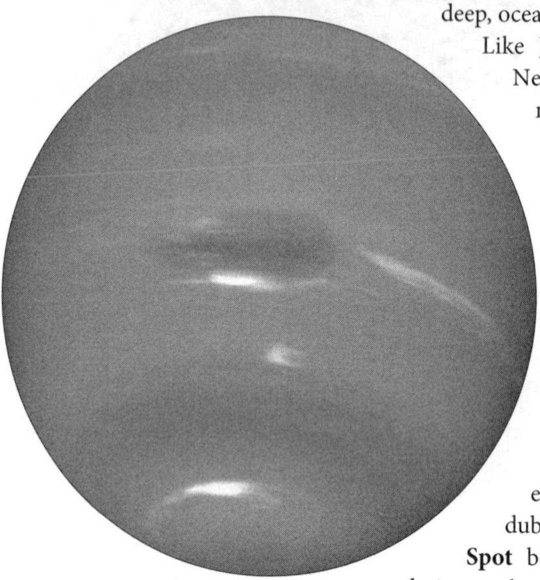

deep, oceanic blue appearance. Like Jupiter and Saturn, Neptune produces more **energy** internally than it takes in from the Sun, and like them has large storm systems. When **Voyager 2** visited the planet in 1989, it took pictures of an immense storm system which scientists immediately dubbed the **Great Dark Spot** because of its similarity to the Great Red Spot on Jupiter (even down to its location in the planet's southern hemisphere).

The wind speeds in the Great Dark Spot measured up to 1500 miles/2400km per hour, the fastest on record, but unlike the Great Red Spot, it faded out over time, and when the Hubble telescope looked at Neptune in 1994, it was gone – replaced by a similar storm in the northern hemisphere. That there were storms on Neptune at all surprised astronomers. Because of its great distance from the Sun, it receives only three percent of the solar energy Jupiter receives, and half that of the much calmer Uranus. The ferocity of Neptune's storms is a testament to the heat generated in Neptune's insides.

Even such a "normal" planet as Neptune still has a few surprises up its sleeve. For one thing, it's the only one of the gas giants on which **clouds** are seen to cast shadows – high, dispersed clouds of **methane ice** floating some fifty miles above the main cloud deck encompassing Neptune's globe. Another oddity is the exceedingly unusual **magnetic field** that is tipped 47 degrees from the rotational axis, and off-centre from the planet's core by more than half the radius of the planet. In other words, the centre of Neptune's magnetic field is actually closer to the planet's cloud tops than it is to the planet's core. Scientists believe the field is being generated not in the core, but in the layers between it and the planet's atmosphere. While Neptune is the smallest of the gas giants, it is also the **densest**, so while Uranus is physically larger than Neptune, Neptune is more massive than its "bigger" twin.

The rings of Uranus

Uranus's rings were discovered in 1977, once again by accident. Astronomers on Earth were observing a star as Uranus passed in front of it, hoping to use information from the event to determine more precisely the size of the planet and the composition of its atmosphere. To their surprise, the star dimmed five times before it passed behind Uranus, and again after Uranus had moved away from the star. There they were – Uranus's rings. Later scientists discovered that there were actually **ten rings** orbiting the planet.

Uranus's rings are not the glittering beauty show Saturn's are; in fact, the most notable thing about them is how hard they are to see at all. They don't spread out, thousands of miles wide, like Saturn's do, and the densest, **Epsilon**, is a mere 60 miles/96km wide. Not only are the rings **thin**, they're also extremely difficult to see because the particles in each ring reflect only one or two percent of the light that shines on them. The rings also exhibit a tendency to "wobble" slightly – a motion scientists believe is a result of interactions of the rings' material with Uranus's moons and with the planet itself. In between the major rings lie thin spreads of **ring material**.

Uranus's rings were discovered by accident when a star passing behind the planet was seen to dim several times, suggesting their existence to observing astronomers.
NASA

The rings of Neptune

The existence of **Neptune's rings** was hinted at in the mid-1980s by the same occultation observations that discovered the rings of Uranus. But verification had to wait until the **Voyager 2** spacecraft passed by Neptune in 1989, and discovered **four main rings**, similar in brightness to those of Uranus. There is something a little unusual about them, however: the outermost ring, known as **Adams**, is described by astronomers as "clumpy". Within the circle of the Adams ring are three substantial arcs ("Liberty", "Equality" and "Fraternity") of additional ring material. The existence of these arcs is puzzling, since over time ring material should even out. This suggests that the Adams ring may be the remnant of a recently destroyed moon. Scientists also believe that the ring's interactions with nearby moon **Galatea** may also keep the arcs from spreading out through the entire ring.

The moons of Uranus

Like other gas planets, both Uranus and Neptune have an array of moons circling their waists. Uranus has 27, most of which are named after Shakespearean characters. **Titania** is the largest, with a diameter of about 1000 miles/1600km, while **Oberon** is only slightly smaller. All the larger moons (Ariel, Umbriel, Titania, Oberon, Miranda) seem to be a mixture of rock and different types of ice. **Miranda** is the most unusual, with a surface so jumbled in its features (huge cliffs, strange grooves, heavy craters) that scientists originally speculated it had been torn apart and then reassembled by its own gravity. An alternative, less dramatic, theory posits Miranda's appearance as due to the welling up of partially melted ice.

Uranus's moons play a role in maintaining the planet's rings. The two innermost moons, Cordelia and Ophelia, neither quite 20 miles/32km in diameter, orbit on either side of Uranus's outermost ring, Epsilon, and act as "shepherds", keeping material from that ring from floating away.

The moons of Neptune

Neptune has the fewest moons of any gas giant – just thirteen, with only three of any real size. The largest, **Triton**, is 1700 miles/2700km in diameter, making it slightly smaller than our own Moon. It's thought that Triton was not formed in tandem with Neptune, but was instead hauled in by the planet – captured by its gravitational fishing-net. Some of the evidence for this includes the fact that Triton is the only large moon of any planet to orbit in a **retrograde direction**. This will eventually be a problem, since

tidal interactions between Triton and Neptune are sucking energy out of the moon and degrading its orbit. At some point in the future it will either be torn apart by Neptune's tidal interactions (giving the planet a rather more impressive set of rings than it has at the moment), or it will simply fall smack into Neptune.

Neptune's moon Triton is thought to be a "captured" moon, a wandering object trapped by the planet's gravitational field – a field that will eventually pull the moon all the way in, destroying it in the process.
NASA

Triton's eventual doom is not the only notable thing about it. At an estimated temperature of -400°F/-240°C, it is estimated that Triton is the coldest object in the solar system – even colder than Pluto. It also has a very thin atmosphere made up of **nitrogen** and **methane**. Triton's strange orbit is matched by its axis of rotation, which is tilted 157 degrees from Neptune's – meaning that the moon, like Uranus, is "on its side" relative to most other moons and planets in the system.

But perhaps the most intriguing thing about Triton is its "**ice volcanoes**" which spew not lava, but huge plumes of methane or liquid nitrogen when the surface of the moon is warmed, even slightly, by the far distant Sun. On its visit to Neptune in 1989, Voyager 2 spotted one of these eruptions, with a plume of material that stretched nearly a hundred miles "downwind". These eruptions have helped to give Triton an unusual landscape of **ridges** and **valleys** and very few large craters of any sort.

Triton isn't the only one of Neptune's captured moons; **Nereid**, Neptune's outermost moon, is thought to be a captured asteroid. With a distance from Neptune ranging from 840,000 to just under six million miles, Nereid possesses the most eccentric orbit of any moon in the system.

Pluto – the first "dwarf planet"

For three quarters of a century, from its discovery in 1930 until the summer of 2006, Pluto was classified as our solar system's ninth planet – a small and strange planet, but a planet nonetheless. But as of the summer of 2006, the International Astronomical Union has seen fit to classify Pluto as a member of a new planetary category altogether: the "Dwarf Planet". What happened to cause this demotion in status?

What happened was **Eris**, a new Pluto-like object discovered floating well beyond Pluto's orbit, not only reaching Pluto's size and mass but exceeding it. If Pluto was a planet, then Eris, being larger, would quite necessarily have to be a planet as well. This was a prospect that alternately caused great joy and consternation to the world's astronomers, to the extent that, at the International Astronomical Union's 2006 meeting, the scientists had to grapple with the fundamental question of what a "planet" really was.

Eventually it was concluded that a planet had three qualities: it must circle the sun; it must collapse into a sphere under its own gravity (a phenomenon called **hydrostatic equilibrium**); and it must have enough mass to clear its surroundings of other astronomical objects. Pluto achieved

FACT FILE

PLUTO

Average Distance from Sun 3.647 billion miles (5.869 billion km)

Diameter 1485 miles (2274km)

Rotation 6.38 days

Orbital Period 248 years

Average Temperature -375°F (-195°C)

Axial Tilt 122.5 degrees

Gravity 1.9ft/s^2 (0.6m/s^2)

Moons 3

two of these qualities, but not the third, and on August 24, 2006, was therefore dropped from the family of planets into the newly created Dwarf Planet category, which it now shares with Eris and Ceres (which is located within the solar system's asteroid belt). As a result of its demotion into the Dwarf Planet category, Pluto has also been saddled with a numerical categorization for astronomical purposes. It is now officially 134340 Pluto.

This controversy regarding Pluto, however, was not new. Even when it was still defined as a planet, Pluto was distinctly unusual. For one thing, it was very small. Its radius is just over 2250km, which makes it smaller than several moons, including Triton (which it closely resembles) and our own. Comprised largely of ice and rock, its composition is different from that of the terrestrial planets, which are made of rock and iron, and of the gas planets, which are primarily hydrogen and helium. In fact, the object Pluto most closely resembles is a comet, even though it is far larger than any comet that has ever been discovered. Beginning in 1992, however, dozens of other icy objects – larger than comets but smaller than Pluto – had been found lurking in space between the orbits of Neptune and Pluto. These are called Trans-Neptunian Objects, or Kupier Belt Objects

Enough Trans-Neptunian Objects have been found that, even before the discovery of Eris, some astronomers were calling into question Pluto's planet status. Indeed, in the early 2000s the scientists at the Rose Centre for Earth and Space at New York's American Museum of Natural History pre-emptively demoted Pluto from the planetary line-up, declaring it "the king of the Kupier Belt comets". Apart from its size and composition, there was the additional matter of the planet's **orbit**. Pluto is traditionally thought of as the most distant planet, with an orbit that averages a distance of 3.647 billion miles/5.869 billion km from the Sun. But Pluto's orbit was actually the most **elliptical** of all the planets; so elliptical that for twenty years out of its 248-year-long orbit, Pluto is actually inside the orbit of Neptune. (This most recently happened between 1979 and 1999.)

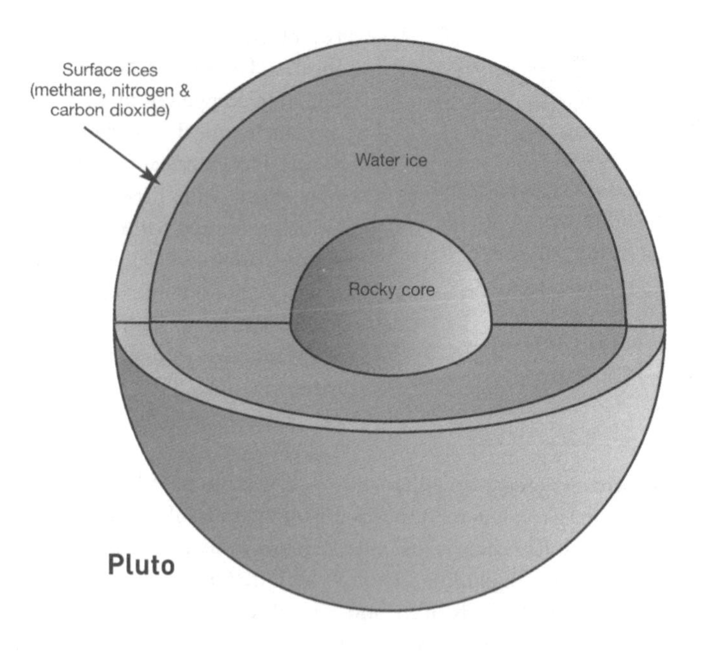

This would seem to set up the chance of a collision between Neptune and tiny Pluto, but it's not going to happen – Neptune and Pluto are locked in a 3:2 orbital resonance. That's to say, Neptune orbits the Sun three times for every two times Pluto does, and that keeps the two of them from ever having any inconvenient close encounters. Pluto's orbit is also steeply inclined from the other planets in the system, with a **seventeen-degree inclination** from the ecliptic, which also helps when it comes to avoiding messy planetary collisions.

Not content to have an odd orbit, size and composition, Pluto is also oddly **tilted**, tipped over from its orbital plane by more than 120 degrees, enough for it to be said that its north pole is pointing south. On Pluto, the Sun rises in the west and sets in the east, and the planet rotates once every six days, ten hours – the third longest rotation period in the solar system after Mercury and Venus. As for **atmosphere**, Pluto does have one, but only just. When closest to the Sun, its atmosphere is thin, probably comprised of **nitrogen**, **methane** and **carbon dioxide**. However, as it recedes from the Sun in its orbit, the surface temperature of the planet drops (the average temperature on Pluto is a brisk -375°F/-226°C) and the atmosphere literally freezes out and drops to the ground.

As mentioned earlier in the chapter, even though astronomers were looking for a ninth planet, Pluto was discovered by accident. One of the reasons it was found at all is that it is fairly **bright**, reflecting thirty percent to fifty percent of what little sunlight it receives, bright enough for astronomers to assume originally that Pluto was Earth-sized (roughly 8000 miles/12875km in diameter). That size estimation shrank over the years, most radically in 1978, when it was discovered that Pluto (tiny though it was) actually had a moon.

Pluto's moon

Pluto's primary moon is called **Charon**, after the boatman who ferried the dead across the River Styx into Pluto's underworld. At just under 800 miles/1280km in diameter, it's more than half the size of the planet it orbits and about twenty percent of its mass. This makes it the single largest moon out there – relative to the size of the planet (Dwarf or otherwise) it orbits. Our own Moon held this distinction before Charon was discovered. It's large enough that while it is **tidally locked** to Pluto (always showing the same face to that planet, just as our own Moon does), Pluto is also tidally locked to Charon. Pluto's rotational period and Charon's orbital period take exactly the same amount of time. Charon hangs motionless in Pluto's sky – well, half the sky, since there's an entire hemisphere of Pluto that never sees Charon at all.

Both Charon and Pluto are covered with **ice** – Charon with frozen water, Pluto with methane ices. Charon orbits just 11,400 miles/18,350km

Pluto and its moon, Charon, viewed from the Hubble Space Telescope. Pluto is the only planet in our solar system not to have been visited by a spacecraft.
NASA

from Pluto and is less bright – both factors that explain why its discovery took such a long time. Once Charon was discovered, however, it allowed scientists to make more precise calculations of Pluto's size. This was in no small part due to a series of **eclipses** scientists observed as Charon's orbital plane synced up with Earth's. It's from these eclipses that we were able to (fairly) accurately determine Pluto's true size (and Charon's too).

In 2005, it was revealed that Charon has some neighbours, in the form of two tiny additional moons, discovered by the Hubble Space Telescope and named **Hydra** and **Nix**. Very little is known about the moons and we can currently only estimate their size, which is somewhere in the range of between 44 and 130km in diameter for each. Nix orbits at about 48,700km, Hydra at about 64,800km, but they don't orbit Pluto specifically. Charon is so large relative to Pluto that the center of gravity between them is outside the surface of Pluto, and they both rotate around that spot, known as the **barycentre**. It's this barycentre that Nix and Hydra orbit (technically speaking, the barycentre also takes into consideration the masses of Nix and Hydra, but they are small enough that the real action, gravitationally, is between Pluto and Charon).

What we don't know

Far off and tiny, Pluto is the (dwarf) planet we know the least about. It is the only planet that has not been visited by a spacecraft, although that will change when NASA's New Horizons mission visits Pluto and its retinue of moons beginning in July of 2015, just before Pluto's atmosphere freezes out from the cold. Until then, a good view of Pluto is difficult even with the best ground-based telescopes, and even the mighty Hubble has only been able to tease out the most general views of it's surface. So while we know some very basic things about Pluto today, real knowledge of the little planet is still some years away.

Observing the newcomers

Humans went for thousands of years without spotting these three elusive planets, so finding them yourself makes for an interesting challenge, and in the case of Pluto you will need a fair degree of luck.

Looking at Uranus

Technically speaking, Uranus is visible to the naked eye, but only just. If you're somewhere with almost no light pollution and no bright Moon, it appears in the night sky with a magnitude of around six, right at the cut-off of visibility. It's rather easier to find with a good pair of binoculars. With a larger telescope, you should be able to make out the disc and the colour of the planet, but since it appears as only four arcseconds in diameter, surface details will elude you. Uranus's moons clock in at a magnitude of sixteen, so don't count on spying them through your telescope at home.

What about Eris?

What do we know about the object that caused Pluto to be kicked out of the planetary club? At the moment, not a whole lot, but here are the details:

▶ While the earliest observations of the object date back to 2003, it was formally discovered (that is, recognized as being planet-like) in 2005, with the official announcement taking place on July 29. Its discoverers were Mike Brown, Chad Trujillo and David Rabinowitz, of CalTech and Yale. Prior to its official designation, it was known formally as 2003 UB313 and informally as "Xena," after the TV fantasy heroine.

▶ It shares a number of traits with Pluto, including a wildly eccentric orbit that brings it from about 3.5 billion miles/5.63 billion km from the sun at its closest approach to more than 9 billion miles/14.5 billion km at its furthest distance. It's orbit is also significantly tilted from the ecliptic, even more than Pluto's. Like Pluto it has a hugely long year: it takes five and a half centuries for Xena to go around the Sun once. It also has a satellite, formally known as Dysnomia, but informally known as "Gabrielle" when Eris was Xena.

▶ Eris is larger than Pluto as best we can guess, but not by much: depending on which measurement you use, its diameter may be larger than Pluto's by as little as 60 miles/100km.

▶ Eris is named after the Greek goddess of discord, which, considering the intellectual havoc its discovery caused among the astronomical community, is an all-too-appropriate name.

Viewing Neptune

Neptune comes in with a magnitude of eight, beyond the limit of the naked eye. To find Neptune at all, you'll need a good pair of binoculars. Neptune is only a little over two arcseconds across in the sky, so unless you have a larger-sized telescope, don't expect to make out the planet's disc and, as with Uranus, you won't see anything in the way of features or moons. Neptune is quite difficult to view, so if you get a good glimpse of it, you're doing well.

Finding Pluto

To resolve Pluto at all requires a telescope with an aperture of at least eight inches, and a dark sky with excellent viewing. Even with the best home telescopes, all you'll then see is a point of light. And to be sure you've actually found Pluto, you'll need to view your planetary candidate over the course of time, and to track its movement against the field of stars (taking pictures through your telescope will help).

The task of locating Pluto is made slightly less formidable by using astronomy software or charts downloadable from the Internet, and using the information to guide your telescope to the right place. Finding Pluto, and then confirming it actually is Pluto, is not an easy task, and if you manage to pull it off you'll have definitely joined the ranks of the advanced amateur astronomer.

Comets, asteroids and meteors

Imagine that the solar system is the movie set of an epic motion picture. There's the Sun, which is, of course, the star of the show. After that there are the planets, who comprise the supporting cast. Finally come the bit players, the extras and the second-spear-carriers-to-the-right; these are the minor denizens of the solar system, the comets, asteroids and meteors, who – just occasionally – get their own little piece of the action.

Comets

If comets are the bit players in the solar-system drama, they're less like walk-on players and more like cameo performers who show up and steal the show from everybody else. Streaking through the heavens with their tails smeared behind them, they are the centre of attention just for that brief moment in which they swoop around the inner solar-system. Dazzling cometary displays don't last long (in astronomical terms), and in fact most comets pass unnoticed by all but a handful of astronomers. Every once in a while, however, a comet appears which is bright enough to be seen by all, and while it's there, it can be a truly magical sight.

Dirty snowballs and elliptical orbits

Far away from the Sun, a comet isn't much to look at; it lacks both glow and tail, and its only permanent part is its dark, dusty nucleus (which, on its own, is neither notable nor unique). For an idea of what a comet actually is, think of a snowball with some dirt in it and you're almost there. Not quite, however, because, according to the information gathered by the Giotto spacecraft as it flew by Halley's Comet in 1986, the chemical composition of the comet nucleus varies slightly from a snowball made on Earth. In addition to water ice and dust, it also contains ices made of carbon dioxide, methane, ammonia and other trace gases. As well as this, comet nuclei are dark and not very large; the largest are just a few dozen miles in diameter, and most are considerably smaller. As a consequence, they are almost impossible to see when far away from the Sun.

The reason we can see comets at all is because those we do see are in highly elliptical orbits. So while they spend almost all their lives in the darkness of space, for a very brief amount of time they swoop into the inner portions of the solar system for a close encounter with the Sun. And when that occurs, something remarkable happens to that dark, crusty

Comet Hale-Bopp viewed against a backdrop of the Northern Lights over Alaska's Chugach mountains. Discovered in 1995 by Alan Hale and Thomas Bopp, it was seen in all its glory two years later.
Cary Anderson/CORBIS SYGMA

snowball of a nucleus. First, it heats up as energy from the Sun strikes its surface, and then jets of dust and vaporized gases erupt from the cometary nucleus, forming the parts of the comets we can see from Earth – the coma and the tail.

The coma

The coma is the concentrated gas and dust that forms the "head" of the comet, and it can be impressively large, up to ten times the diameter of the Earth itself. Of course, the coma is extremely tenuous, since it's made up of material cast off from a nucleus just a few miles wide (which is itself hidden somewhere in the coma). Comas usually begin to form when comets are three or four AU out from the Sun (an AU is one earth orbit, about 93 million miles/150 million km), but occasionally they form further out. Comas typically have their greatest diameter before they cross the boundary of Earth's orbit, and as they get closer to the Sun they begin to compact a bit.

The tail

Once material erupts from the nucleus of the comet, the solar wind pushes it away, and the impressive comet tail is born. Tails can stretch for millions of miles – in extreme cases they can reach further than the distance between the Earth and the Sun. With most comets, there are actually two tails. The first of these is a gas tail, made from gases ion-

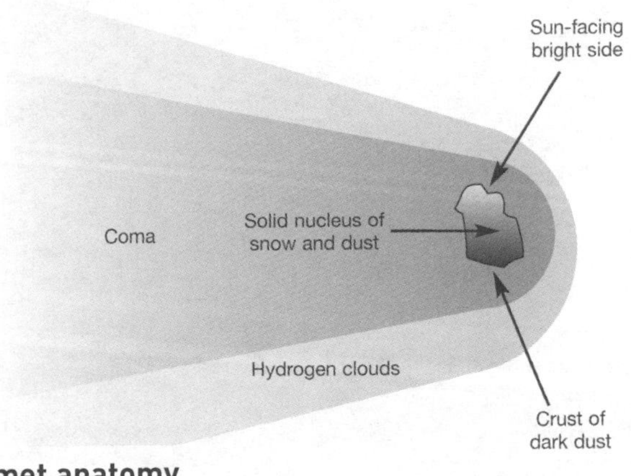

Comet anatomy

ized by interaction with the solar wind. Bluish in colour, it often appears streaky or twisted, and shoots directly out from the comet. The second is a dust tail, which is yellowish, featureless and often curved, as the dust particles lag behind the motion of the comet in its orbit. One exceptional example of a curving dust tail came with **Comet McNaught** (also known as "The Great Comet of 2007"), the brightest comet in decades (in mid-January, it was bright enough to be seen during the day).

Because the tails are formed by the action of the solar wind, they always point away from the Sun. This means that while the tail is behind the comet nucleus as it moves towards the Sun, it's ahead of it as the comet moves back to the far reaches of the solar system. Tails usually form between 2 and 1.5 AU from the Sun, then they grow as the comet streaks around the Sun, before disappearing at about the same distance.

Long orbits and short orbits

How long a comet's cold, dark and unnoticed existence lasts depends on what sort of comet it is. Comets in our solar system are divided into two groups: short orbital period comets and long orbital period comets. The former, as their name suggests, have relatively short orbits around the Sun, less than two hundred years for a single trip. The most famous of these is **Halley's Comet**, which returns to our skies every 76 years. Most

Comet P1 McNaught, taken from Swifts Creek, Victoria, Australia.
Henry Firus/Flagstaffotos

Regular visitors

Here's a list of some short period comets that will be swinging back our way within the next few years, with their orbital periods indicated in years. This isn't anywhere near a complete list, but it gives an idea of the number of short period comets out there.

2008

Tuttle (13.51), Boethin (11.23), Shoemaker-LINEAR (7.88), 2000 WT168 (7.66)

2009

Wolf (8.21), Kearns-Kwee (9.47), Swift-Gehrels (9.21), Klemola (10.82)

2010

IRAS (13.29), Ge-Wang (11.17), Tempel 2 (5.38)

2011

Honda-Mrkos-Pajdušáková (5.252), Crommelin (27.4), Schwassmann-Wachmann (5.36)

2012

Gehrels 2 (7.22), Shoemaker-Levy 7 (6.91), Kowal-LINEAR (7.90)

2013

Helin-Roman-Crockett (8.12), Wiseman-Skiff (6.68), Urata-Niijima (6.67)

2014

Harrington-Abell (7.54), Helin-Roman-Alu 2 (5.77), Comas Solá (8.80)

Halley's Comet will return in 2061.

of these comets originally inhabited the Edgeworth-Kupier Belt (commonly known simply as the Kupier Belt), a band of space between 35 and 1000 AU from the Sun which astronomers estimate holds hundreds of millions of comets. Some astronomers believe that the biggest of all these dirty snowballs is the dwarf planet Pluto (see p.152). It's also been suggested that there may be tens of thousands of Kupier-Belt objects 60 miles/100km in diameter or larger.

Most comets stay in the belt, regardless of size, but some are knocked into elliptical orbits through gravitational interactions with planets and other objects, and become the comets we see in our sky. Once we find short period comets, it's easy for us to determine their orbits, although it's a mistake to assume these orbits are very stable. Comet orbits can be changed by additional gravitational interactions with planets, and by those jets of gas and dust adding minuscule bits of thrust to the comet itself. And, of course, every once in a while a comet will simply smack into a planet, as Shoemaker-Levy 9 did with Jupiter in 1994.

Long orbit comets come from a different place entirely – the **Oort Cloud**. This is a place so distant from the rest of the solar system we can't see it and indeed can only speculate that it's there at all. In theory, the Oort Cloud is a repository of material that was originally part of our solar system but was evicted from the neighbourhood by the forming planets. This cloud of material now extends a light year or more from the Sun. It is thinly populated (each chunk of the cloud would be millions of miles/kilometres from other chunks), but spread out over such a large area that the combined mass of the Oort Cloud bodies would be as much as the planet Jupiter. Objects in the Oort Cloud tumble into the solar system and become comets when they are jostled by passing stars or other interstellar phenomena. Given the distances involved in these comets' orbits, orbital periods can last many thousands of years. The recent comet **Hale-Bopp**, for example, has an orbital period of about 2500 years, long enough for astronomers of the future to forget it had ever come around at all.

Meteors from the Geminid meteor shower flash through the night sky over Arizona's Meteor Crater. Astronomers are unsure whether the source of the shower, 3200 Phaeton, is a comet, an asteroid or something in between.
Jonathan Blair/CORBIS

The Barringer meteor crater in the Arizona desert is named after Daniel Barringer, the first man to prove conclusively that such phenomena were caused by meteors hitting the ground. The impact is thought to have occurred some 50,000 years ago.
Charles O'Rear/CORBIS

Meteors

When comets get close to the Sun, they heat up, causing geysers of **gas** and **dust** to burst from the comet's nucleus. The dust helps to create the coma and tails of the comet, but long after the comet has gone, these cast-off pieces of comet create something else as well – **meteor showers**. Several times a year, the Earth passes through the orbit of long-gone comets, and the tiny dust particles still floating around in the orbit fall into the Earth's atmosphere. Friction burns them up, creating those blazing streaks across the night sky. The meteors in these showers are known by the name of the constellation from which they seem to come; the **Geminids**, for example, appear in mid-December out of the constellation **Gemini**.

The leftover grains of dust from comets make **meteors**, but not all meteors are made up of comet dust grains. Tens of thousands of tons of interplanetary debris of all sorts falls into the Earth's gravitational maw every year. This material originated from comets, asteroids, and other planets or even from our Moon, not to mention the man-made objects

What killed the dinosaurs?

Sixty-five million years ago, the inhabitants of Earth looked up to see a massive **fireball** roaring through the sky. Those creatures that survived the subsequent explosion would soon be dead from the severe aftereffects: fires, floods, huge tsunami waves washing over the land, acid rain, dust in the atmosphere blocking out heat and light from the Sun, followed by the death of plant life and starvation for animals. It was not a good time to be a dinosaur.

How do we know that an asteroid helped to kill off the dinosaurs? Several pieces of evidence contribute to its likelihood. First, there is a thin layer of **iridium** found in rocks dating back 65 million years. This rests in a geologic layer, known as the "K-T boundary", that separates the last age of the dinosaurs (the Cretaceous) from the Tertiary Era which followed it. Iridium is an element that is rare on Earth but is rather more common in asteroids and meteorites, so a layer of it at the K-T boundary suggested it got there by extraordinary means.

All that was missing was a crater to confirm an asteroid hit. In 1978, one was found off the coast of the **Yucatan Peninsula**. It was huge; scientists eventually estimated it to be more than 180 miles/290km in diameter. To make a crater that big, you'd need an object at least 5 miles/8km across – an asteroid large enough to bring down the dinosaurs.

launched into orbit now on their way back down. Most of this stuff is tiny, and so burns up in the atmosphere leaving only a momentary streak of light to note its passage.

Earth-bound meteorites

Some of this leftover material has a bit more heft to it. Any particle that masses more than about 0.04 of an ounce/1g has a good chance of surviving its fiery journey through the Earth's atmosphere to smack into the ground. Once it does that, it becomes a **meteorite**. Unsurprisingly, most meteorites are very small, and the majority of them land in the sea or on unpopulated areas, so their arrival causes little alarm. Impacts large enough to leave an actual crater on the surface of the planet happen very rarely – about once every five thousand years or so. The really big impacts, the ones that wipe out dinosaurs or inspire movies like *Armageddon*, happen once or twice every couple of hundreds of millions of years (although, of course, one can't time these things precisely).

The meteorites we find on Earth are of one of three types: **stony**, **stony-irons**, and **irons**. Stony meteorites come in two versions: **chondrites**, which have small, rounded, formerly melted objects called chondrules, and **achondrites**, which resemble igneous rocks here on earth. Chondrites are interesting because they are probable leftovers from the beginning of the solar system, giving us a glimpse of what things were like billions of years ago. Achondrites seem to have been chipped off larger objects, like asteroids, moons or planets. Stony-iron and iron meteorites were also originally part of larger asteroids or planets, ripped free through collisions and other celestial smash-ups.

Asteroids

Asteroids were ushered into our awareness on the very first day of the nineteenth century, in a manner rather similar to the discovery of Uranus (see p.142). On January 1, 1801, **Father Giuseppe Piazzi** was busily mapping star positions when he noticed that one of his "stars" was moving from a position between the orbits of Mars and Jupiter. It turned out to be **Ceres**, the first (and, so far, largest) of the asteroids, of which more than ten thousand have been discovered.

Solar system leftovers

As more and more asteroids popped up, scientists speculated that these minor planets were the residue of a **larger planet** between Mars and Jupiter that had somehow come to a bad end. But even if you were to assemble all the known asteroids together, the resulting chunk of rock and metal would

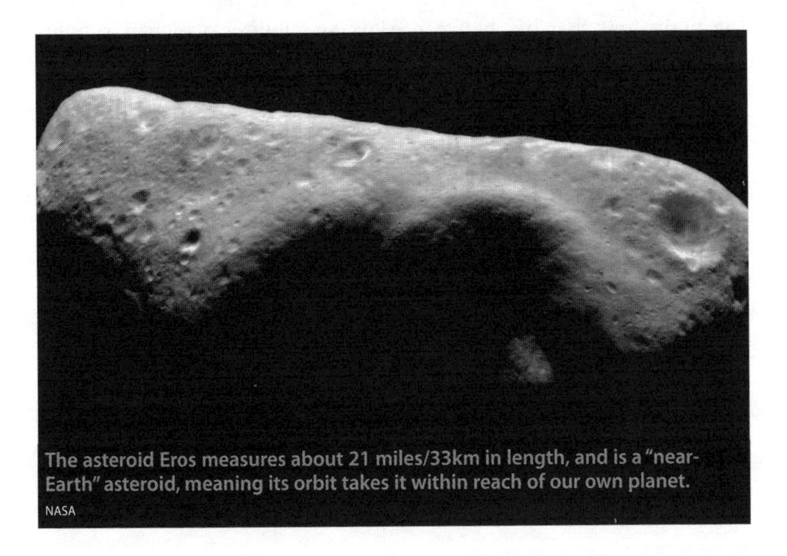

The asteroid Eros measures about 21 miles/33km in length, and is a "near-Earth" asteroid, meaning its orbit takes it within reach of our own planet.
NASA

have a mass less than 1/1000 of the Earth. Asteroids aren't likely to be the fragments of a previous planet; rather, like the dust of flour that remains after you make a cake, they're simply the **leftover fragments** from the early formation of the solar system.

Asteroids are mainly **stony** objects, although some are a mixture of rock and metal, and some are almost entirely metal (these are the ones deep thinkers dream of one day mining for their resources). **Ceres** used to be the largest of the asteroids but was recently elevated to the status of "dwarf planet" – the same designation that Pluto was demoted to (see p.152). There are thirty or so asteroids with diameters of 200km or larger, 250 are 100km or more in diameter, and there are an estimated one million asteroids no smaller than 1km in diameter.

There are so many asteroids that astronomers **naming** them quickly switched from using classical names (Ceres, Juno, Pallas, Vesta) to a more populist approach. There are asteroids named after astronauts (Armstrong), science fiction witers (Heinlein), TV characters (Mr Spock), painters (O'Keeffe), computers (NORC), Egyptian pharaohs (Imhotep) and even pop groups – the Beatles give their name to a comet (Beatles) and the group's individual members all get an asteroid each (Lennon, McCartney, Harrison, Starr).

Within the main belt

About 95 percent of the asteroids we know about orbit the Sun in the **main belt** that lies between Mars and Jupiter, which is roughly between 200 and 305 million miles/320 and 490 million km from the Sun. If you have seen the asteroid-chasing sequence in the movie *The Empire Strikes Back*, you will be disappointed to learn that the real asteroid belt is not nearly so crowded as the one the *Millennium Falcon* ducks and weaves its way through. Your average asteroid floats alone, unencumbered by the sight of another asteroid, although there are a few that are a bit more sociable. **Antiope**, for example, is actually two asteroids orbiting a central point 100 miles/160 km from each other. Other asteroids have small **moonlets** orbiting their larger bodies. There are also large groupings of asteroids, known as **families**, which share similar orbits. Many of these may actually be shattered chunks of larger asteroids, because while asteroids may be solitary, they still occasionally collide.

Asteroids in the belt are not evenly distributed. There are **spaces** in the belt, like grooves in a vinyl record, where asteroids don't go (or at least, don't stay long). These are called **Kirkwood Gaps**, and they exist at spaces within the belt in which an asteroid's orbit would be in a direct ratio to Jupiter in its orbit. Jupiter's immense gravity works on asteroids in the resonant orbits, nudging them over time into different orbits.

Tunguska, 1908

In the morning of 30 June, 1908, an asteroid (or perhaps a comet nucleus) estimated at about 500 feet/150m in diameter, ploughed through the atmosphere above central **Siberia**, exploding above the **Tunguska River**. It was exactly as if a massive atomic bomb had detonated, right down (or up, as the case may be) to the mushroom cloud. Miles of forest were flattened. Farmers dozens of miles away reported being knocked off their porches by the blast wave and windows were shattered for hundreds of miles. Dust sucked into the atmosphere by the blast spread out over Europe and Asia, reflected far-off sunlight and lit the night sky; a letter-writer to the London *Times* reported being able to read books from the light of the night sky.

Twenty-two years after the event, Soviet scientist **Leonid Kulik** was still able to see the vivid effects of the explosion, and wrote, "One has an uncanny feeling when one sees giant trees snapped across like twigs, and their tops hurled many metres to the south." It was lucky that the asteroid (or comet) exploded in such a sparsely populated region. As far as we know no humans died in that event. Had the same explosion happened over London, the blast would have levelled much of the city, and taken most Londoners with it.

The forest at Tunguska photographed shortly after the 1908 asteroid hit.
Smithsonian Institution

Outsiders

Asteroids do exist outside of the main belt. The largest groups are the **Trojan asteroids**, two groupings of asteroids that precede and follow the planet Jupiter in its orbit. Beyond Jupiter lie the **Centaurs**, icy objects that are rather more like comets than other asteroids. Of more interest to us, however, are **Near Earth Objects**, asteroids that have orbits that cross Earth's orbit from time to time. Some come closer to the Earth than the orbit of the Moon, and nearly 1000 of them are larger than half a mile/1km in diameter (more than enough to cause problems on a global scale).

Getting hit by one isn't simply likely, it's a near certainty, and, of course, it's happened before. The meteor that is believed to have wiped out the dinosaurs was a near-Earth asteroid which got a little too near. The good news is that (at the moment) there are no Near Earth Objects (that we know of) which threaten life on Earth. Astronomers peg the odds of one of these planet-busters hitting Earth this century at a relatively comforting 5000 to 1 (so it's still safe to go ahead and make plans for your retirement).

Searching for comets

Two of the most famous comets of recent times were discovered by amateurs. **Comet Shoemaker-Levy 9**, which impacted into Jupiter in 1994, was co-discovered by David Levy (a celebrated amateur astronomer who has found a total of 21 comets) and the husband-and-wife team of Eugene and Carolyn Shoemaker. **Comet Hyakutake**, one of the brightest comets of the past fifty years, was discovered by **Yuji Hyakutake** in 1996 using only a pair of binoculars. Comet finding is one area of astronomy where amateurs can and do make a real contribution; it can even lead to a kind of immortality, since the first person who reports on a discovery has the comet named after them. There's even an award given to amateur astronomers who locate new comets: the Edgar Wilson Award, which has a cash payout. You probably won't get rich off it, but it's a nice bonus.

To discover a comet, it helps to know what it would look like in your telescope or binoculars, and obviously the best way of doing that is to look at one that has already been located. There should be several such comets in action at any one time, and their current positions can be located through the Internet. Harvard University has a good site for this purpose that's worth exploring (**http://cfa-www.harvard.edu/iau/Ephemerides/Comets**).

It also helps to have a reasonably detailed **sky atlas**, which will help you eliminate a great number of objects you might otherwise misidentify as comets. One of the first catalogues of non-stellar objects, created by eighteenth-century comet enthusiast **Charles Messier**, was made specifically to identify fuzzy, comet-like objects that weren't in fact comets at all. For your part, you don't want to get excited over the prospect of finding a new comet, only to discover it's, say, the Andromeda Galaxy (also known as M31 in Messier's catalogue).

When to look

The best time to search for comets is in the **early morning**, before the Sun rises, and in the **evening**, after the Sun has set. The presence of the Moon is a factor to consider, since its glare can hinder your efforts. You'll need to work around its position in the sky. In the mornings, the few days just before and just after the new Moon are a good time for searching. If it's just before a new Moon, you'll need to get up quite early to get in some viewing time before the Moon cracks the horizon. In the evenings, the few days after a full Moon are best as you'll have time to search before the Moon rises. More comets are found in morning observations, so if you're really keen to find a comet, set that alarm clock.

How to look

Comets become more visible the closer in they are to the Sun, so your search should start in the space above where the Sun will rise (or has

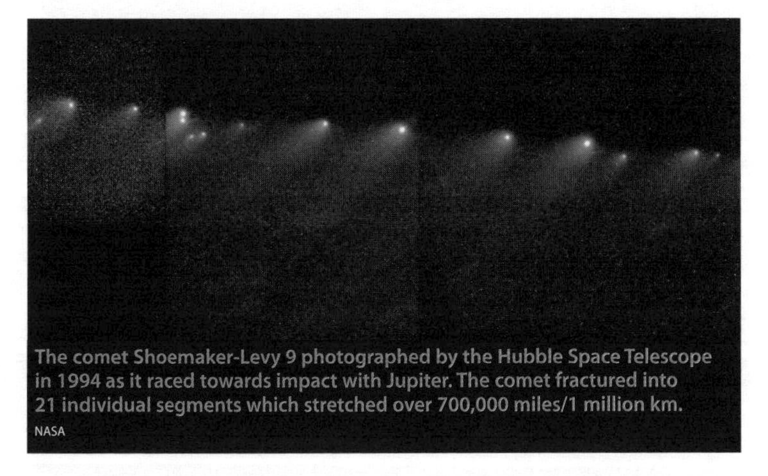

The comet Shoemaker-Levy 9 photographed by the Hubble Space Telescope in 1994 as it raced towards impact with Jupiter. The comet fractured into 21 individual segments which stretched over 700,000 miles/1 million km.
NASA

just set), and should encompass a "rectangle" that starts at the horizon and reaches a point about 45 degrees into the sky. What happens next is pretty simple: all you do is slowly sweep the sky with your binoculars or telescope.

For an **evening search**, start your observations in a lower corner of your rectangle (let's say the left corner), observing the objects in your telescope's or binoculars' field of view. This should take as little as a few seconds (if you're sure there's nothing to see) although it can take longer if you're looking through a particularly dense bit of sky. After you're sure you haven't seen a comet, move your field of view slightly to the right, allowing for a bit of overlap from your previous position. Repeat until you reach the edge of your rectangle and then move your field of view up slightly (again, allowing for some overlap). You then start another horizontal sweep, this time moving from right to left. Keep at it until you reach the top of your rectangle (or you get bored – this is a laborious process). For a **morning search**, you do exactly the same thing, but starting from the top of the rectangle and moving downwards.

The key to comet hunting is patience. Even successful comet-hunters log literally hundreds, and sometimes even thousands, of hours of skywatching before they bag their first comet. It's a time-consuming process.

What to do if you find one

So, there you are, patiently scanning the sky, when suddenly a fuzzy, glowing patch shows up in your field of view. You've found a **comet**! Or have you? What should you do next?

- **Check your star atlas** Make sure what you're seeing isn't a nebula, or galaxy, or star cluster, or whatever. You may also be experiencing "ghost images", caused by light from nearby stars. These are all common occurrences for newer comet-hunters, so don't feel bad if your comet turns out to be something else entirely. It happens to everyone.
- **Check your telescope** Make sure what you're seeing isn't a flaw in your telescope optics or a star near the limits of your telescope's resolving power.
- **Check its movement** To do this, come back to the comet an hour or so later to see if the object you're looking at has moved relative to the stars; ideally, if you're sure you won't crack under the pressure, check the next evening. Comets move, so if you don't see movement in what you're looking at over the space of an evening, and definitely over the space of two evenings, then it's probably not a comet.
- **Enlist help** Get other star-watchers (who you know and trust) to check with their own telescopes, or privately contact a local observatory for confirmation. Try to be discreet about your enquiries; posting an "Is this a comet?" question on an

Internet newsgroup or website is the quickest way to enable someone else to get credit for your work.

- **Be as certain as possible** that what you've got is a comet. If you're going to report your discovery as a comet, it's going to be checked and verified by professionals, who won't appreciate following it up only to find it's a nebula or a galaxy that's well known. For every comet verified, there are five false alarms. Fire off enough false alarms and you'll soon develop a not-so-good reputation.
- **If you're convinced** beyond reasonable doubt that you've got a comet on your hands, then you can report it to the **Central Bureau for Astronomical Telegrams** (**CBAT**), which, despite its name, accepts reports through email and regular mail (its telex number, ironically enough, was disconnected because of disuse). CBAT requires you to supply as much of the following information as possible:

  The position of the comet, as accurately as possible (the star atlas comes in handy again here). CBAT prefers you to position the comet to at least 1' in Declination and 0.1 minute of time in Right Ascension, so they don't have to waste a whole lot of time finding it again. Note which epoch you're using: 1950 and 2000 are the most common (your star atlas should tell you which epoch it is based on).

  The time you noted the position of the comet, in Universal Time (UT), not in your local time. UT (also known as Greenwich Mean Time) is determined through Greenwich, England, and is given on a 24-hour clock. The US Naval Observatory can tell you the Universal Time (http://tycho.usno.navy.mil/cgi-bin/timer.pl)

  As many details about the comet as you can: shape, brightness, whether it has a tail, and so on. Also include location on your observing position, the kind of equipment you used to spot the comet, and also what you did to make sure your "comet" was, indeed, a comet (the star atlases you used, for example).

  Your full name, mailing address (include your email) and a telephone number where you can be reached.

Once you've collected all this information, email it to "cbat@cfa .harvard.edu" or send it by post to CBAT, 60 Garden St, Cambridge, MA 02138, USA. If you're the first to find the comet, they'll let you know.

Observing asteroids

Asteroids in the belt between Mars and Jupiter are a challenge for amateur astronomers for two very good reasons: they're small (none larger than 700 miles/1100km in diameter), and they're dim (usually eighth magnitude or less, even at opposition). So sighting one is generally regarded as a feather in your cap.

The easiest way to find an asteroid is simply to look where they are. There are plenty of online lists of locations of many of the most prominent asteroids, including **Near Earth Objects**. CBAT, the same organiza-

Meteor showers through the year

Here are some of the most common meteor showers, with the dates when you can see them at the peak of their display. You should also be able to see smaller displays of meteor activity on the few days on either side of these dates. Just point yourself in the direction of the constellation provided and enjoy the show.

Name	Constellation	Date
Quadrantids	Boöes	January 4
Lyrids	Lyra	April 21
Eta Aquarids	Aquarius	May 5
Delta Aquarids (S)	Aquarius	July 28
Delta Aquarids (N)	Aquarius	August 6
Perseids	Perseus	August 12
Orionids	Orion	October 21
Taurids	Taurus	November 5
Leonids	Leo	November 17
Geminids	Gemini	December 14
Ursids	Ursa Minor	December 23

tion that confirms comet finds, also has a useful website that lists which asteroids come into opposition in any given year, along with the date, the magnitude and the location in the sky (**http://cfa-www.harvard.edu/iau/Ephemerides/index.html**). Once you've pointed your telescope in the right place, check what you see through it with a handy **star atlas**. If you find a faint star where one should not be, it might be your asteroid. You can confirm this by checking back an hour or two later to see if the "star" has moved (asteroids at opposition move relatively quickly).

If you're really gung-ho about asteroids, try spotting an **asteroid occultation**, which occurs when an asteroid passes in front of a star. When this happens, the magnitude of the star will appear to drop while the asteroid cruises by. Be warned, however, that these occultations are almost literally over in the blink of an eye, and it's a rare asteroid occultation that lasts more than a minute. If you want a list of asteroid occultations go to the website of the **International Occultation Timing Association** (**http://www.lunar-occultations.com/ iota/iotandx.htm**).

Observing meteors

Here's one astronomical phenomenon that is best seen with the naked eye alone. There are two types of meteors: those that appear as part of a **meteor shower**, and those that appear as their own, spontaneous one-off events called **sporadic meteors**. Meteor showers can allow you to see up to dozens of meteors in a single hour, while sporadic meteors are truly catch-as-catch-can, since they can happen at any time and from any corner of the sky. Meteor showers are best viewed when their radiants (the area of space which they appear to come from) are high in the night sky – there's less atmosphere to look through in order to see the show.

Most meteors you observe will be momentary streaks of light, but a few will put on a bigger display. Those meteors called **fireballs** can get to be brighter than the planet Venus; they can also leave what is called a train, which is just that – a train of smoke (more modest meteors can also do this). Some fireballs will explode in the air, becoming what's known as **bolides**. All of which should add a bit of excitement to your viewing.

Stars and their lives

Stars are huge collections of gas: primarily hydrogen with some helium and trace amounts of other gases, brought together by gravity and ignited into nuclear fusion by the immense, crushing heat and pressure at their centre. Of course, that's not quite all there is to them.

There are hundreds of billions of stars in our galaxy alone, encompassing a vast range of sizes, masses and ages – from tiny red dwarf stars, whose natural lives are longer than the current age of the universe, to massive stars dozens of times larger than the Sun, which consume their fuel and then die violently in the blink of an eye. There are bright, hot stars blazing in the heavens still in their stellar swaddling-clothes, and there are small, dark corpses of dead stars, surrounded by nebulas of their own gases hung around them like mourning crepe. Everywhere you look in the heavens, you see the stages of the stars' lives – birth, life, death and rebirth, as another generation of stars forms from what is left over from a previous generation.

The sextant is a navigational instrument used by sailors to determine their latitude on the sea by calculating the angle between the horizon and the stars.
Private collection

Plotting the stars

Every star in the sky is placed in a position on a chart known as the **Hertzsprung-Russell diagram**, named after two early-twentieth-century astronomers, Enjar Hertzsprung of Denmark and Norris Russell of the US. The H-R diagram features two axes, one charting the star's **surface temperature** (which is intimately related to the visible colour of the star), and the other charting the star's **luminosity** (how bright it is in absolute terms). Stars in the top left corner of the H-R diagram are the hottest and most luminous stars in the sky, while those down in the bottom right corner are the dimmest, coolest stars.

The vast majority of all stars are found on a mostly diagonal swath of the H-R diagram called the **main sequence**. This is where stars are for the main portion of their lives, burning hydrogen in their stellar furnaces. For stars on the main sequence, luminosity and surface temperature are linked; the hotter you are on the main sequence, the more luminous you are as well. Outside the main sequence are very large but very cool stars whose luminosity is a by-product of their immense size; these are the **red giants** and **red supergiants**. You'll find these near the top right corner of the H-R diagram. Near the bottom left corner are the stars that are hot but rather dim because they're so small – the **white dwarfs**. Both red giants and white dwarfs represent stars that have left the main sequence at the end of their natural stellar lives.

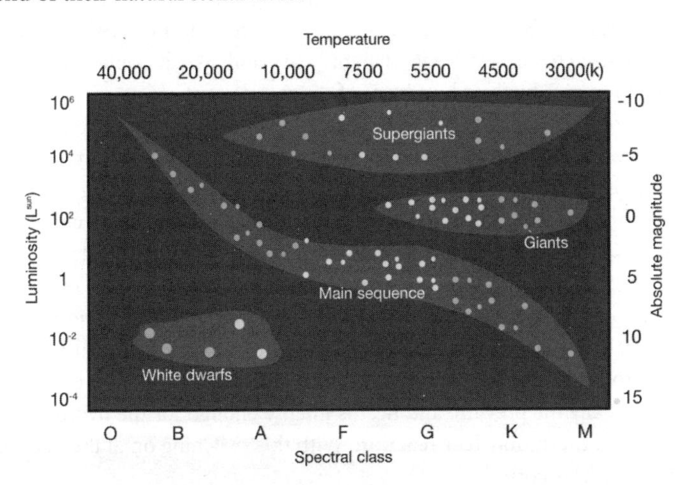

Simplified Hertzsprung-Russell diagram

Stars on the H-R diagram are collated into **spectral classes**, arranged by surface temperature (and therefore by colour). Stars of spectral class O are the hottest at 50,50,000°F/27,760°C (or more), followed by B, A, F, G, K and M, whose members generate a relatively cool 5000°F/2780°C (or more), followed by B, A, F, G, K and M, whose members generate a relatively cool 5000°F. The letters are left over from an earlier classification system that was alphabetical but was subsequently rearranged. The mnemonic which many astronomers use to remember the order of the spectral classes is "**O**h **B**e **A** **F**ine **G**irl, **K**iss **M**e".

The spectral classes are further subdivided by astronomers for more accuracy: by surface temperature numbered 0 to 9 (with 0 representing the hottest), and by luminosity with Roman numerals indicating a star's class. All main sequence stars are class V, and are called by astronomers the "main sequence dwarfs". Class I and II stars are supergiants, class III are giants, and class IV are "subgiants". White dwarfs are class D (which is the Roman numeral for the number 500). The Sun is a type G2V, which means it's a G-type star (yellow, and between about 8500 and 10,300°F, or 4700 and 5700°C) in the main sequence.

A star is born

For a star to be formed, the first thing that's needed is the raw material – **hydrogen**, and lots of it. Fortunately, hydrogen is in plentiful supply: it floats around, with other trace elements, in huge, cold **molecular clouds** that can stretch for hundreds of light years and which contain enough hydrogen to give birth to hundreds of thousands of stars.

Every so often something will cause the **cloud** to become agitated (a shockwave from a nearby exploding star, for example) or parts of the cloud will simply become so massive that they begin to collapse, drawn together by an ever-increasing tug of **gravity**. These collapsing parts of the cloud are known as **protostars**, and after about 100,000 years of pulling material into themselves, they begin to generate serious **heat** – not from nuclear reactions in their core but from the **compression** and **gravitational contraction** of the gas itself. Protostars in this stage would appear to us to glow faintly. Deep in the protostar's **core**, however, things are getting really hot and cramped, and eventually the pressure and heat is intense enough for the hydrogen to ignite into a **thermonuclear reaction**. With this switching on of the nuclear furnace, a star is born.

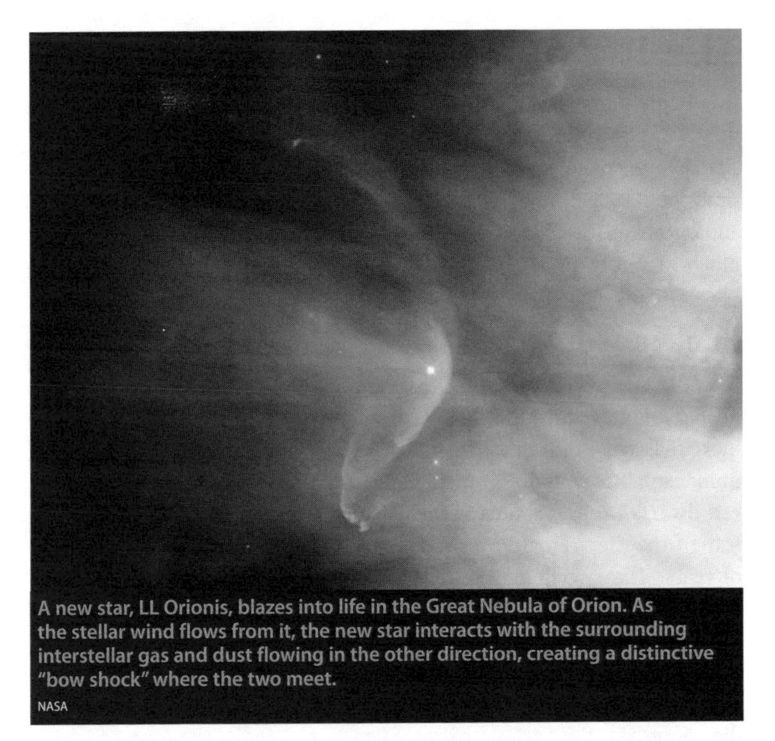

A new star, LL Orionis, blazes into life in the Great Nebula of Orion. As the stellar wind flows from it, the new star interacts with the surrounding interstellar gas and dust flowing in the other direction, creating a distinctive "bow shock" where the two meet.

NASA

These new stars are hot and bright, and they're still swaddled by the gas from which they're born. Astronomers call these brand-new stars **T Tauri** stars, although they're not just found in the Taurus constellation, just named after a particular star there. Because the molecular cloud that gave birth to this new star is far larger than the star itself, a new star will usually find itself just one among many recently created stars in a single "stellar nursery". Depending on whether the new stars are large or small, these nurseries are known as **OB associations** (for the large, bright new stars) or **T associations** (for smaller, less-massive new stars). As these stars switch on their engines, their heat and power blow the remaining gas and dust away from them, and also **ionize** the cloud they're sitting in, creating dazzling **emission nebulae**. One famous example of an emission nebula is the **Orion nebula**, home of many newborn stars and a place where stars are still being born.

The life of a star

Eventually, stars leave the nest – migrating out of the nurseries in which they were born and starting their adult life on the **main sequence**. All stars live the majority of their lives on the main sequence, and during this time the star maintains a more or less constant temperature, luminosity and output as it works its way through the **hydrogen fuel** at its core. How much time it takes depends on how large the star is.

Fuel consumption

It would seem to make sense that **smaller stars**, with less hydrogen to burn, would get through their nuclear fuel faster than stars many times more massive. But it doesn't work that way. Small, low-mass stars are not unlike small cars; they sip their fuel and go further on less. Some of the **red dwarfs** use their nuclear fuel so sparingly that their life expectancy – the time they remain on the main sequence – is hundreds of billions of years. That's far longer than the universe itself has been in existence. On the other end of the scale are massive, **blue-hot stars**, dozens of times more massive than the Sun. Like expensive sports cars, these stars guzzle their fuel like there's no tomorrow, and will use every drop of it in just a few million years. They are the stellar exemplars of the slogan "Live fast, die young".

(Where does the Sun fit in to all this? Well, it's rather on the smallish side, has a main-sequence life expectancy of about ten billion years, and is about halfway through that lifespan at the moment.)

One reason that more massive stars burn their fuel faster is that they burn it differently from smaller-mass stars. **Small-mass stars** produce nuclear fusion in their cores through what is known as a **proton-proton chain**. In these reactions (trillions of which occur every second), **hydrogen nuclei** slam together, eventually forming **helium nuclei** and releasing a small amount of energy in the process. This produces a sure and steady burn of the nuclear fuel over a long period of time.

High-mass stars, on the other hand, burn their hydrogen through a process called the **CNO cycle**, in which nuclei of the heavier elements – carbon, nitrogen and oxygen – are used as catalysts for the reaction. To go back to our car analogy, this is very much like a "nitro injection" right into the engine of the star: the star burns its fuel hotter, faster and with rather astounding output of **energy**. A star with just fifteen times the solar mass of the Sun will crank out tens of thousands of times more energy

This image of nebula NGC 3603 offers a view of all the life stages of stars. The gaseous nebula to the right is a birthplace of new stars, which ignite and begin the main sequence portion of their lives. Stars then age and grow old, as in the star cluster in the centre of the image, and begin their death processes, shown by the star venting gas at upper centre left.
NASA

than our modest yellow star, and live a correspondingly shorter life. The CNO cycle does happen in stars the size of our Sun, but comprises only the tiniest fraction of their energy output.

Rotational speed

There's another difference between high-mass and low-mass stars during their lives in the main sequence – the **speed** at which they **rotate**. All stars arrive on the main sequence rotating rapidly, a side effect of the **angular momentum** generated during formation. Small stars eventually lose this angular momentum and slow down (our Sun, for example, rotates about once every 25 days at the equator). This loss of angular momentum is related to the small stars' **magnetic fields**, generated in part by **convective motion** in their outer layers.

Very large stars lack convective motion, have weak magnetic fields, and rotate very quickly – the bigger they are, the faster they spin. Some of the very largest stars rotate on their axis once a day. The speeds at which these stars rotate can be so fast – given their size – that parts can literally fly off, forming **rings of gas** girdling the star.

Death of a star

(the low-mass version)

Sooner or later, stars burn through all their hydrogen, an event that signals the end to their long life on the main sequence. What happens next depends on how big – or how small – the star was during the course of its normal life.

For small- and medium-sized stars, with a mass equivalent of eight times the Sun (or smaller), the beginning of the end comes when the hydrogen in the core has been entirely converted into helium. When that happens, the burning of hydrogen continues but it moves from the core of the star into a thin "shell" surrounding what is now the helium core. The outer layers of the star then expand enormously, turning it into a **red giant**, and the surface temperature drops to about 6500°F/3600°C. (When, in the distant future, the Sun turns into a red giant, its expansion will almost certainly engulf the orbit of our own planet.)

As the surface of the star **expands**, the core of the star **contracts**, becoming ever smaller and hotter. When it reaches a temperature of about 360 million°F/200 million°C and has a density of over 600,000 pounds per cubic foot/9.6 million kg per cubic metre, the **helium fuses** (just as hydrogen did before it) in a sudden event known as a **helium flash**. Where before the star converted hydrogen into helium, now the star converts helium into **carbon**. And with a new source of nuclear energy, the star settles down again… but not for long.

The **helium fuel** of a star burns for considerably less time than the hydrogen, and soon (in astronomical terms) the star is once again left with a core of material it's not fusing, and with a thin shell of material (in this case helium) that is burning away. Again, the outer layers expand as the core contracts, but this time the outer layers **pulsate**, causing the size and surface temperature of the star to jerk back and forth – the star

The planetary nebula Mz3 (also known as the "Ant Nebula") at the end of its life cycle. Stars the size of our own Sun do not explode as a supernova but nevertheless shed layers of gas, which form dazzling nebulae such as this.
NASA

becoming three times larger at its most expansive stage than it is when it's most contracted. Stars in this phase of life are known as **Mira variables** (see p.187).

Eventually, the **outer layers** of the star are blasted away, leaving the star core exposed and surrounded by an envelope of the gases that used to be its outer layers. The exposed core is incredibly hot – more than 250,000°F/139,000°C – and the gas surrounding the core is illuminated by the **ultraviolet light** that flows off what's left of the star. Here on Earth, we see this illuminated gas as a **planetary nebula**, so called because many are disc-shaped in telescopes (not because of any relation to real planets).

White dwarfs

Because low-mass stars have insufficient mass to turn carbon into nuclear fuel, the remaining cinder (which used to be the star's core) simply cools down, slowly, for the rest of eternity. These stellar corpses are called **white dwarfs** (although they can be any colour, depending on how hot they are). Despite not being massive enough to contract further, they're still incredibly dense: a white dwarf can have the same circumference as the Earth, but the mass of the Sun (which is 330,000 times more massive than the Earth). And there it remains, small, dense, and slowly going black against the backdrop of other stars.

Death of a star
(the high-mass version)

High-mass stars don't just "live fast, die young", they also seem to conform to the belief that "it's better to burn out than fade away". Burn out is exactly what these stars do – brilliantly, violently and spectacularly.

The first part of a high-mass star's endgame is like that of its lower-mass cousins, only much, much faster. High-mass stars complete their hydrogen-burning stage and swell up, not into red giants but **red supergiants** – stars so large that if they were placed in our solar system, their outer surface would reach to the orbit of Jupiter. Like small-mass stars, they crunch the

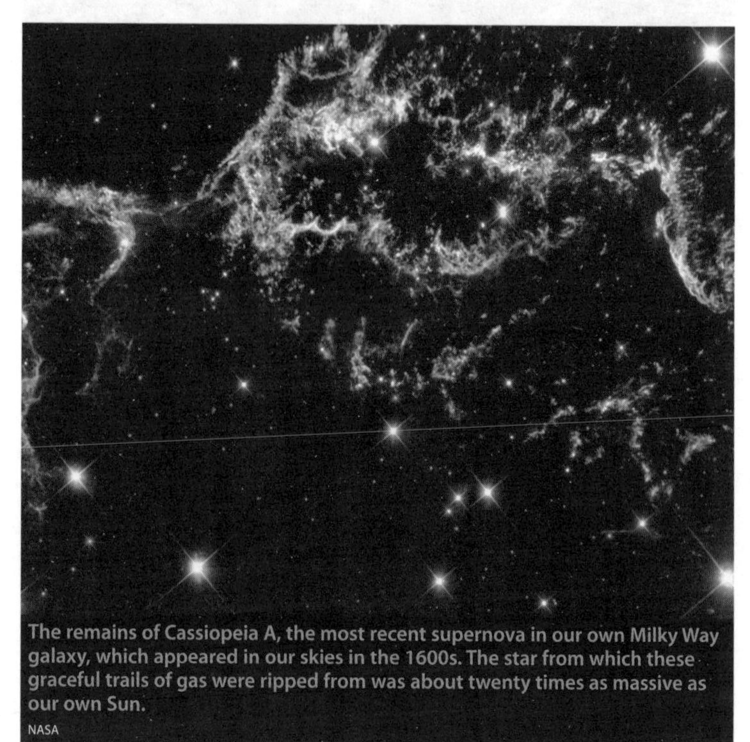

The remains of Cassiopeia A, the most recent supernova in our own Milky Way galaxy, which appeared in our skies in the 1600s. The star from which these graceful trails of gas were ripped from was about twenty times as massive as our own Sun.
NASA

helium in their cores and it begins to burn, fusing into **carbon**. Unlike smaller stars, however, these giants retain enough mass in their cores to reignite their nuclear engines to burn carbon. The star is back in business.

Not for long, however. Stars burn helium for far shorter than they burn hydrogen, and they burn carbon even quicker. Once the carbon has been burned through, the star continues to fuse progressively heavier elements in its core: **neon**, **magnesium**, **silicon** and **sulphur**, each cycle of fusion taking a substantially shorter amount of time. Finally, what's left in the star's core is a compact ball of **iron**, an element whose atomic structure doesn't allow it to fuse in order to generate any more energy.

What happens next is quite simple – the star goes kaboom! The iron core at its centre, no longer supported by the power of nuclear fusion, collapses under its own mass. As it **implodes**, the temperature in its core reaches unimaginable levels – tens of billions of degrees. The iron disintegrates and the suddenly free **protons** and **electrons** fuse together to form **neutrons**. Waves of neutrinos blast through the star, carrying off excess energy and escaping towards space. A **shockwave** from the collapsing core rips through the rest of the star.

This shockwave does two things. First, it creates heavy elements (gold, uranium, platinum and so on) in the remaining layers of the star, through a process called **rapid neutron capture** (the r-process for short). Second, it blows the star apart, hurling the outer layers into space at speeds reaching more than 12,000 miles per hour/19,000km per hour. This exploding star is called a **supernova** (a **type II** supernova, to be precise), and it can be as luminous as the equivalent of ten billion Suns. The star goes out in a blaze of glory, leaving behind a nebula of exploded gases and what remains of the stellar core.

Neutron stars and black holes

Exactly what remains of the core depends on how massive it was when it imploded. If the core of the star is less than three solar masses (three times as massive as our Sun), then the core stops collapsing once the neutrinos formed in the implosion can no longer be squeezed together any more tightly by **gravity**. What then remains is a **neutron star**, a tiny object – just a few miles in diameter – a million times denser than a white dwarf (which is already incredibly dense). If you tried to extract a tablespoon of **neutron star**, you'd need some heavy lifting equipment, because that tablespoon would have about a billion tons of densely packed neutrons in it.

Neutron stars spin incredibly fast and have immense magnetic fields, from which **energy** escapes at the magnetic poles. As the neutron star's rotation whips it around (up to hundreds of times per second), we are able to pick up the energy here on earth as pulses – which is why these stars are called **pulsars**. These energy pulses are often radio, gamma or X-ray waves, but a few of them pulse visibly. One such can be found in the **Crab Nebula**, which is a remnant of a supernova that exploded in 1056, less than a thousand years ago.

What if the core is greater than three solar masses? Well, then, gravity compresses it even further – to such an extent, in fact, that the core collapses into an infinitely dense, dimensionless point with such crushingly high gravity that not even light can escape its clutches. It becomes the celestial monster known simply as the **black hole**, also known as a **singularity** (see p.209). Everything that was in the stellar core is now trapped in the black hole forever; any other material falling into the black hole is similarly doomed.

Variable stars

While most stars live their lives with some measure of stability, at the end of their lives most will also experience some degree of **variability** in terms of energy output and luminosity. Some stars, while actually stable, will appear to be variable because they are **binary stars**: two or more stars that orbit each other. This makes for some interesting twinkling in the night sky. Listed below are some of the different types of variable stars (with the reasons why they vary).

Cepheids

Cepheids are part of a group of variable stars known as **pulsating variables** – stars that actually expand and contract in size over a period of time. This expansion and contraction is caused by a layer of **ionized hydrogen** and **helium** in the deep levels of the star, which makes the layer opaque and keeps energy in. The star expands against this layer until its opacity drops and energy can escape once more, at which point the star contracts and the process starts all over again. Cepheids are **yellow supergiants** and their expansion and contraction is dramatic; their diameter can vary by as much as ten percent their temperature by as much as 1800°F/1000°C, and their luminosity by as much as two magnitudes.

What makes Cepheids particularly interesting to astronomers is that their **period of variability** – the time it takes to expand and contract – is directly tied into the star's luminosity. While the period of a Cepheid can range from one day to a month, all Cepheids with the same period have the same intrinsic luminosity, whether they are near or far from us. Since astronomers can calculate the distance to a star by comparing its absolute brightness against its apparent brightness, this makes Cepheids extremely useful in determining the distances of faraway galaxies and star clusters.

RR Lyrae stars and Delta Scuti stars

Another form of pulsating variable, **RR Lyrae stars** are smaller and dimmer than Cepheids. They also have shorter periods of variability, from a few hours to a day. The variability of their brightness is also less dramatic, usually one order of magnitude or less. Like Cepheids, RR Lyrae stars can be used to measure distances.

Delta Scuti stars have a period of variability of just a few short hours, and their variance in magnitude is so small that you wouldn't notice it without the help of specialized equipment.

Mira variables

The star **Mira**, in the constellation **Cetus**, was the first star known to be variable. The Babylonians were the first to suggest its variable appearance, but it wasn't until 1638 that the Dutch astronomer **Phocylides Holwarda** managed to chart Mira's period of 330 days. Astronomers were so impressed by the star they gave it a name which means "miraculous".

The star also gives its name to an entire class of long-period variable stars, all of which are **red giants** or **supergiants**. With these **Mira variables** the difference in luminosity from one end of the cycle to the other can be immense; a star can be a hundred times more luminous at the top of its cycle than it is at the bottom. Unlike Cepheids, Mira variables also vary in brightness from cycle to cycle, while some even brighten and dim irregularly with no discernable cycle. Mira variables are the most common type of variable star.

R Coronae Borealis stars

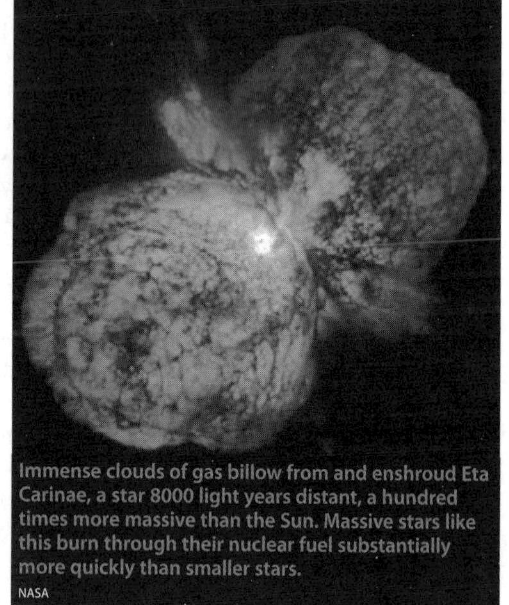
Immense clouds of gas billow from and enshroud Eta Carinae, a star 8000 light years distant, a hundred times more massive than the Sun. Massive stars like this burn through their nuclear fuel substantially more quickly than smaller stars.
NASA

R Coronae Borealis stars are rich in **carbon**, and every so often they blow great clouds of the stuff into space. Once the carbon cools down it condenses into soot – so much soot that it blocks out light from the star itself! As a result the star can lose several orders of apparent magnitude, and take months to get back to its original apparent luminosity.

Luminous blue variables

The very largest stars belong to spectral type O, and when they skip off the main sequence, they become more unstable than other stars. The result is that every few hundred or thousand years, huge amounts of **gas** blow off a star, dramatically increasing its luminosity. As the cloud of ejected gas cools over time, luminosity declines no less dramatically. The most famous example of a **luminous blue variable** is the star **Eta Carinae**, which in the 1840s was briefly the second brightest star in the sky after Sirius, only to drop down past the limits of naked-eye observability a few years later. It's now about fifth magnitude. Eta Carinae's eventual fate is to become a supernova.

Flare stars

Flare stars are exactly what they sound like, stars that flare up as much as one full order of magnitude in a matter of minutes. How do they do it? Flares explode off the surface of the star itself, just like the immense flares that sometimes arc off our own Sun. All flare stars are **red dwarfs**, which are the coolest and dimmest stars out there.

Binary stars

Some stars travel through the universe alone, others have company in the form of another star or stars. These stars circle around a centre of gravity in orbits that range from mere hours to hundreds, or even millions, of years. Such stars are called **binaries**, and they are common, particularly for larger, hotter stars (small stars, like red dwarfs, are usually single stars). **Proxima Centauri**, our closest celestial neighbour at 4.3 light years away, is a binary star, part of the triple-star system of **Alpha Centauri**.

Visual binaries

Some binary stars can be easily observed in binoculars and telescopes as two or more separate stars. **Alpha Centauri** is an example of these so-called **visual binaries**: Alpha Centauri A & B, two bright stars, can be viewed apart with a small telescope, while dim **Proxima Centauri** floats off to the side. In

Could life exist in a binary star system?

We know that single stars can have planetary systems that support life because ours does. But could a system with two stars support life? Or would the heat and energy of two stars simply fry any emerging life? The good news for believers in alien life forms is that a binary system could harbour life, but the stars would have to be a fair distance from each other, in a stable orbit, on the main sequence and not too variable.

As it happens, there is a binary system that fits the bill. Our near neighbours in the Alpha Centauri system are a good example. The two bright stars in the system, Alpha Centauri A and B, are solidly middle-aged (about six billion years old) and move around in an orbit that takes them as close as 11 AU from each other (roughly the distance from the Sun to Saturn) and as distant as 36 AU (as far as Neptune). Each star could have its own system of planets, but they'd need to be in relatively close so as not to be disturbed by the gravitational pull of the other star – roughly 3 AU. In our own solar system, four planets reside within 3 AU of the Sun – Mercury, Venus, Earth and Mars.

Alpha Centauri A is a G2V star very much like our own, while Alpha Centauri B is a slightly cooler, orange K0V star. And in either case, a more distant star would be far enough away that the amount of energy received from it by an Earth-sized planet orbiting the other star would be small indeed. In both cases, not a bad place for life to try its chances.

addition to visual binaries, observers can also see **visual double stars**, stars not in an orbital relationship, but which appear to be close to one another because they lie along a similar line of sight to Earth.

Not all binary stars are visual binaries; their binary nature has to be discovered by other means. One way to do this is to look at the star's **spectrum**. Every star absorbs certain wavelengths of light, which show up as dark bands in the star's spectrum. If a star's spectrum lines are seen to vary their position over a regular period of time, this is evidence that the star is a binary star. These types of binary stars are known as **spectroscopic binaries**.

Eclipsing binaries

There are times when the orbits of binary stars are lined up with Earth in such a way that they can be seen to eclipse each other as they pass in front of and behind each other. When this happens, the apparent brightness of the star dims, sometimes by a full order of magnitude or more. These are called **eclipsing binaries**, and the most famous example is the star **Algol** (in the constellation **Perseus**), whose apparent magnitude drops significantly every three days when the smaller star of the binary pair crosses in front of its larger, brighter partner.

Novas and supernovas

Stars that spend their lives together naturally influence how each other functions. Stars that orbit each other from close distances can even feed off each other, as the more massive star pulls **gas** off its less-massive partner. This can happen, for example, when a **white dwarf** star has a binary partner – although physically smaller, a white dwarf can be more massive, and will suck down gas from the other star.

What's interesting about this process (other than its creepy, vampiric nature) is that the white dwarf's gravity compresses and heats this captured gas, and as more gas accretes, the hydrogen finally erupts, blowing a huge amount of the gas back into space, but not destroying the white dwarf itself. As a result of this, the brightness of the binary star system can increase by hundreds of times; these eruptions are called **novas** (or, to be Latin about it, novae). Novas are rare – they occur on the same star so infrequently that no one's ever seen it happen twice, and here on Earth, we can only see them happen about once a decade or so. However, **dwarf novas**, in which smaller amounts of gas erupt (with the

Some famous supernovas

Supernovas may be simply explosions, but they're celestial events so infrequently observed in our skies that when they occur, people sit up and take notice. Here are some of the most memorable that have occurred over the years:

1054

On July 4 the **Chinese** witnessed a magnificent display of celestial fireworks. The emergence of a supernova in the constellation of **Taurus** was so bright it could be seen during the day for three weeks, and at night for nearly two whole years (although there is no record of the supernova being recorded anywhere else). The wreckage of this particular supernova eventually became the **Crab Nebula**, discovered by English astronomer **John Bevis** in 1731. The Crab Nebula is still expanding, at the rate of about 700 miles/1100km per second.

1572 and 1604

The famed astronomer **Tycho Brahe** observed the first of these supernovas in the constellation **Cassiopeia**. It was visible for 16 months and could be seen in daylight for about two weeks. Brahe's colleague **Johannes Kepler** spotted another in **Ophiuchus** in 1604 that was slightly dimmer but which stayed in the sky for about 18 months. Astronomers today suspect it was a type Ia supernova. It's also the last supernova recorded in the Milky Way galaxy.

1987

Located in the **Large Magellanic Cloud**, a small companion-galaxy to our own Milky Way located some 170,000 light years away, this supernova (named **1987a**) is the closest recent explosion. Even so far away, it was visible to the naked eye, achieving a magnitude of three. The Hubble Space Telescope recently viewed its remains, which displayed gorgeous rings of expanding gas.

star's brightness increasing in a commensurately smaller fashion), are seen rather more frequently.

Sometimes a white dwarf will pull down so much material from its neighbour that the hydrogen reaches **critical mass** and explodes, blowing up the white dwarf in the process. This kind of explosion is a **supernova** called a **type Ia**, and it's even brighter than the type II supernova described earlier in the chapter. The intrinsic brightness of a type Ia supernova can be gauged by how quickly it fades from sight. Astronomers use type Ia supernovas to determine the distances of galaxies that are immensely far from us in space, and in the process learn about the age of the universe. Even in their deaths, stars shed light on the nature of the universe itself.

The Milky Way

and other galaxies

The Sun and all the stars we can see in the night sky are connected by proximity and gravity, part of a very much larger collection of stars known as a galaxy. Stars in their billions make up a galaxy, along with nebulae, dust and interstellar gas. Just like stars, galaxies crowd together to form clusters, and these form even larger clusters, which in turn form vast walls and structures that literally span the universe. But immense as galaxies are, they are merely the cells that make up the body of the cosmos, of which our own galaxy, the Milky Way, is just one small part.

The Milky Way – our home galaxy

When our ancient ancestors looked up into the night sky, they saw a faint glowing band of light streaming across the constellations, becoming particularly bright somewhere near the constellation **Sagittarius**. In order to explain what they saw, they made up stories. The **Seminole Indians** regarded it as a road the dead took to a glorious city in the sky; the **Vikings** likewise saw it as a path to **Valhalla**. For the **Chinese** it was a vast river that separated two lovers, represented by the stars **Altair** (in the constellation Aquila) and **Vega** (in Lyra). The ancient **Egyptians**, whose lives were dominated by the Nile, thought of the Milky Way as a heavenly **Nile**. But it was the ancient Greeks who called it the **Milky Way**, since, for them, it was created by the Goddess Hera spraying the heavens with her milk. Indeed, the very word "galaxy" comes from gala, the ancient Greek word for milk.

In reality, the glowing haze derives from billions of stars, arrayed in a disc across the sky. But the Milky Way is so immense that the vast number of stars we can see are merely a tiny fraction of the total number, which is estimated at between 200 billion and twice that amount. Hanging together by gravity in a space that is 100,000 light years wide and possibly

© NASA

The Earth rising above the surface of the Moon. Photographed during the Apollo 8 Moon mission in 1968.

A lunar eclipse, showing the Earth's shadow cast across the Moon's surface.

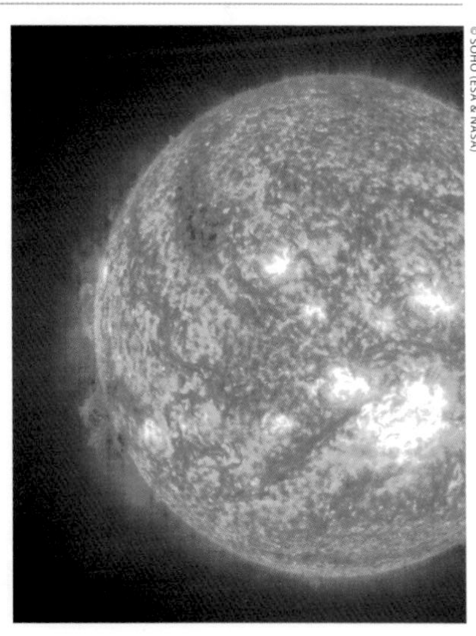

© SOHO (ESA & NASA)

The photosphere, the visible surface of the Sun, is a raging sea of luminous gas.

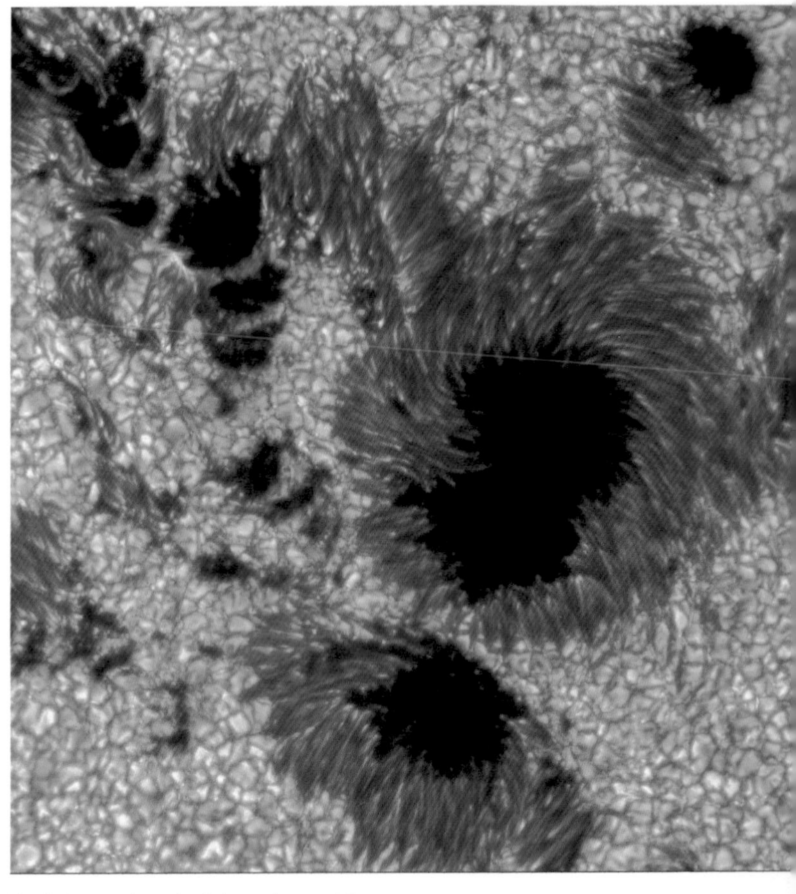

The darker area, the umbra, is the cooler part of the sunspot.

© SOHO (ESA & NASA)

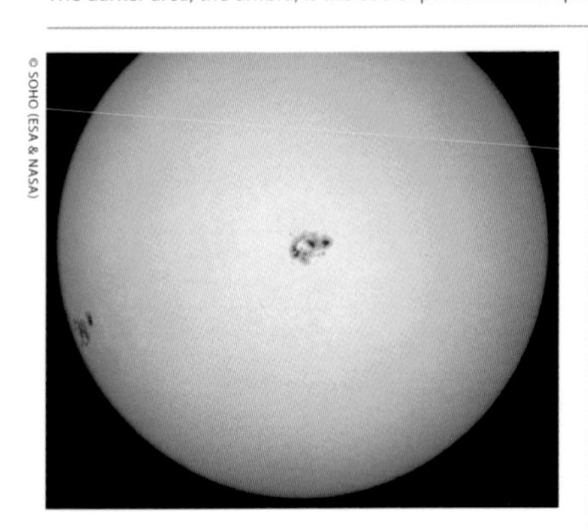

The black marks are an array of sunspots many times the size of the Earth.

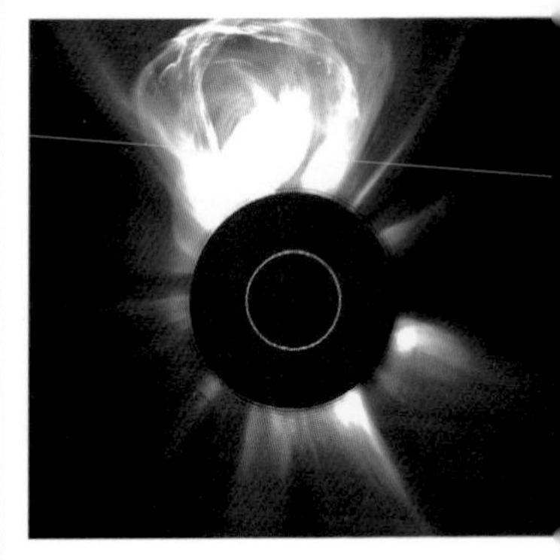

A coronal mass ejection from the Sun, blasting billions of gas particles into space.

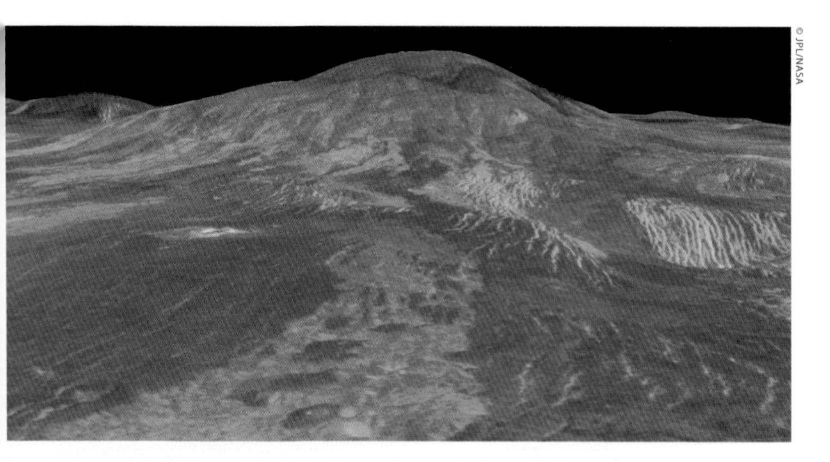

© JPL/NASA

A computer-generated 3D image of Venus showing lava flows at the base of Sif Mons.

© NASA

A composite image of Mercury from photographs taken by Mariner 10. The pale strip is due to an absence of data.

Radar has enabled scientists to penetrate the layers of cloud that surround Venus and produce this image of the planet.

A section of the Valles Marineris (Valley of the Mariners), Mars's great canyon which is over 3000km long.

© NASA/JPL-CALTECH

A panoramic view of Mars's bleak and dusty surface.

Channels within a Martian crater provide evidence of recent liquid water.

© NASA/JPL

A view of Jupiter taken by Voyager 1 in 1979.

© NASA/JPL

A close-up view of Jupiter's 300-year-old storm, the Great Red Spot.

© NASA/JPL

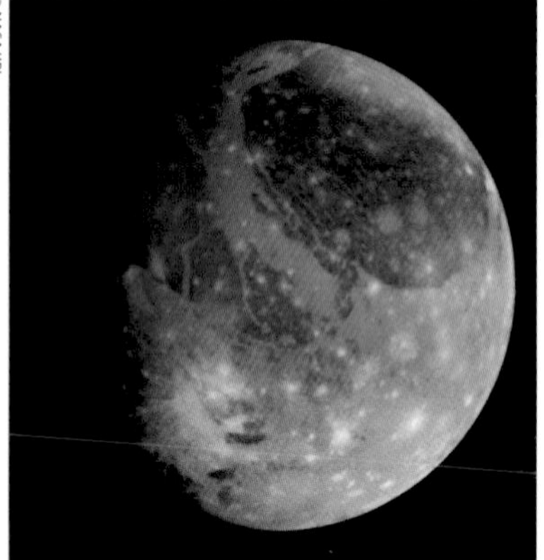

Ganymede, the largest of Jupiter's moons.

© NASA/JPL

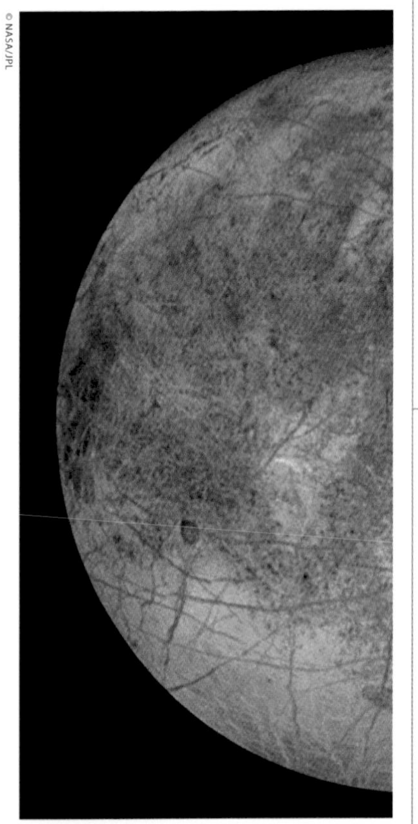

The red lines on Europa are cracks in the icy surface of this Jovian moon.

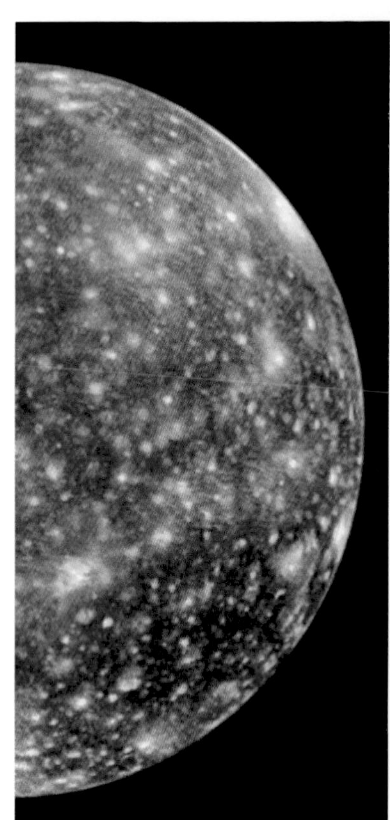

Callisto is the most heavily cratered of Jupiter's moons.

Io, another Jovian moon, is the most volcanically active body in our solar system.

© NASA

A magnificent view of Saturn captured by the Cassini orbiter.

Saturn's B and C rings.

© NASA/JPL

Saturn's A ring and Encke Division.

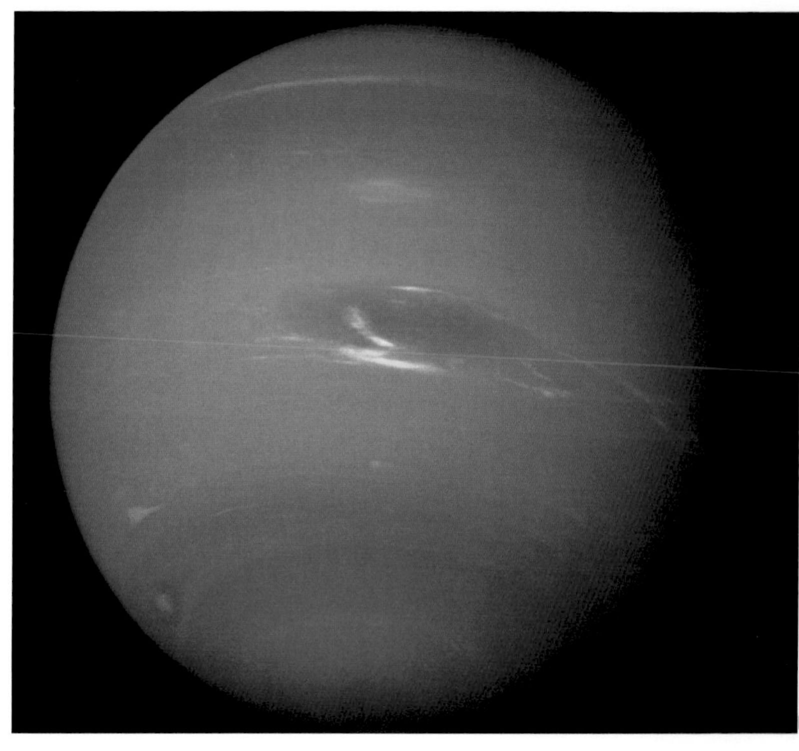

The planet Neptune with its massive storm, the Great Dark Spot, at the centre.

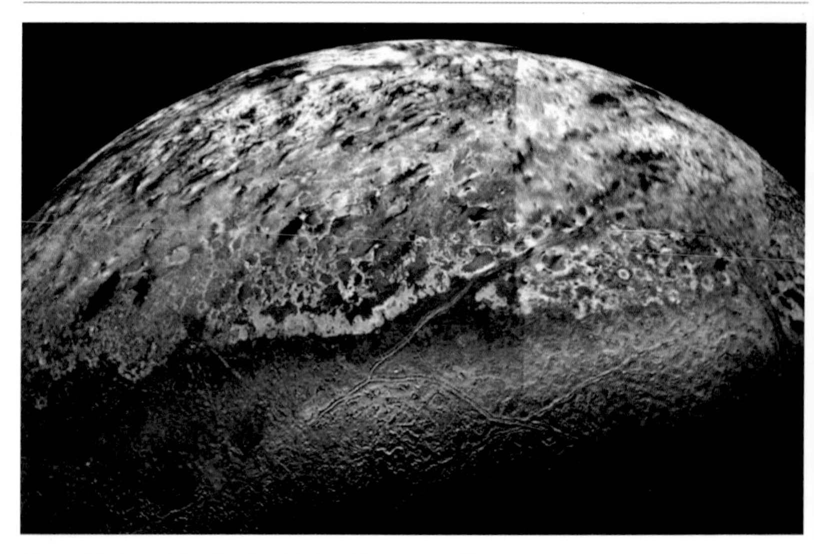

A composite image of the polar cap of Neptune's largest moon, Triton.

© HERMAN MIKUZ, CRNI VRH OBSERVATORY

The tail of Comet Hyakutake grows as it nears the Sun: the heat changes the material on its icy surface.

© NASA/JPL

Hale-Bopp, visible in 1997, was one of the largest comets ever observed.

Radar image of part of the impact crater in the Yucatan peninsula caused by an asteroid hit some 65 million years ago.

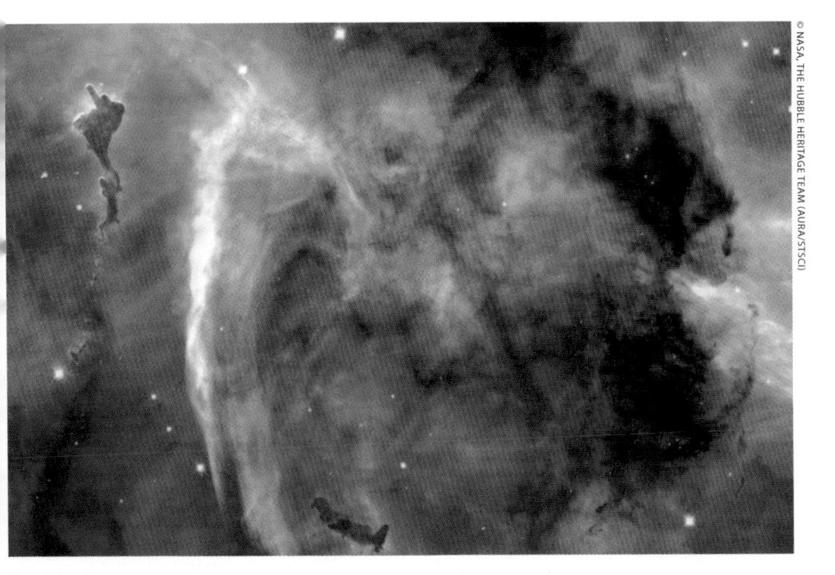

© NASA, THE HUBBLE HERITAGE TEAM (AURA/STSCI)

Part of the Keyhole Nebula photographed by the Hubble Space Telescope.

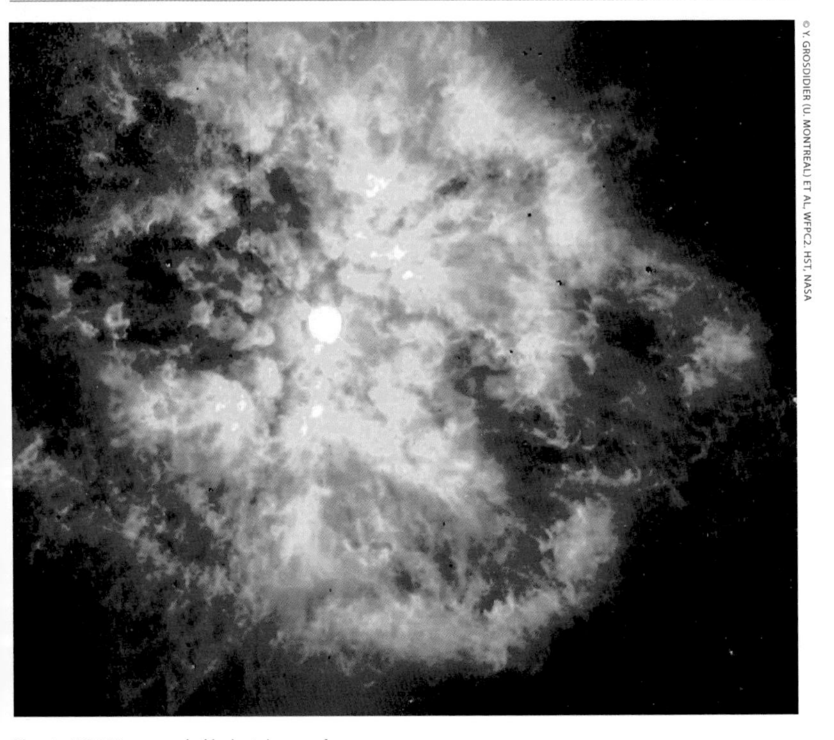

© Y. GROSDIDIER (U. MONTREAL) ET AL, WFPC2, HST, NASA

The star WR 124 surrounded by hot clumps of gas.

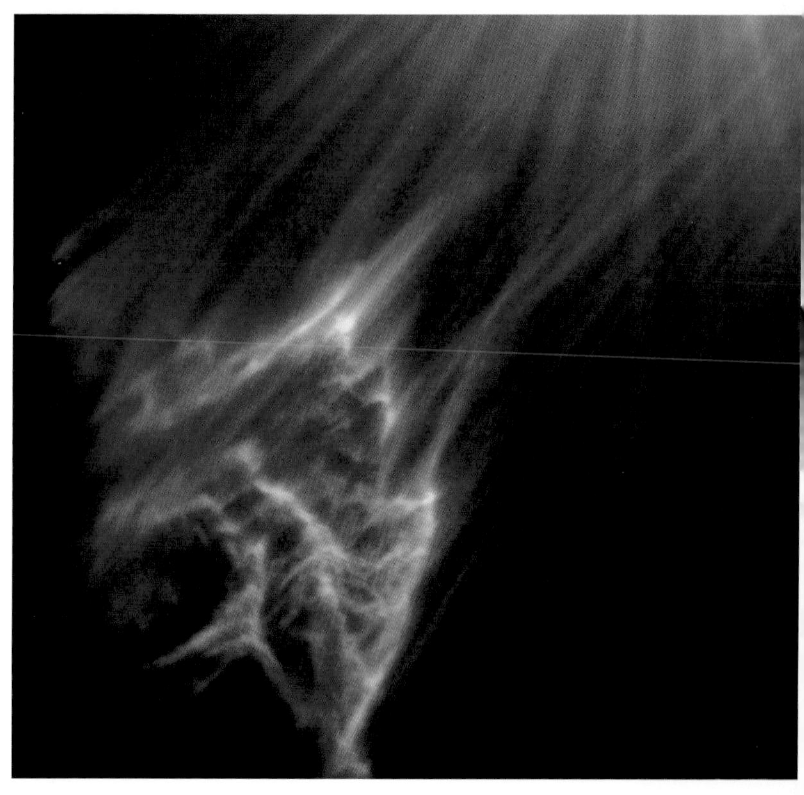

Light from a star in the Pleiades cluster reflected off black clouds of gas and dust.

Dense, opaque dust clouds, called globules, silhouetted against bright stars of the region IC2944.

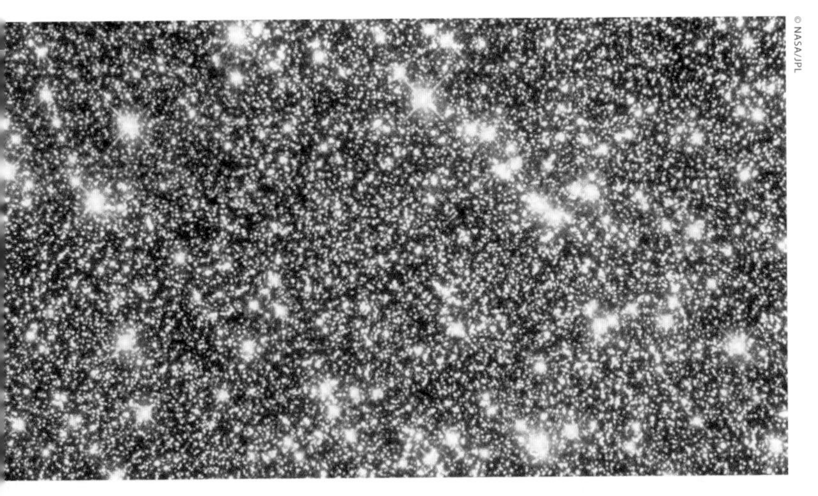

© NASA/JPL

A section of the massive globular star cluster Omega Centauri.

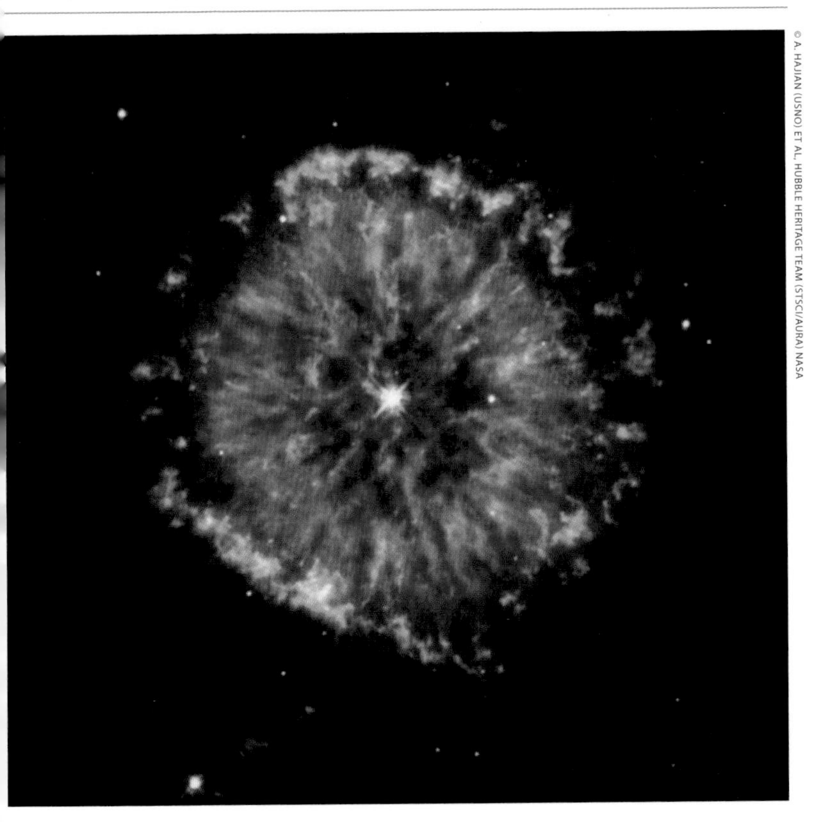

© A. HAJIAN (USNO) ET AL, HUBBLE HERITAGE TEAM (STSCI/AURA) NASA

Hot clouds of gas make up the nebula NGC 6751.

A view of the 30 Doradus Nebula, a vast star-forming region.

An unusual edge-on galaxy, showing details of its warped dusty disc.

© NASA/JPL

NGC4650A is one of the 100 known polar-ring galaxies.

Many galaxies (including our own) are just as thin as the lenticular galaxy NGC 5866, viewed here edge-on.

The gravitational force of a large spiral galaxy draws in a smaller neighbouring galaxy.

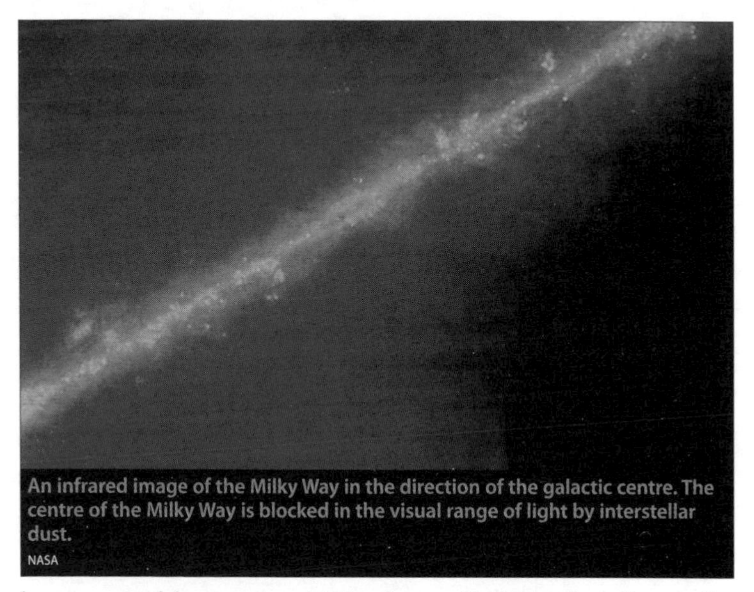

An infrared image of the Milky Way in the direction of the galactic centre. The centre of the Milky Way is blocked in the visual range of light by interstellar dust.
NASA

larger, most of these stars are hidden from us by thick veins of interstellar dust that block visible light. Our own Sun is located somewhere between 25,000 and 28,000 light years from the centre of the galaxy.

Spiral arms

The shape of the Milky Way is a familiar one: a graceful **spiral** that most of us associate with galaxies (although not every galaxy has such a shape). Recent observations, including 2005 data from the Spitzer Space Telescope, suggest that the Milky Way's form is actually a **barred spiral**, which means that the spiral arms of the galaxy emanate out from a huge, bar-shaped structure at its heart. Our Sun rests on one of these spiral arms, in a fragmentary chunk called the **Orion arm**, nestled between the larger arms of **Sagittarius** and **Perseus** (named after the constellations in which they appear).

Galactic bulge

A spiral galaxy like the Milky Way has two distinct portions to it: the **galactic bulge**, the great mass of stars resting in the centre of the galaxy, and the **galactic disc**, which features the spiral arms as well as masses of gas and dust. Each of these areas has a substantially different feel to it, which is largely defined by the type of stars that exist in each. Astronomers have divided galactic star populations into two large categories: **Population I stars**, which are young, and **Population II stars**, which are rather older (some as old as the galaxy itself). The Milky Way's galactic bulge is a huge, dense sphere filled with aging Population II stars that glow orange and red, and with ancient **globular clusters** – smaller collections of similarly constituted stars. In this region of the Milky Way, there's not much in the way of the free-floating gas that helps to create new stars; consequently, within the galactic bulge, star formation has come to a standstill.

Galactic disc

The **galactic disc** is a flattened region spreading out from the middle of the galactic bulge where huge amounts of gas and dust pile up in the spiral arms of the galaxy. There are numerous opportunities for stellar creation in this environment, as well as for the display of star-making nebulae and open clusters of young, hot stars just starting their lives on the main sequence. The Sun is one of these **Population I** stars and is about half the age of the Milky Way galaxy.

Just as planets circle around the Sun, so the Sun orbits the galactic centre. It makes good speed at 140 miles/220km per second, but the galaxy is so large that even at this speed a single orbit takes about 225 million years. Interestingly, although the Sun currently rests within one of the Milky Way's spiral arms, it won't stay there forever but will pass through the arm in its journey around the galaxy (as will other stars). The galaxy's spiral arms, therefore, are not a permanent collection of stars locked in position relative to each other. Although the galactic disc is flattened, it is not perfectly so; astronomers have observed warping of the disc. The origin and extent of the warp is still being determined, although this sort of warping is not uncommon in spiral galaxies.

The size of the Milky Way's disc is a matter of some recent discussion among astronomers. Recent observations of the Andromeda galaxy, our close neighbour, have revealed its galactic disc to extend further than previously thought, and the recent observation of certain structures in the Milky Way, including a new spiral arm, suggest the Milky Way's disc may do the same. The disc of the Milky Way may also be encircled by a proposed **torus** (a doughnut shape) of stars known as the Monoceros Ring; this ring may be the remains of a small galaxy being torn apart by the gravity of the Milky Way, or may just be part of the warp of the galactic disc.

Density waves

Stars, dust and interstellar gases slow down and collect into the spiral arms as a result of **density waves**. What originally causes these density waves, and how long they last, are mysteries that are still unsolved. In the meantime, however, be glad it happens, since the increased density of gas, dust and stars in the spiral arms is one of the primary causes of stellar development, including most probably the birth of our own Sun. Without the arms of the Milky Way acting as a stellar cradle, we probably wouldn't be here.

Stellar halo

As well as the stars within the galactic bulge and the galactic disc, the Milky Way also sports a **stellar halo** of stars that float above the bulge and below the disc. Many of these stars are bound together in over 150 **globular clusters**, and are as old as the galaxy itself. The stars and clusters in the galaxy proper have mostly regular, near-circular orbits around the galactic centre, but the stars and clusters in the stellar halo do not; their orbits are highly **elliptical** and they swoop in and out of the galactic disc at wild angles.

Galactic neighbours

Beyond the confines of the Milky Way are several small **satellite galaxies**, chained to it by gravity. The two best known to us are the **Large Magellanic Cloud**, which sits 170,000 light years away, and its smaller neighbour, the **Small Magellanic Cloud**, at 200,000 light years. For a long time it was believed that these were the Milky Way's closest neighbours, but recently even closer ones have been discovered – the Sagittarius Dwarf, at 80,000 light years away, and the Canis Major Dwarf, which at 25,000 light years distance from our solar system is currently the closest observed dwarf galaxy.

Over time the Milky Way is going to consume these satellite galaxies, merging them into its own massive bulk. Its gravitational pull is already pulling apart the Sagittarius Dwarf, and astronomers expect it will swallow the Magellanic Clouds several billion years from now. The two have already had at least one close call: the path of gas that exists between them, called the **Magellanic Stream**, suggests they were torn from each other

When galaxies collide

The Andromeda spiral galaxy is more than two million light years away, but it's gaining on our galaxy at the rate of more than seven million miles/eleven million km a day. At some point in the future, there's likely to be an intergalactic fender-bender as these two massive entities collide, spilling gas and stars everywhere in a vast cosmic crack-up.

As bad as this sounds, anyone around for the collision won't have to worry about a star sweeping up out of nowhere and suddenly frying the solar system. For all the hundreds of billions of stars in both galaxies, most of the space in the galaxies is just that – space; and individual stars will simply slide by each other. What will collide is gas, and in colliding the merging galaxies will compress these vast, cold fields of gas, thereby experiencing an eruption of star formation unlike any either galaxy has ever seen. Among the new stars will be a spate of supermassive stars that will blaze up and then blaze out in supernova blasts, all in the space of a relative eyeblink.

Once the two galaxies complete their collision, what will remain is a new, larger elliptical galaxy (the delicate spiral arms of both the Milky Way and the Andromeda galaxies being casualties in the collision). An observer from where the Earth is today would no longer see the faint, glowing band of light we see today; instead, he or she would be able to look down all the way into the centre of this new galaxy.

during a close approach. As for the Canis Major Dwarf, it's a likely source for the stars found in the Monoceros Ring.

Dark matter and worse

Even without the added bulk of these satellites, the Milky Way is massive – somewhere between 750 billion and one trillion solar masses. What is interesting about this fact is that most of that mass is not tied up in stars: an estimated ninety percent of it is made up of mysterious **dark matter** that envelops the visible portion of the galaxy in a dark halo. Since no one is quite sure what "dark matter" is, other aspects of the dark halo, including its actual shape and size, are also somewhat murky.

If you think dark matter sounds weird, check out the large, mysterious entity lurking at the exact centre of the Milky Way. This is **Sagittarius A***, roughly the size of the Earth's orbit and millions of times as massive as the Sun. Sagittarius A* emits radio signals, generated by gas and other material falling into its gravitationally gaping maw. In September of 2002, astronomers using the **Chandra X-Ray telescope** confirmed that this mysterious object was in fact a supermassive **black hole**.

A dazzling shot of the Whirlpool galaxy (M51), located over 30 million light years away in the constellation Canes Venatici. It was discovererd in 1773 by Charles Messier who described it as a "very faint nebula".
NASA

The Local Group

The Milky Way and its satellites are not alone: they are part of a small, dumbbell cluster of galaxies, spread out over three million light years, known as the **Local Group**. Included in this cluster are two other spiral galaxies: the **Andromeda Galaxy**, physically larger than the Milky Way, though probably not as massive, and the **Triangulum galaxy**, much smaller than either Andromeda or the Milky Way, but closer in size to the "average" spiral galaxy. The Local Group is also home to nearly three dozen smaller galaxies, most of which are located relatively close to either the Milky Way or Andromeda galaxies and create "satellite systems" for those larger galaxies.

The Local Group isn't isolated either: other nearby galactic clusters are gravitationally influenced by it (and it by them). Over time, the members of the Local Group will change as galaxies wander in and out of the cluster or merge with other galaxies. Even the Milky Way itself won't last forever; there's a good chance that within the next few billion years, it will collide with the Andromeda galaxy to form one huge galaxy (see above).

How galaxies form

Galaxies are some of the oldest individual objects in the universe – nearly all the galaxies we know of were originally formed about a billion years after the Big Bang. While there's a general consensus about the timing of the birth of galaxies, there's some dispute as to how the galaxies formed. Of the two major theories, **monolithic formation** posits they were formed very much like stars, with immense clouds of **hydrogen** and **helium** slowly condensing into huge galactic structures. In this formulation, galaxies gained their familiar spiral shape (or not) depending on their rotation as they condensed. They also left behind relatively small clumps of gas as they contracted, which became the **globular clusters** of stars we find surrounding galaxies.

The other theory, and the one that's gaining in popularity among astronomers, is called **hierarchical formation**. This suggests that many of the large galaxies we see today weren't born but grew incrementally, as groupings of smaller galaxies merged together, and then merged with other groupings, and so on. Some of these ever-larger groupings would eventually manifest some of the characteristic galactic features, like a **central bulge** of stars – and, if their gravity were to attract enough gas, they would also develop the familiar **spiral arms**.

More and more astronomers are favouring this view of galactic formation for several reasons, not least because they can observe galaxies merging together today. Indeed, the Milky Way's cannibalism of the Sagittarius Dwarf is just one example of this process taking place.

Galaxy types

Just as there is more than one type of star – there is also more than one type of galaxy. They range in size from dwarfs of a few million stars to

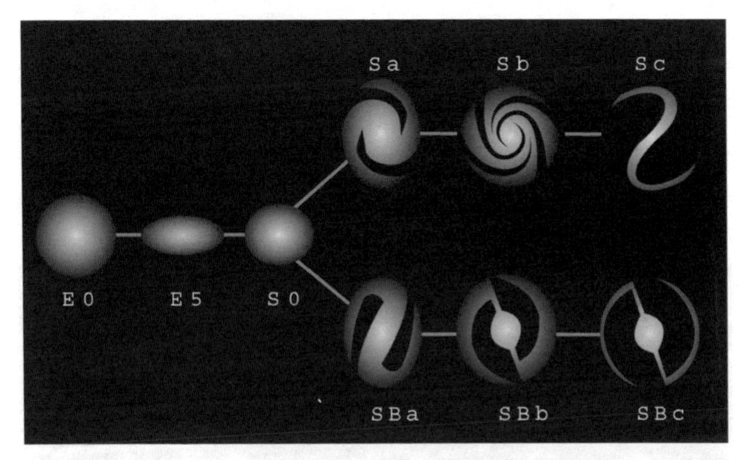

Hubble classification of galaxy types

massive systems of hundreds of billions of stars. And despite the popular image of a spiral galaxy, galaxies take on many different forms as well, all of which are classified according to a system created by the noted astronomer **Edwin Hubble**.

Spiral galaxies

Other **spiral galaxies** share much in common with our own spiral galaxy, the Milky Way: there's the bright, dense **galactic bulge**, filled with aging Population II stars and globular clusters, and the **spiral arms** filled with dust, gas, young stars and open clusters (although the spiral arms are usually not very thick, relative to the width of the galactic bulge). Spiral galaxies range in size from "mere" billion-star galaxies to trillion-star monsters, and their diameters range anywhere from 10,000 to 200,000 light years or more.

Spiral galaxies are classified first by the details of their construction. **Simple spiral galaxies** have spiral arms that appear to branch off directly from the centre of the galaxy. **Barred spirals** feature one or more column-like structures of gas, dust and stars, jammed through the centre of the galaxy, from which the spiral arms flow. About half of all spiral galaxies are barred spirals. Simple spirals are classified by an "S", barred spirals by an "SB".

The second criterion for classifying spiral galaxies focuses on the size of the central bulge and the shape of the spiral arms. A spiral with a large central bulge and tightly wound arms is indicated by a small letter "a", a galaxy with a small bulge and well-developed arms by "c", while one with an appearance somewhere in the middle gets a "b". If a galaxy falls somewhere between the accepted standard for each letter, it will often have two letters in its classification. All of which means that the Milky Way, which is a barred spiral galaxy with a reasonably well-developed system of arms, is classified "SBbc".

A dazzling shot of the Whirlpool galaxy (M51), located over 30 million light years away in the constellation Canes Venatici. It was discovererd in 1773 by Charles Messier who described it as a "very faint nebula".
NASA

Galaxies are haphazardly oriented, so in observations we see spiral galaxies at all angles. These range from **head on**, which shows the classic spiral shape, to **on edge** (for example the Sombrero galaxy), which reveals thick lanes of dust in the spiral-arm structure, with every angle in between.

Elliptical galaxies

Elliptical galaxies are the spiral galaxies' dowdy relatives. These galaxies lack the glamorous arms that grace spiral galaxies. They also lack much of the interstellar gas that is the basis for new star formation, so what you have is an immense elliptical collection of stars, slowly growing older, with no new stars to take their place. There's a very large range of sizes and luminosities in elliptical galaxies, from **dim dwarfs** of a few million stars (like overgrown globular clusters) to supergiants several times larger and more luminous than our own galaxy. Elliptical galaxies are classified by shape on a scale from 0 to 7. An elliptical galaxy which is nearly spherical is an E0 galaxy, while one that's thin and elongated is an E7.

Lenticular galaxies

Lenticular galaxies are in between ellipticals and spirals: they feature a galactic bulge and a disc and may even have a bar like a barred spiral. What they lack, however, is the spiral's arms. It's possible that lenticular galaxies – classified as "S0" galaxies in the Hubble system – represent an actual evolutionary stage between spirals and elliptical galaxies, but no one knows for sure. We do know that most stars in these galaxies tend to be older, and that it is usual to find lenticular galaxies in the company of other galaxies.

Irregular galaxies

Spiral galaxies have a spiral shape, and elliptical galaxies are elliptical, but there are some galaxies that have no organized shape at all; they're just ragged collections of stars and dust, hanging out in the cosmos. These are the **irregular galaxies**, some of which, like the Magellanic Clouds, are close neighbours. Although irregulars don't share the symmetry of spirals or ellipicals, they can share some other galactic features. For example, the Large Magellanic Cloud has a distinct bar shape within it, which means it's known as a **barred irregular**.

Dwarf spheroidials

These small, spherical galaxies have just a few million stars in them. They appear to be related to elliptical galaxies, in that they share a similar shape (though much less dense) and are usually conglomerations of older stars with no ongoing stellar formation. Small they may be, but **dwarf spheroidials** are the most common type of galaxy in the universe, and most of the members

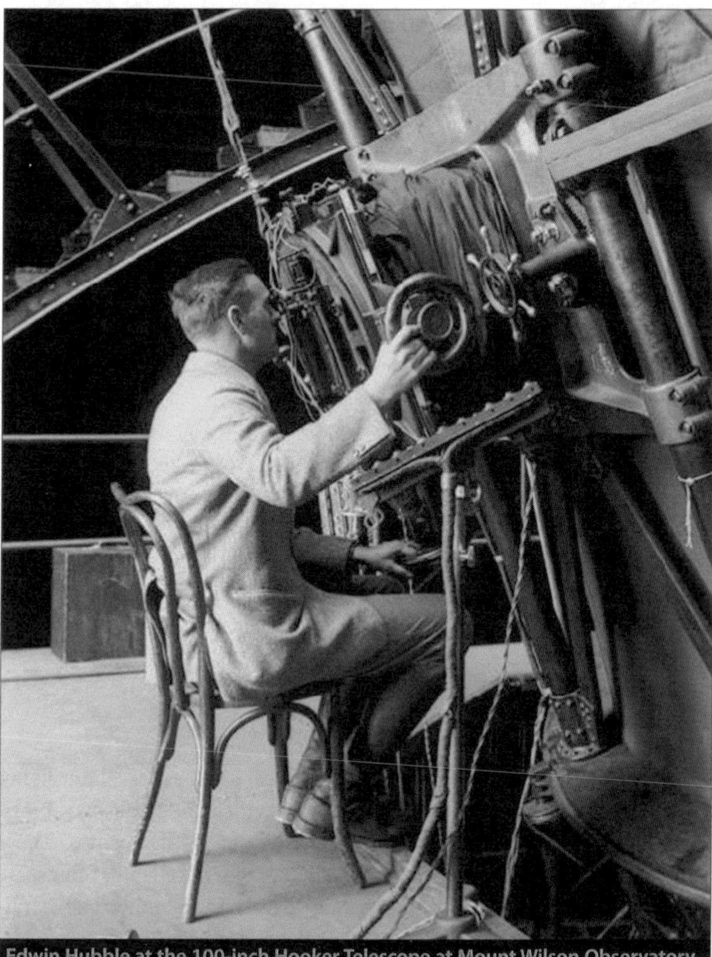

Edwin Hubble at the 100-inch Hooker Telescope at Mount Wilson Observatory in California. Hubble was the first to realize that the Milky Way was not the only galaxy in the universe.

of our own Local Group of galaxies are dwarf spheroidials. Dwarf galaxies also come in other shapes as well, including ellipticals and irregulars.

Low surface brightness galaxies

Low surface brightness galaxies have galaxy-sized collections of gas, but not very many stars at all (almost the opposite of elliptical galaxies which have many stars but little gas). Without many stars, this type of galaxy is extremely difficult to see, and amateur observers will fail to observe them in the night sky.

Active galactic nuclei

The Milky Way is not the only galaxy to have a massive **black hole** at its centre; evidence suggests that most galaxies have one. Much of the time, these black holes sit there quietly (because there's nothing around to get sucked into them), but if a star comes by too closely, or other material somehow spirals into their grasp, then a huge amount of energy is generated all across the electromagnetic spectrum. If there's a lot of material falling into a black hole, the energy created can outshine the galaxy itself, sometimes by a factor of 1000. When this happens, the galaxy is said to have an active galactic nucleus or **AGN**. These can be huge – larger in size than our entire solar system – and they can hide black holes billions of

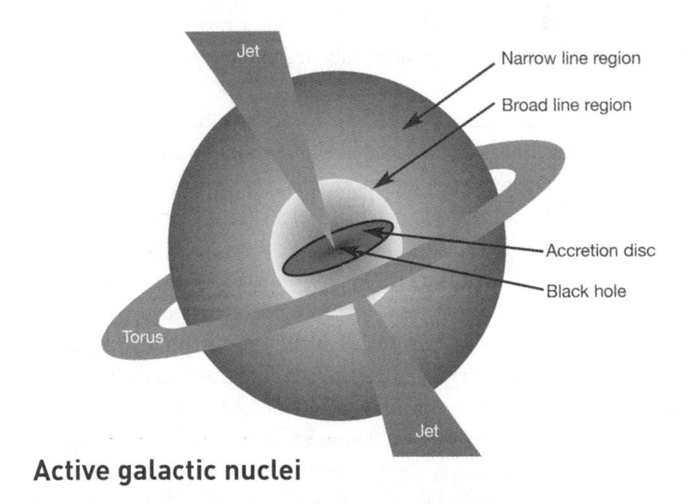

Active galactic nuclei

How did black holes get there?

Astronomers are pretty certain that at the centre of most galaxies lurks a black hole, but how did it get there in the first place? While no one is entirely sure, two main theories have emerged. The first suggests that black holes have always been there. Created in the aftermath of the Big Bang, before even galaxies had formed, their intense gravitational pull acted as the initial "push" which caused gas to begin to accrete around them, eventually forming the galaxies we see today.

The other theory has it that black holes were created after the galaxies, and arose as the result of interactions between the galaxy's stars. If two stars closely approach each other, one of them may lose so much energy that it will "fall" right down to the centre of the galaxy. Get enough of those stars together, and you'll have enough mass in one place to collapse into a black hole, with any later arrivals helping to feed the black hole's mass.

times more massive than our Sun. Some of the different types of AGNs we can observe include the following.

Quasars

Quasars were the first AGNs to be discovered, identified in the 1960s with the help of radio telescopes. Billions of light years away (and thus, billions of years in the past), these objects were exceptionally luminous in the observable spectrum of light, as well as luminously variable, brightening and dimming over the space of less than a day. They also poured out impressive amounts of radio waves and other radiation. Not knowing what to make of them, astronomers labelled them "quasi-stellar objects", or quasars for short.

The true nature of quasars was a controversial subject for many years. It wasn't until the 1990s and the use of the **Hubble Space Telescope** that it was verified that quasars were located in the centre of galaxies substantially dimmer than the quasars themselves (and therefore previously unobservable). Though quasars can be huge, they are very small in relation to their galaxies, and the visible light created by them is generated in that small area. However, the radio emissions created by quasars can stretch out in "lobes" for vast distances outside the quasar's own galaxy.

Since their first discovery, the definition of quasars has been extended to include objects of similar luminosity and variability, but which are not actively putting out radio signals. These are called **radio-quiet quasars** and account for about ninety percent of the quasars we know about. In addition to pouring out light, radio waves and other forms of energy, quasars can also blast tremendous jets of **gas** at high percentages of the speed of light. Indeed, because of a quirk of line-of-sight viewing, some quasars appear to be shooting out gas fragments faster than the speed of light itself (for instance the quasar known as 3C 273, located in Virgo). However, it's all a trick of perspective, and in reality, Einstein's speed limit for light is still in effect.

None of the nearby galaxies we observe have quasar-like objects in them, even though they almost certainly have black holes at their centres. All the quasars we can see are billions of light years away, and therefore seem to be a product of the early history of our universe.

Blazars

Because galaxies are oriented at all angles to the Earth's point of view, sometimes the jet of hot, fast-moving gas bursting off a quasar will appear to be pointed directly at the Earth. When that happens, we get a **blazar** – a special class of quasar that appears to us to be both more luminous and

This unusual spiral galaxy, which lies in the southern constellation of Circinus, is a "Seyfert galaxy" – a type of galaxy with a galactic centre substantially brighter than usual.
NASA

more luminously variable than your standard quasar. Blazars come in two main types: **optically violently variable quasars**, which as the name suggests exhibit wild swings in luminosity, and **BL Lacs**, which in addition to being variable also emit more polarized radio waves than other quasars.

Radio galaxies

Radio galaxies generate the same sort of radio emissions as quasars and have many other characteristics in common, but they aren't as optically bright as quasars typically are. Think of them as quasars with the volume turned down. Some astronomers believe that radio galaxies actually are quasars, positioned relative to our point of view so that the disc of material which has accreted around the black hole blocks the quasar's brightness.

Seyfert galaxies

Seyfert galaxies are named after astronomer Carl Seyfert who spotted the first galaxy of this kind in 1944. These galaxies have optically luminous nuclei, but are not nearly as energetic as quasars – in other words the galactic nucleus doesn't drown out the rest of the galaxy with its glare. About one percent of all spiral galaxies we see are Seyfert galaxies, and there's some suggestion that all spiral galaxies exhibit Seyfert-like tendencies at one point or another.

Clusters

Galaxies are sociable things; it's a rare galaxy that's out there on its own. Galaxies conglomerate into **clusters** – groups of galaxies bound together by gravity. Our **Local Group** of three dozen galaxies is one such cluster, but other nearby clusters include the **M81 Group** (13 galaxies, 12 million light years away), the **M96 Group** (9 galaxies, about 38 million light years away) and the **M51 Group** (8 galaxies, 37 million light years distant).

Clusters are not fixed entities; individual galaxies will sometimes leave one grouping to join another and entire clusters can merge or split. An example of this is the small cluster of galaxies known as the Maffei 1 Group, which features five galaxies about ten million light years away. It's likely that this group was once part of our own Local Group but then broke away.

Regular and irregular clusters

Sometimes clusters have **structure** and sometimes they don't. Those that do are called **regular clusters**. These are spherical in shape and have a core densely packed with galaxies – usually large elliptical galaxies formed from the merging of smaller galaxies. Then there are **irregular clusters** – galaxies that hang together gravitationally but have no well-defined centre. These include the **Virgo Cluster**, which is sixty million light years away and houses some two thousand galaxies. As one of the largest clusters nearest to our Local Group, there's evidence to suggest that in the distant future the Virgo Cluster will eventually pull the Local Group into itself.

Superclusters

As large as the Virgo Cluster is, it is just part of a much larger collection of galaxies known as the **Virgo Supercluster** (or the Local Supercluster). This **supercluster** of galaxies spans more than a hundred million light years, and includes 160 galaxy clusters containing within them more than 25,000 galaxies of all sizes and classifications. This sounds impressive (how could it not?) but as superclusters go the Virgo Supercluster is on the small size. If you want a really large supercluster, the **Horologium supercluster** is almost a billion light years away and spans 550 million light years.

The Great Wall

These superclusters are not randomly distributed across the universe. Detailed mapping of the cosmos shows that galaxies, clusters and superclusters form **walls** and **filaments** of matter that span hundreds of millions of light years. The largest of these structures, so far as we know, is a single **Great Wall** of galaxies that is some five hundred million light years long, two hundred million light years wide, and fifteen million light years deep. This makes it the single largest object in the known universe.

But unimaginably large as the Great Wall may be, consider this: in between these walls and filaments are **enormous voids** in which almost nothing exists – yawning gaps into which galaxies fear to tread. These voids account for ninety percent of the area of the universe. Or to put it another way, the single largest component in the universe is nothing at all.

Our strange universe

"Not only is the universe stranger than we imagine, it is stranger than we *can* imagine."

Sir Arthur Eddington

The more we learn about the universe in which we exist, the more the truth of the above comment is borne out. Every question we answer about our universe ends up provoking several new questions – an endless cycle of discovery. As our knowledge of the universe grows, we know one thing for sure: it's a very weird and wonderful place.

Black Holes – deep, dark mysteries

Perhaps the strangest entities we know about in our universe are **black holes**, those infinitely dense, infinitely small objects created when the cores of large, **burned-out stars** collapse in on themselves through sheer force of gravity. We have already looked in the previous chapter at how black holes come about (see p.195); now let's focus on some of the more bizarre aspects of the phenomena.

Black holes can't be seen

The reason for this is simple: the **gravitational pull** of the black hole is so strong that at a certain distance from it, not even light can escape from it. That distance is known as the **event horizon**, and the more massive the black hole, the larger the distance to the event horizon. The distance itself is called the **Schwarzwald radius**, after German physicist Karl Schwarzwald, who discovered its formula. The event horizon is often thought of as the "surface" of a black hole, but that's not quite true, since there is no physical surface but simply a **boundary**, inside of which nothing can be seen. At the centre

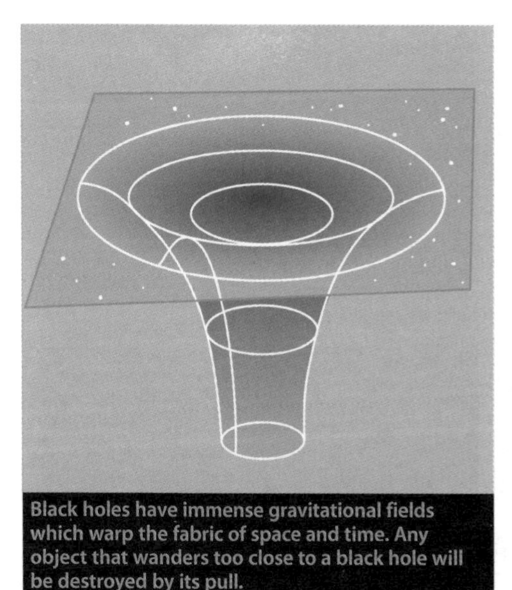
Black holes have immense gravitational fields which warp the fabric of space and time. Any object that wanders too close to a black hole will be destroyed by its pull.

of the event horizon is the black hole itself, the **singularity**, a place where all laws of physics – as we currently understand them – are put on hold.

But if you can't see a black hole, how can you know it exists? Well, you can see the **gas and dust** the black hole pulls into itself, especially if it's feeding off a nearby star. As the gas and dust accumulates, it rotates around the black hole forming an **accretion disc**. This accretion disc heats up to over 1 billion degrees Fahrenheit/555 million degrees Centigrade, causing the matter to emit powerful and detectible waves of **X-ray energy**. One of the first cosmic sources of X-rays ever to be discovered was **Cygnus X-1**, a very small and very hot object which is tearing the surface off its companion star, a **blue supergiant**. We can also see the effects of super-massive black holes in the centre of galaxies, in the intense radiation of quasars, blazars and other **active galactic nuclei**.

There's another way to "see" a black hole, and that's by observing the distortion its gravity creates on a galaxy or star (or other background object) passing behind it. A black hole's **gravitational field** is so intense that light visibly warps around the passing object.

Time acts strangely near a black hole

A black hole's immense gravity causes **time** to slow down the closer one gets to the black hole's event horizon – a strange quirk explained by Einstein's **theory of relativity.** One way to understand this would be to imagine two people synchronizing their watches, before one of them was launched towards the nearest black hole. The closer the space traveller got to the black hole, the slower his or her watch would seem to tick, from the point of view of the person left on Earth, until finally it would seem hardly to move at all. But, from the point of view of the traveller, there's been no slowing down – it's simply that the person on Earth has sped up to a nearly infinite speed.

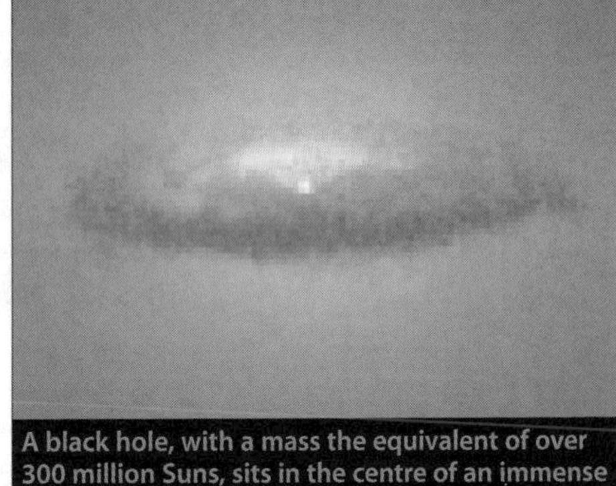

A black hole, with a mass the equivalent of over 300 million Suns, sits in the centre of an immense dust cloud at the heart of elliptical galaxy NGC 7052. All of the dust will eventually find its way into the black hole's gaping gravitational maw.
NASA

Alternative black holes

Scientists theorize that black holes of all different masses could have been created in the incredibly hot and dense universe that existed just after the **Big Bang.** These could have ranged in mass from the unbelievably tiny (the mass of an atom) to some that would have had more mass than an entire galaxy, or even a cluster of galaxies. As was mentioned in the previous chapter, there's some speculation that these sorts of **early black holes** could have been the kernels around which the galaxies themselves formed.

It is hard to believe, but scientists are working towards creating **artificial black holes**: "optical" black holes that can be created theoretically by slowing down light to a few miles per hour in a field of particles called a "Bose-Einstein condensate" and then trapping that light in a swirling vortex of the condensate. These artificial black holes are not a threat (they won't cause the Earth to be sucked up in their maw) because they're not created through gravity; on the other hand they might help physicists answer some thorny questions about the sort of black holes you find in space.

Evaporating black holes

Black holes may **evaporate** one day by **releasing energy** back into the cosmos. You may wonder how this will happen since a black hole's gravitational field is so strong that nothing – not even light – can escape it. Well, according to **quantum physics**, particles of virtual matter can appear spontaneously, seemingly out of nothing at all, and when they appear they are always in pairs. If these particles happen to appear just outside

Physicist Stephen Hawking, who, in 1974, came up with the theory that virtual particle/anti-particle pairs are sometimes created outside the event horizon of a black hole.
Michael S. Yamahita/CORBIS

a black hole's event horizon, one of them might get sucked in, while the other virtual particle might be ejected. When these virtual particles part company, their parting causes energy to be extracted from the black hole's gravitational field. This energy is called **Hawking radiation**, after Stephen Hawking, who dreamed this all up in 1974.

If this happens enough times, then the black hole loses **energy** and thus loses **mass** – since mass and energy are fundamentally the same thing in different forms. And the less massive a black hole is, the more likely it is to create Hawking radiation, and thus lose more mass. Eventually the black hole loses so much mass that it simply **evaporates**, going out in a final, furious flash of **radiation**. This sort of evaporation process is astoundingly slow, especially for larger black holes – trillions of times longer than the universe has actually existed. However, some of the smaller black holes created near the beginning of the universe (those with a mass of less than a billion or so tons) may be just about ready to experience evaporation. So far, no one has observed it happening.

Black holes or "wormholes"

While we picture black holes as dimensionless points in space, sitting there devouring anything in their path, there's a possibility that they are actually **rotating**. And if this is the case, then the singularity at the bottom of their gravity well is not a single point but an almost dimensionless **ring** of extremely **high density**. If such rings exist, it would be theoretically possible to launch a spacecraft (or any type of matter) right through the centre of such a ring, which would then travel unimaginable distances to an entirely different part of the universe (or even another universe altogether). Such black holes are referred to as **wormholes**, or more formally as "**Einstein-Rosen bridges**".

Science fiction fans, of course, know that writers continually use wormholes to shuttle people all over the universe. In the real universe, however, even those physicists who believe in wormholes are sceptical that they could have any real use. While it is theoretically possible that wormholes exist, it's equally possible that they are wildly unstable, or that the intense radiation firing off around one would fry anyone unwary enough to get close.

Dark matter

Every planet, star, galaxy and supercluster we see in the universe is merely the tip of the iceberg as far as matter is concerned. Just as the substantial part of an iceberg lies unseen below the waterline, so more than ninety percent of all the matter in the universe most probably exists as **dark matter** – material we can infer exists, but which we can't see directly.

Astronomers and physicists believe dark matter exists because of the influence it seems to have on the matter we can see. An example of this is the way **stars orbit** in **spiral galaxies** much faster than they should if the observable matter was the total amount of matter in the galaxy – at such high speeds they should be shooting right out of the galaxy. The reason they don't is because the galaxies are far more massive than can be accounted for by the visible matter (the stars, gas and dust), and that this higher mass allows stars to maintain their high speeds and still stay bound to the galaxy.

Dark matter may also account for the clusters and superclusters of **galaxies** that form the **immense filaments** and **walls** we see in our deep-space surveys. When the universe began, the distribution of matter was almost entirely uniform. Today, our charting of the universe shows that **visible matter** (in the form of galaxies) is no longer evenly distributed but clumped together in a far more "lumpy" fashion. To explain this "lumpiness", physicists have postulated a certain type of dark matter that tends to gather into clumps. Such dark matter attracted ordinary visible matter, which then formed the stars and galaxies (and the filaments and walls) we see today.

Astronomers have divided dark matter into two types and, in their wisdom, they have named them **MACHOs** and **WIMPs**. MACHOs stands for Massive Compact Halo Objects, WIMPs for Weakly Interacting Massive Particles.

MACHOs

Also called **baryonic dark matter**, this is really just another name for matter as we understand it, but matter that is dark (i.e. not radiating any energy we can see). Some of these **MACHOs** might include cooled-down **white dwarfs**, **brown dwarfs** (huge balls of gas that are yet too small to become stars), **dust**, **asteroids** or far-flung **comets**. This material could be what makes up the vast "dark halo" surrounding the Milky Way and other galaxies (thus the "Halo Objects" portion of the name).

We can't see MACHOs, but they can give themselves away because of their gravity. If a MACHO wanders in front of a **light source** (such as a star), its gravity can bend the path of that light, concentrating it (from our point of view) and making the star appear to be brighter for a short time. Astronomers have seen stars in the **Magellanic Clouds** brighten briefly, suggesting the presence of MACHOs between the clouds and us. From these observations, they've estimated that there may be hundreds of billions of MACHOs surrounding the Milky Way (although even that number can't account for all the mass in the universe as we see it today).

Some MACHOs are losing their "dark matter" status, as ever more sensitive equipment ferrets them out and observes them directly. In 2001, for example, nearly thirty cold white dwarfs were directly observed, as a result of combing through some of the data of earlier MACHO-hunting observations. All of these white dwarfs were within 450 light years, but most dark matter of this type will remain dark to us for a long time to come.

WIMPs

WIMPs (Weakly Interacting Massive Particles) are a class of subatomic particles that are, for the most part, entirely new and theoretical, although one of them, the **Neutrino**, definitely exists. Neutrinos were long thought to have no mass, but at least one type is now believed to possess a minuscule amount (less than two percent of the mass of an electron), which is enough for it to be thought of as a component of this sort of dark matter. **Axions**, theoretical particles with about a trillionth of the mass of a proton, are also potential WIMPs, and may be so numerous that they make up in numbers what they lack in mass. Other WIMPs that may be heavier than other subatomic particles have been given names such as "photinos" and "squarks" – variants on the names of more familiar subatomic particles.

If WIMPs do exist, they don't interact with "normal" matter very well (which is why they're called "weakly interacting"); they only affect normal matter through **gravity**, or if they smash into it – which doesn't occur very often. This makes verifying their existence extremely difficult. Current attempts to locate them involve using highly sensitive **sensors** which have been cooled to near absolute-zero temperatures in order to help isolate potential impacts from the WIMPs. So far, there have been no verifiable observations.

Dark energy?

In addition to dark matter, astronomers have been thinking on something called dark energy, which is a hypothetical form of energy postulated to explain the fact that the rate of expansion of the universe appears to be increasing over time, which suggests something is "pushing," or exerting negative pressure, on the universe as it expands. Right now, what dark energy is – or even if it actually exists – is a mystery, although there is increasing evidence for dark energy to exist, much of it through observation of type Ia supernovas, using their apparent brightness to measure the history of the expansion of the universe. If the best-understood model of dark energy turns out to be correct, then dark energy would be the predominate form of mass or energy, accounting for nearly three-quarters of the total mass-energy in the universe.

One of the vaguely depressing side-effects of dark energy as we currently understand it is that as the universe continues to accelerate as it expands, the less of the universe we would be able to observe; objects outside our local cluster of galaxies would eventually fall beyond our "cosmic horizon" – they would be literally too far away for their light to reach us. Our "universe" – or the universe of our descendents, billions of years from now, would effectively be a vastly smaller place.

The exploration of dark energy, however, is at its very earliest stages; expect it to be a focus of controversy and wild speculation (i.e. the scientific process in its rawest form) for some time to come.

Antimatter – not just science fiction

The idea of **antimatter** – familiar to all *Star Trek* fans – is very simple: it's like normal matter, but the opposite. So, for example, instead of **electrons**, which have a negative electrical charge, antimatter has **positrons**, which have a positive charge. And while the proton has a positive charge, its antimatter version, called an **antiproton**, has a negative charge.

Antimatter is extremely rare, primarily because when matter and antimatter meet, they annihilate each other, transforming all their matter into

BESS about to be launched on one of its 36-hour searches for antimatter.
NASA

energy in one single blast. Despite the obvious difficulties of detection, positrons were discovered in 1932, just a few years after physicist **Paul Dirac** predicted the existence of antimatter. Then in 1995, scientists in Switzerland managed to create nine atoms of **antihydrogen**, each of which existed for a few billionths of a second before colliding with atoms of normal matter, causing the annihilation of both. More recently, larger batches of antimatter have been created – more than 50,000 antihydrogen atoms were whipped up in September 2002.

One of the big mysteries of the universe is why there is so much matter around, but hardly any antimatter. Physicists hypothesize that when the universe was formed, equal amounts of matter and antimatter were created along with it. Of course, it could be that matter predominates locally and that antimatter predominates elsewhere; there could be entire galaxies with antimatter stars, nebulae, planets and even people. In the 1990s, a joint NASA–Japan project called BESS (Balloon-borne Experiment with a Superconducting Solenoidal magnet) launched sensors into the upper atmosphere to look for evidence of faraway antimatter (specifically, cosmic ray antihelium), but found nothing conclusive.

Extra-solar planets

Until recently, astronomers had no proof that our solar system was not the only one to exist around a main sequence star. Then in October 1995, **Michel Mayor** and **Didier Queloz** of the Geneva observatory announced that they had spotted a planet, about half the mass of Jupiter, orbiting the star **51 Pegasi** in the Pegasus constellation. Since that initial discovery, astronomers all around the world have spotted more than 260 planets around other stars. In some instances they have discovered more than one planet around a single star, indicating the existence of complete planet systems like our own.

Observing the wobble

Although a very small number of extra-solar planets have been directly observed (beginning with a massive planet orbiting the star 2M1207 in the Centaurus constellation, sighted in 2004), most extra-solar planets were not discovered by direct observation. The planets we detect around other stars are tremendously small relative to the star, and those in close orbit are also affected by the glare of the star. As a result, planets are detected indirectly, by observing the star itself. When a planet orbits a star, the two of them actually orbit a **common centre of gravity**, known as a **barycentre**. Given the relative masses of the star and the planet, this centre of gravity is far closer to the star than it is to the planet.

For example, the barycentre between the Sun and Jupiter lies just barely outside the surface of the Sun. As the star and the planet orbit, we can observe the star's "**wobble**" around the barycentre, and from that wobble, we can determine the size of the planet and its orbit around the star. Currently, we can track this wobble in two ways: by radial velocity and by astrometry. **Radial velocity** is derived from the fact that when a star wobbles, it moves very slightly closer to us, and then ever-so-slightly further away. When it moves away from us, distinctive absorption lines in its spectrum shift towards the red side of the spectrum; when it moves towards us, they shift towards the blue. By measuring how the spectrum lines shift over time, we know the size of a star's wobble. The second method, **astrometry**, measures a star's movement against other, more distant stars (whose distance makes them effectively stationary). By measuring how much the star wobbles relative to the other stars, we again can figure out both size and orbit.

Detecting the planets

Astronomers have ways to detect the planets themselves – not just the wobbles they cause their stars. The aforementioned planet around 2M1207 (known as 2M1207b) was discovered through **infrared imaging**. Another method for spotting planets is called **optical interferometry**. This technique trains more than one telescope on a star and then uses the interaction of light between the two telescopes to "cancel out" the light waves coming from the star directly. This allows that which is usually hidden in the glare of the star – such as a planet – to be seen. Optical interferometry has already been used to locate planet-like objects orbiting faraway stars, and newer instruments, such as the twin telescopes of Hawaii's **Keck Observatory**, are beginning their own planetary explorations. Another technique is microlensing, in which the bending of light caused by a massive object – like a planet – is used to verify its existence; how much the light bends tells astronomers how massive the object is. This technique has been used in the discovery of several planets.

One of the drawbacks of the current progress of planet-detecting technology is that it can only locate planets considerably more massive than the Earth. The smallest confirmed planets have masses more than five times that of Earth, and many others are the size of Jupiter and even larger.

Pulsar planets

51 Pegasi was the first main sequence star, other than our own, to be found with a planet orbiting it. Then in 1991, radio astronomer **Alexander Wolszczan** of Pennsylvania State University found planets orbiting a **pulsar** – the rapidly spinning, burned-out corpse of an exploded star. Unlike the planets orbiting main sequence stars, these planets and the pulsar (known as **PSR 1257+12**) were relatively small: two planets with about three times the mass of Earth, and one with the mass of our Moon (a possible fourth planet, about the size of Saturn, was found later). Wolszczan discovered the planets when he realized that the signals coming from the pulsar were being "pulled" – stretched at some points, while being compressed at others. This suggested that something (or somethings) were making the pulsar move.

The planets around PSR 1257+12 are extremely intriguing: first of all they are the smallest planets we know of outside the solar system, and then there's the fact that the planets exist at all – after all, a pulsar is what's left of a star that has exploded. Since the orbits of the Earth-mass planets are similar to those of the planet Mercury around our Sun, there is no way, at that distance, that they could have survived the star going supernova. It is therefore possible that these planets were created after the original star went supernova. This is good news for astronomers, since it suggests that planet formation is actually common. This in turn means more planets to discover and a greater possibility of discovering life elsewhere in the universe. Not that you will find it on the planets of PSR 1257+12, since these are bathed in lethal radiation which is being hurled out from their dead star.

Hot Jupiters and habitable zones

Many of the giant planets we've discovered exist in orbits that are incredibly close to their stars. For example, the planet around **51 Pegasi**, the first main sequence planet detected, orbits its star from less than five million miles/eight million kilometres and makes a complete orbit in just over four days. Dozens of other giant planets are also jammed in close to their stars; enough for astronomers to have given them their own category – **hot Jupiters**. The existence of so many of these planets has puzzled astronomers, because in the one previous example of a planetary system (our own), the larger, gaseous planets all orbit far away from the Sun, while smaller, terrestrial planets orbit close in. Further discoveries have uncovered more "conventional" systems, however, including one discovered in June 2002 around the star **55 Cancri**, which appears to be much closer in design to our own system.

Another system that looks a lot like ours orbits the star **Gliese 581**, located in the Libra constellation; this system includes three planets with relatively low masses (that is to say, only a small multiple more massive than Earth), including two that are within the "habitable zone" for that star, just as the Earth is in the habitable zone for the Sun. It doesn't mean life will be found on these planets (or even if it were, that it would be like ours), but for astronomers it's still an exciting discovery.

The Kepler mission

Will we ever "see" Earth-sized planets? The "low-mass" planets around Gliese 581 are examples of how our ability to detect small planets is increasing, and there is more to come as spacecraft designed to detect planets start their work. One of these spacecraft is already on the job: COROT (short for COnvection ROtation and planetary Transits), launched by the French Space Agency and the ESA in December of 2006. It's capable of finding planets a few times larger than Earth. NASA has a spacecraft specifically designed to locate Earth-like planets in habitable zones; it's called the Kepler mission. It was originally scheduled for launch in 2007 but delays pushed it back to 2009. Kepler will use a specialized telescope to hunt out terrestrial-sized planets around Sun-like stars. Because the light from a star is dimmed briefly (when a planet crosses in front of the star's disc), by measuring the amount stars dim over the course of the mission life, Kepler should be able to spot planets down to the size of Mercury. And by noting the frequency of the transits, it should also be able to provide information about the planet's orbits and their temperatures. Advances in **optical interferometry** (see p.000) are also bringing our capabilities closer to the point when we may be able to spot "near-Earths", and open a new chapter of speculation about the possibility of life out there.

The search for extraterrestrial life

Given our history, it's not unreasonable to assume that one day we'll find out that life, rare as we think it to be, is just another common feature of the universe.

But as much as the average astronomer (or person in the street) may believe or hope in life "out there", the actual scientific search for other life in the universe has largely been relegated to sideshow status compared to other astronomical endeavours. "Little green men" make for good television and movies, but apart from the film *Contact* (based on a novel by astronomer Carl Sagan), it's all much more "fiction" than "science", and the popular ideas of UFOs, crop circles and big-headed aliens mad for bizarre probing techniques doesn't help to give the search for **extraterrestrial intelligence** a serious or respectable sheen.

Speed limit

Nevertheless, the search for extraterrestrial intelligence is grounded in the reality of physics; as much as we like to picture warp drives and hyperspace, in the real universe we are (so far as we know) bounded by one very real constraint – the **speed of light**. At 186,000 miles/300,000km per second, light moves almost unimaginably fast, but because of the immense distances between stars, it can take years, centuries, millennia for light (or any other energy) to reach us from another star.

The fastest spaceship created so far has been the **Cassini spacecraft**, which sped towards Saturn at 43,000 miles per hour – the tiniest fraction of the speed of light. Practically speaking, it's not very likely that humans

Faster than light

The problem with space travel is that you can only go so fast. The upper speed-limit is the speed of light, and even getting to that takes effort. Thanks to the nature of the universe, as described by Einstein, the closer matter gets to the speed of light, the more energy it requires to go faster. In fact, actually achieving the speed of light requires an infinite amount of energy, which we're obviously not going to be able to provide. So if we're going to move around the entire universe at speed, we have to find a way to do it without actually breaking the speed of light. It sounds counterintuitive, but theoretically there may be ways to make it happen.

One way might be through wormholes, which can be understood as tunnels through space. Rather than taking the long way round, the tunnel connects far-distant points, just as folding a sheet of paper can connect opposite ends of the paper. Assuming wormholes exist, one of the practical problems about using them to get around the universe is that you wouldn't have any idea where exactly in the universe you were coming out (or, for that matter, when).

Another way might employ something like the famous "warp drives" of Star Trek fame. Interestingly enough, there is a theory that is reminiscent of a warp drive which involves detaching a "bubble" of normal space from the rest of space. While everything within the bubble and normal space moves at the speed of light or less, the bubble itself can travel faster than the speed of light, and can, presumably, reattach itself to space when it's reached where it wants to go. This is called an Alcubierre Warp Bubble. Unfortunately for us potential Starfleet captains, such a warp bubble requires the use of exotic matter that is, at this point, purely theoretical.

will travel to other stars in the near future. But it's also a good bet that other life forms, should they exist, are constrained by the speed of light just as we are.

Listening for the signals

So how will we find any extraterrestrials (or how will they find us)? Quite simply by listening. Since the 1930s, humans have been announcing our presence to the universe in the form of the **radio** and **TV broadcasts** that we've been putting out over the air. These signals have been travelling into space all that time, and the earliest transmissions are now more than sixty light years distant. So, while the more refined among us may recoil in horror over the idea that the first contact extraterrestrials might have of us is a rerun of *I Love Lucy* or *The Flintstones*, it's a bit too late to worry about it now.

By the same token, if intelligent life forms exist out there in the universe, it's not unreasonable to assume that at some point in their development they may also be broadcasting their existence to the cosmos – either intentionally or unintentionally like us. If that's the case, what we humans need to do is put our ears to the sky and listen for signals which are clearly non-natural.

The one drawback to this approach is that if we pick up a signal from very far away, we won't actually be able to communicate, just receive. A signal picked up from a star five hundred light years away would have started on its way to us around the same time that Magellan was circumnavigating the Earth. It wouldn't be a conversation so much as archeology. But it'd still be interesting to hear.

Radio telescopes

How do we put our ears to the sky? By using radio telescopes and scanning for signals that indicate intelligence. The very first attempt to do this was in 1960, when radio astronomer **Frank Drake** trained the radio telescopes at Green Bank, West Virginia, at two stars similar to our Sun, **Tau Ceti** and **Epsilon Eridani**, to hear what he could hear. Apart from one false alarm caused by a military project, Drake heard nothing but static. More recently, an extensive search project called **Project Phoenix** used the largest radio telescope in the world (at **Aricebo,** in Puerto Rico) to hunt for signals from space. Nothing definite turned up.

The radio telescope at Green Bank, West Virginia, the site of several searches for extraterrestrial life
Scott T. Smith/CORBIS

Those who have been searching for signals have long wished for a telescope they could use full-time, and in October 2007, they got one in the form of the **Allen Telescope Array**. The ATA (named after Paul Allen, a founder of Microsoft and the man who is primarily funding the project) chains together more than forty small radio telescopes in a single array, and there are plans to eventually chain 350. The ATA's current SETI mission includes surveying the galactic centre; as more telescopes come online the array will expand and refine its survey to target specific stars. In addition to searching for extraterrestrial life, the ATA is simultaneously carrying out other related research.

The Drake Equation

How many new civilizations can we expect to discover during the course of our search for extraterrestrial intelligence? No one can say for sure (or even if we'll find any at all). But Frank Drake, who did the first systematic search for extraterrestrial life, has come up with an equation to provide an estimate of how many civilizations could be capable of sending signals into space. It's called the **Drake Equation**, and it looks like this:

$$N = R^* \cdot fp \cdot ne \cdot fl \cdot fi \cdot fc \cdot L$$

"N" is the total number of civilizations in our galaxy we can detect. You get "N" by multiplying all the factors on the other end of the equation together. Here's what they are:

R* This variable considers the rate of formation of stars suitable for the development of life – some stars are too massive and burn out too quickly, while others are too small and don't generate enough energy.
fp Of those stars, how many of them have planets around them? That is what this factor stands for.
ne This represents the number of those planets actually capable of supporting life. In our solar system, only one planet out of nine (so far as we know) is capable of that.
fl Of those planets which can support life, how many do? That's the number this variable represents.
fi Mere life is not good enough – it has to be intelligent life. This variable tracks that number.
fc It's also not enough to be intelligent – the intelligent life on other planets has to develop technology that we can actually perceive; radio signals

Messages into space

Humans aren't only sending reruns and rap music into space. On occasion we actually make the effort to send messages that actually tell a little bit more about ourselves. Here are two messages we've sent into the great dark universe:

▶ **The Pioneer Plaque** This was a plaque attached to the **Pioneer 10** spacecraft, launched in 1972, which headed out into the Milky Way after a rendezvous with Jupiter. Pioneer 10's path beyond our solar system doesn't take it near any stars in particular, but just in case someone sees it floating by, the plaque is designed to provide some basic information about who we are, and where we are. Who we are is represented by a picture of a **human male** and **female** (the male raising his hand in greeting), whose size is represented by a unit of measurement based on a radio wavelength put out by the hydrogen atom. The plaque also shows a model of our solar system, with the path Pioneer 10 took out of it, and the location of the Sun, relative to the centre of the galaxy and 14 nearby pulsars. The plaque was designed by astronomer **Carl Sagan** and artist **Jon Lomberg**.

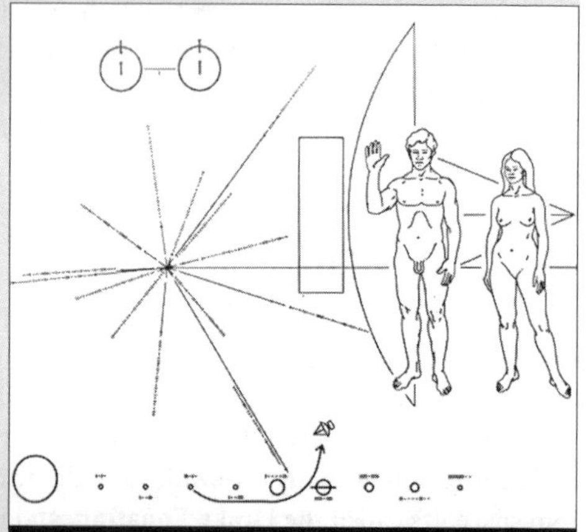

An image of the plaque attached to the Pioneer spacecraft. It's unlikely that it will ever be examined by extraterrestrial life, which given its rather cryptic and soulless character, might well be a good thing.
NASA

▶ **The Voyager Plaques** When the twin spacecraft Voyager 1 and Voyager 2 were launched in 1977, Carl Sagan was again involved. This time he decided to be a little more ambitious in his choice of material, which was to be stored in a kind of time capsule. Each capsule contained a golden record (a stylus and cartridge was thoughtfully included, and instructions on how to play the record were etched onto the record itself). On the record were more than one hundred different images from Earth (in the form of recorded signals) plus a collection of natural sounds like rainfall and chimpanzees, as well as several dozen audio greetings in Earth's many languages. There's also a 90-minute selection of music which includes Navajo chants, Glenn Gould playing Bach, and Chuck Berry singing "Johnny B Goode" (the last of which prompted one comedian to joke that the first message we'll ever get from space may well be "Send more Chuck Berry!"). The records were designed to last a billion years. Since it'll be at least 60,000 years before either Voyager spacecraft comes near another star, that is probably a good thing.

are the most obvious, but there may be others. This variable takes that into consideration.

L Finally, for how long does a civilization which can broadcast its existence so that we can perceive it actually do so? Time is of the essence, and that's what this variable is concerned with.

What "N" finally ends up being depends on the numbers for the variables, most of which we have no certain way of knowing. But consider that there are probably two hundred billion stars in the galaxy. Let's suppose that only one percent of those stars could support life, and half of those had planets, and half of those had planets that supported life. Of those planets, let's say one in five actually does support life, and only one in a hundred of those support intelligent life. Among those intelligent species, one half are capable of sending signals into space, but only one in fifty is actually doing so right now.

Even after all that winnowing, we're still left with **ten thousand planets** with intelligent life shouting out into the void. Now all that's left to do is find them. Maybe one day we will.

STAR CHARTS AND HOW TO READ THEM

Star charts and how to read them

The night sky holds thousands of stars and dozens of nebulae, clusters and galaxies visible to the naked eye – and the numbers expand exponentially with the use of binoculars and telescopes. Since it would be impossible to keep track of them all without some sort of system, several methods of classification have been devised over the centuries, the most important of which groups them by constellation.

Constellations

The **constellations** began simply as collections of stars that our ancestors grouped together. Often this was simply because the patterns the stars formed seemed to resemble something, whether it be animals (bulls, lions, birds, snakes), people (heroes, hunters, virgins) or objects (arrows, ships, scientific equipment). Nearly every culture devised their own constellations, but most of the ones we now recognize as "official" come from a few particular sources.

The largest proportion come from the Almagest, a collection of astronomical and mathematical texts by the Greek astronomer **Ptolemy of Alexandria**, who was active in the middle of the second century AD. Among the 48 constellations he noted are the signs of the **Zodiac** (which lie on the path of the **ecliptic** and through which the Sun moves over the course of the year) and nearly all of the prominent, well-known constellations visible from the northern hemisphere. Ptolemy didn't create these constellations, he simply gathered information which had been accumulated over the years.

About 1400 years passed before any new constellations were observed that still survive to this day. In the 1590s Dutch navigator **Pieter**

Dirkszoon Keyser, assisted by **Frederick de Houtman**, added a dozen constellations to the southern sky, which had until that time been largely unseen by Europeans. After Keyser's death, his charts were passed on to Dutch astronomer **Peter Plancius**, who then added several of his own constellations, three of which are still recognized. Seven more constellations were added in the 1680s by Polish astronomer **Johannes Hevelius**, while French astronomer **Nicolas Louis de Lacaille** added fourteen more southern constellations in the mid-1700s. The constellations were finally "fixed" in the early twentieth century by the **International Astronomical Union**, which recognized 88 official constellations and set each constellation within specific boundaries.

Naming the stars

Nearly all the most prominent stars have traditional names, and many of these use **Arabic words** – a reflection of the importance of Arab astronomy between the ninth and eleventh centuries AD. Stars beginning with the letters AL (Aldebaran, Algol, Altair) are nearly always of Arab origin since Al is the Arab definite article. The exception to this rule is, of course, stars with Alpha in their name – Alpha being the first letter of the **Greek alphabet**. This follows a system introduced by astronomer **Johannes Bayer** in 1603, whereby stars are named after a **Greek letter** followed by the genitive (possessive) case of the constellation's **Latin name** (e.g. Alpha Centauri, Sigma Orionis, Mu Scorpii). The ordering of the letters usually – but not always – reflects the brightness of the star: so Alpha Centauri is the brightest in the Centaurus constellation, followed by Beta Centauri, which is brighter than Gamma Centauri and so on. Bayer's system is very useful for locating a star. If someone refers to the star Theta Tauri, for example, you know that the star lies within the boundaries of the constellation Taurus but that it's not likely to be very bright. Other star-naming schemes use numbers and letters in association with the constellation name (T Tauri, 19 Piscium). Just to confuse the matter still further, named stars are also known by Greek letters so that Aldebaran, for example, is the same star as Alpha Tauri.

When a star is named using a letter of the Greek alphabet (followed by the constellation name), the letter is usually spelt out when it appears in the text, but written just as the Greek character itself when it appears on a star chart e.g. Alpha Centauri (in text) = τ (on the Centaurus constellation chart).

Beneath is the full Greek alphabet (lower case) followed by the spelling of each character.

α Alpha	ϵ Epsilon	ι Iota	ν Nu	ρ Rho	φ Phi
β Beta	ζ Zeta	κ Kappa	ξ Xi	σ Sigma	χ Chi
γ Gamma	η Eta	λ Lambda	ο Omicron	τ Tau	ψ Psi
δ Delta	θ Theta	μ Mu	π Pi	υ Upsilon	ω Omega

It's important to note that stars which appear related in constellations very often have nothing to do with each other in reality. Stars we see as being close together in our night sky may in fact be hundreds or even thousands of light years apart in actual space. Their "relationship" is merely due to the fact that we see the night sky as a **two-dimensional backdrop** – all stars are so far away from us that in effect they all seem equidistant. However, if we were to look at the night sky from a position several light years distant to our own position in space, most of the constellations we recognize would be stretched and twisted as the stars' positions changed relative to our own.

In addition to the official constellations, there are several prominent **asterisms**: patterns that aren't recognized as constellations in their own right, but which are nevertheless recognizable objects in themselves. The most famous asterism is the **Plough** (also known as the Big Dipper), located in the constellation Ursa Major; others include the **Teapot**, in Sagittarius, and the **Circlet**, in Pisces.

Nebulae, clusters and galaxies

Non-stellar or "deep-sky" objects like nebulae, clusters and galaxies aren't typically known by their constellation, although there are some notable exceptions, like the **Orion nebula** or the **Andromeda galaxy**. Some objects are given proper names that describe them in some way, like the Crab nebula and the Sombrero gGalaxy, the names of which both fancifully describe their shape.

Most, however, have been catalogued over the years in one of the several catalogues of deep-sky objects. The **Messier Catalogue**, the most well known, was begun in 1760 by the French astronomer Charles Messier. Messier listed over a hundred objects, all denoted by the letter "M" plus a number. Other prominent catalogues include the **New General Catalogue** (NGC), first compiled in 1888, and the **Index Catalogue** (IC), first compiled in 1895 (with a second volume added in 1908). The result of this profusion of catalogues is that deep-sky objects can have several different names (or catalogue listings); for example the Andromeda Galaxy is known as both M31 and NGC 224, and the Crab Nebula as M1 and NGC 1952.

Using the star charts

Most of the individual charts cover a single constellation, although several charts have combined two of the smaller or more undistinguished constellations into one entry. In each chart the most prominent stars and deep-sky objects are labelled (along with the neighbouring constellations). Stars with proper names are shown with those names; likewise deep-sky objects with popular names, like the Crab Nebula or the Pinwheel Galaxy. Those stars that are labelled also have their **apparent magnitude** listed, so you can tell how bright a star is supposed to be; bear in mind that variable stars will visibly fluctuate in brightness over time. Remember that with apparent magnitude, the higher the negative number, the brighter the object in the sky, and the higher the positive number, the dimmer the object. The full Moon, for example, has an apparent magnitude of -12.7, making it easily the brightest object in the sky after the Sun. The brightest star in the sky is **Sirius**, with an apparent magnitude of -1.5.

Generally speaking, the stars that are named in the charts are of 6th magnitude or brighter, which means that they will be visible to the naked eye in good conditions, although, inevitably, stargazers in urban areas will see fewer stars. Deep-sky objects are typically of 10th magnitude (or brighter), which means that only the very brightest of them is visible to the naked eye, so you'll need binoculars or a telescope to get the most out of them. Each chart is oriented so that **north** is at the top of the page, and every constellation is set on the grid of coordinates (**right ascension** and **declination**) that indicates its general position on the celestial globe (see p.17). In order to help locate the named stars, their precise coordinates are given in the text accompanying each chart.

Best time to view the constellations

The best time to see a constellation is when it is **on meridian** – in other words, when it reaches its highest possible point in the night sky. At this point the whole of the constellation is visible, and you have the least amount of atmosphere between you and it.

Below are the dates on which the constellation will be on meridian at 9pm local time (10pm when it's daylight-saving time). The constellation will be on meridian on these dates and at this time, no matter where you are on the globe. The specific date follows the constellation.

January Auriga (10), Caelum (15), Columba (30), Dorado (20), Eridanus (5), Lepus (25), Mensa (30), Orion (25), Pictor (20), Taurus (15)
February Camelopardalis (1), Canis Major (15), Gemini (20), Monoceros (20), Puppis (25)
March Cancer (15), Canis Minor (1), Carina (15), Lynx (5), Pyxis (15), Vela (25), Volans (1)
April Antlia (5), Chamaeleon (15), Crater (25), Hydra (20), Leo (10), Leo Minor (10), Sextans (5), Ursa Major (20)
May Canes Venatici (20), Centaurus (20), Coma Berenices (15), Corvus (10), Crux (10), Musca (10), Virgo (25)
June Boötes (15), Circinus (20), Libra (20), Lupus (20), Serpens Caput (30) Ursa Minor (25)
July Apus (10), Ara (20), Corona Borealis (30), Draco (20), Hercules (25), Norma (5), Ophiuchus (25), Scorpius (20), Triangulum Australe (5)
August Aquila (30), Corona Australis (15), Lyra (15), Sagittarius (20), Scutum (15), Serpens Cauda (5), Telescopium (25),
September Capricornus (20), Cygnus (20), Delphinus (15), Equuleus (20), Indus (25), Microscopium (20), Pavo (25), Sagitta (30), Vulpecula (10)
October Aquarius (10), Cepheus (15), Grus (10), Lacerta (10), Octans (20), Pegasus (20), Piscis Austrinus (10)
November Andromeda (10), Cassiopeia (20), Cetus (30), Phoenix (20), Pisces (10), Sculptor (10), Tucana, (5)
December Aries (10), Fornax (15), Horologium (25), Hydrus (10), Perseus (25), Reticulum (30), Triangulum (5)

Not all constellations can be seen from everywhere on Earth. A sky-watcher in London or New York, for instance, will never see the constellations Crux, Pavo or Octans because they lie too far south and never rise above the horizon, and other constellations are only partially visible.

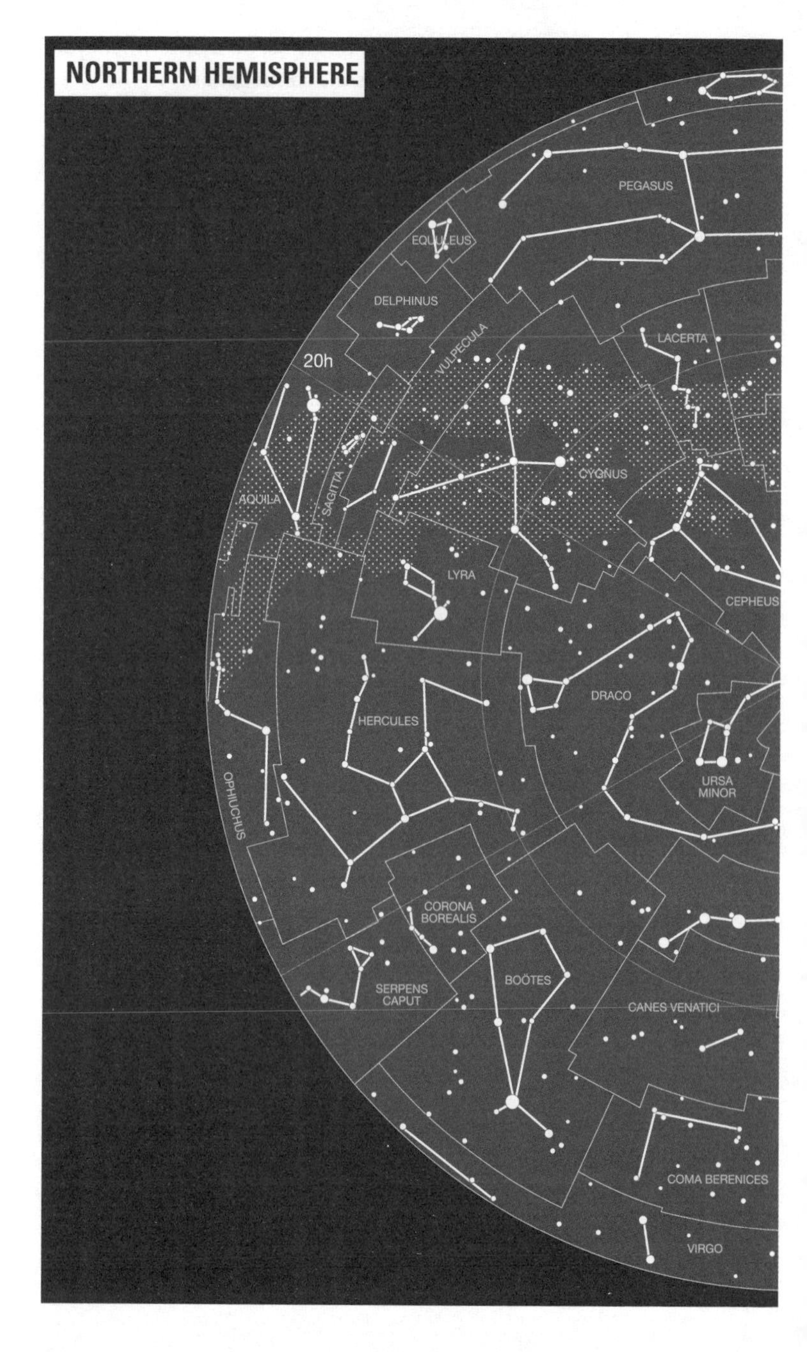
NORTHERN HEMISPHERE
PEGASUS
EQUULEUS
DELPHINUS
VULPECULA
LACERTA
20h
SAGITTA
CYGNUS
AQUILA
LYRA
CEPHEUS
DRACO
HERCULES
URSA MINOR
OPHIUCHUS
CORONA BOREALIS
BOÖTES
SERPENS CAPUT
CANES VENATICI
COMA BERENICES
VIRGO

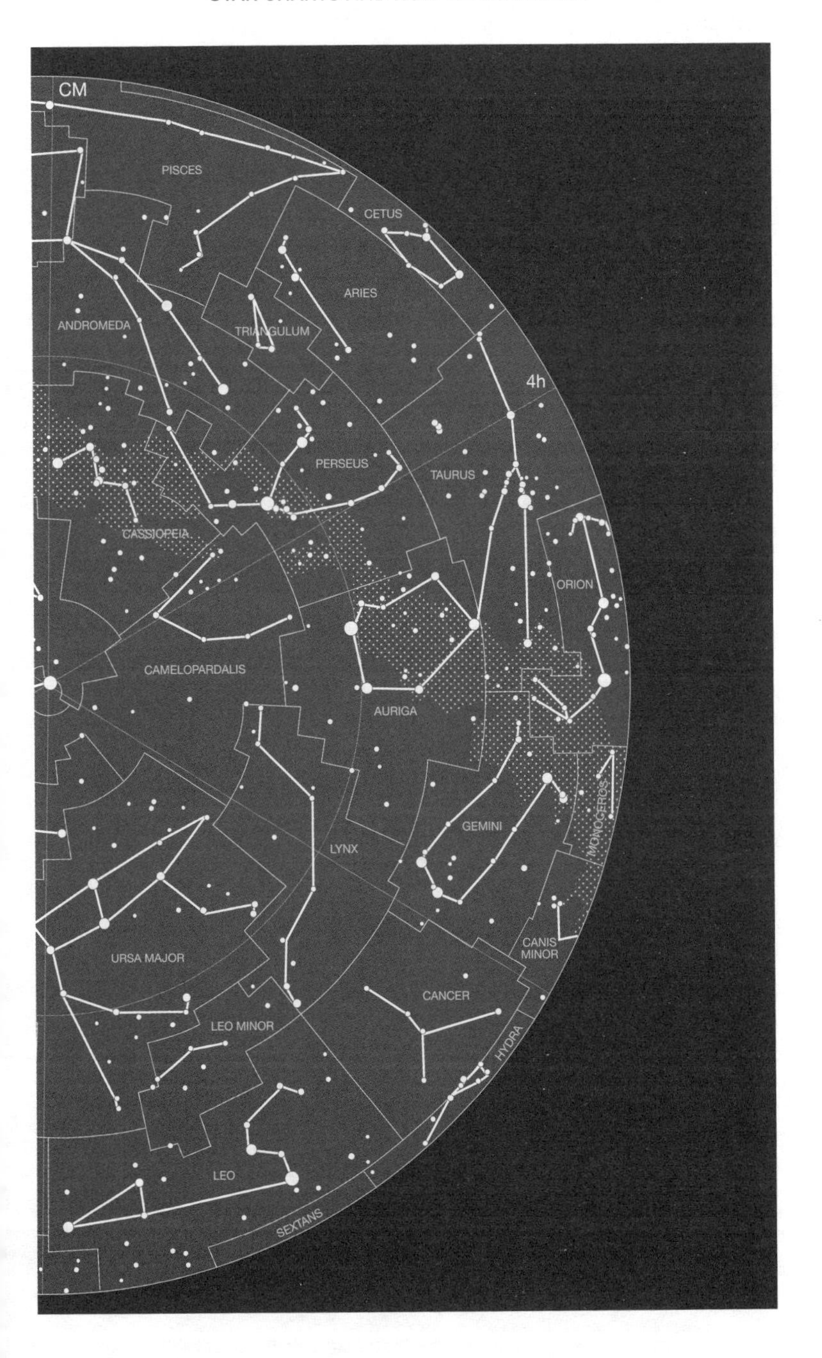
CM
PISCES
CETUS
ARIES
ANDROMEDA
TRIANGULUM
4h
PERSEUS
TAURUS
CASSIOPEIA
ORION
CAMELOPARDALIS
AURIGA
GEMINI
MONOCEROS
LYNX
CANIS MINOR
URSA MAJOR
CANCER
LEO MINOR
HYDRA
LEO
SEXTANS

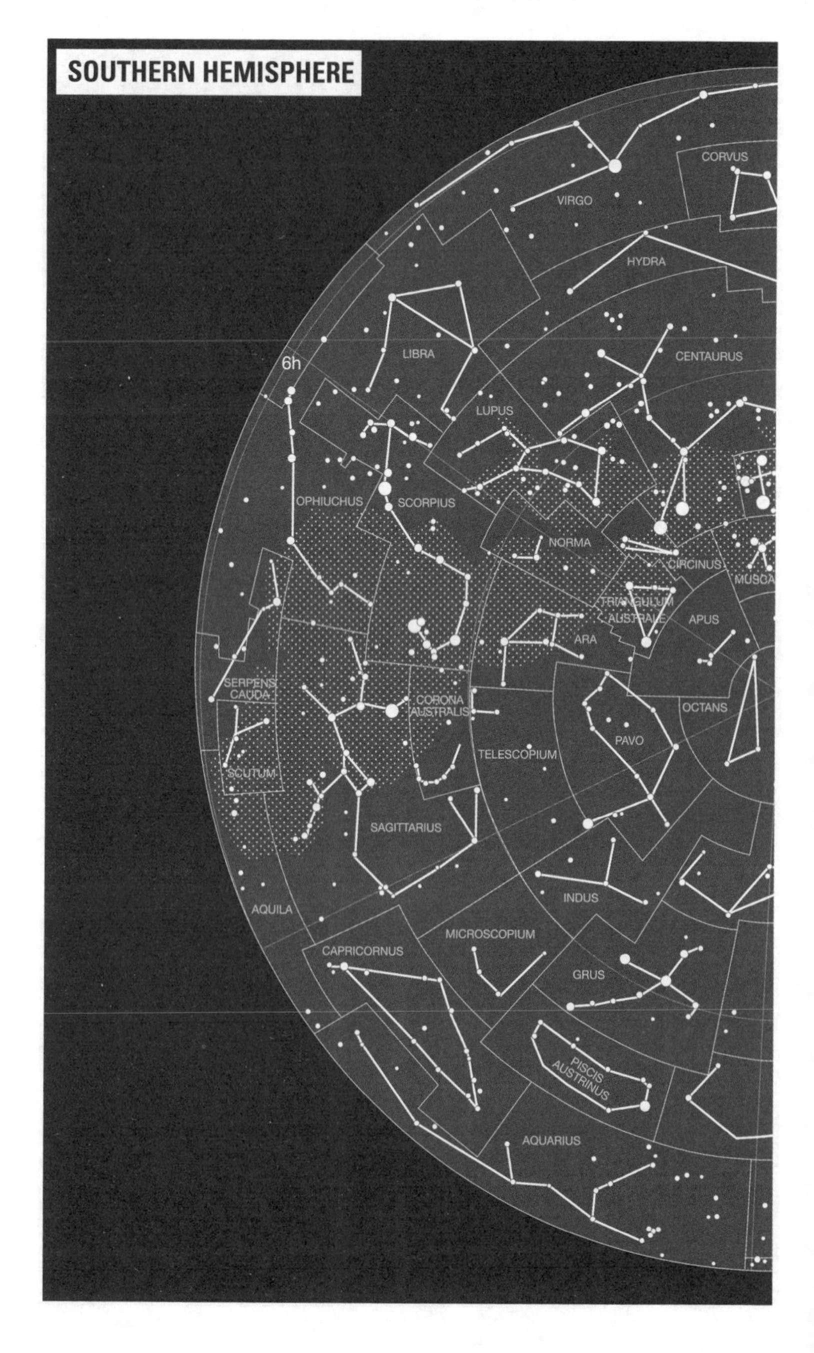
SOUTHERN HEMISPHERE
CORVUS
VIRGO
HYDRA
LIBRA
CENTAURUS
6h
LUPUS
OPHIUCHUS
SCORPIUS
NORMA
CIRCINUS
MUSCA
TRIANGULUM AUSTRALE
APUS
ARA
SERPENS CAUDA
CORONA AUSTRALIS
OCTANS
PAVO
TELESCOPIUM
SCUTUM
SAGITTARIUS
INDUS
AQUILA
MICROSCOPIUM
CAPRICORNUS
GRUS
PISCIS AUSTRINUS
AQUARIUS

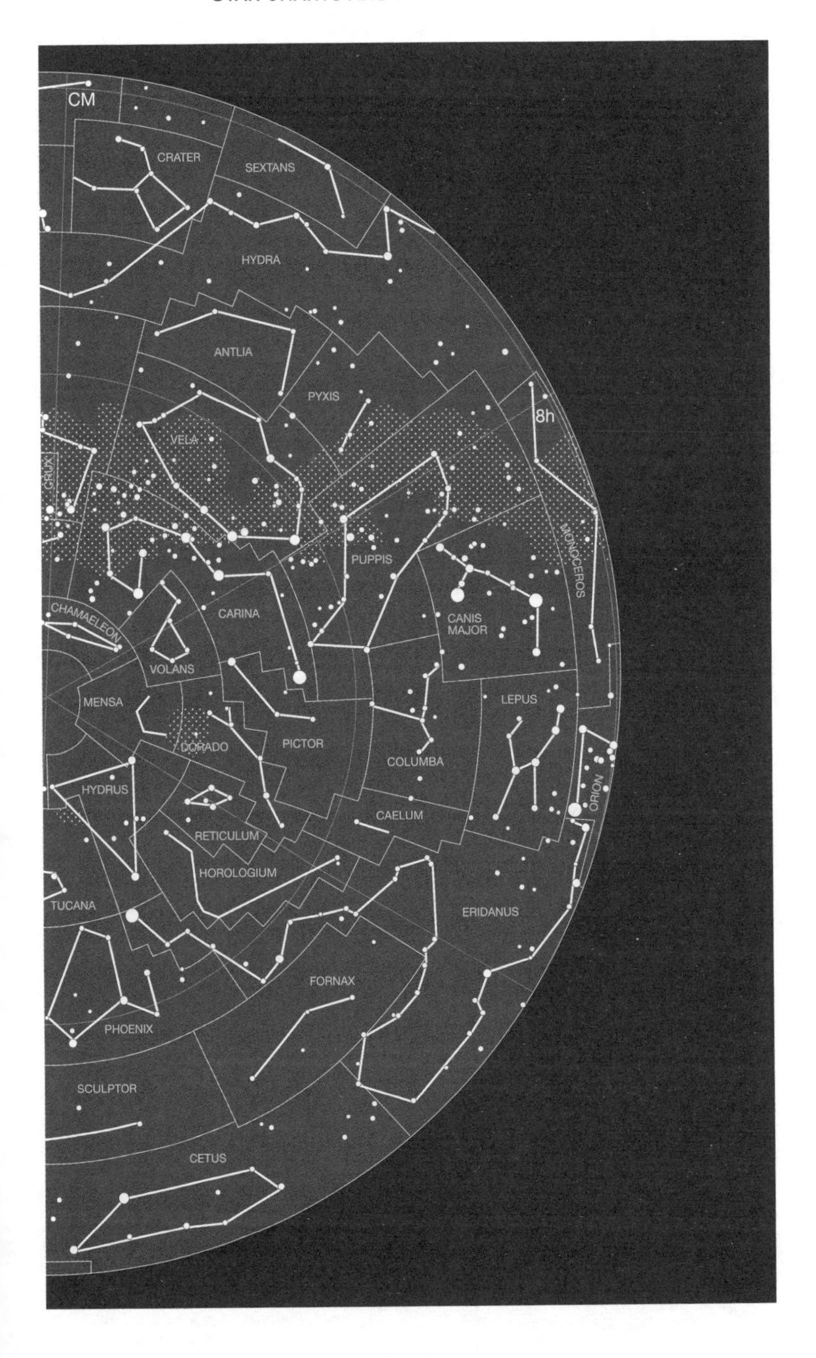
CM
CRATER
SEXTANS
HYDRA
ANTLIA
PYXIS
VELA
8h
CRUX
PUPPIS
MONOCEROS
CHAMAELEON
CARINA
CANIS
MAJOR
VOLANS
MENSA
LEPUS
DORADO
PICTOR
COLUMBA
HYDRUS
ORION
CAELUM
RETICULUM
HOROLOGIUM
TUCANA
ERIDANUS
FORNAX
PHOENIX
SCULPTOR
CETUS

Andromeda

Coordinates of Brightest Stars Alpheratz (Mag 2.06): RA 00h 08m 23.2s, Dec +29° 05' 26"; Mirach (Mag 2.06): RA 01h 09m 43.8s, Dec +35° 37' 14"

Size of Constellation Boundaries 722 square degrees

Bounded by Cassiopeia, Lacerta, Pegasus, Pisces, Triangulum, Perseus

Constellation Legend In Greek mythology, Andromeda was the daughter of King Cepheus and Queen Cassiopeia (whose constellations lie directly to the north). When Cassiopeia boasted that she was more beautiful than the sea nymphs, the sea god Neptune flooded the land and sent a monster to ravage it. To placate him, Andromeda was chained to a rock and was about to be devoured by the monster when Perseus (a constellation to the east) rescued her at the last moment and turned the monster to stone. Perseus later married Andromeda.

Constellation Sights The big draw of Andromeda is of course the Andromeda galaxy (also known as M31), the nearest major galaxy to our own. On a clear night, you'll be able to make it out with your eyes as a faint, glowing smear; binoculars will bring out the central bulge, and a telescope will resolve its spiral arms. A telescope will also make out the Andromeda galaxy's smaller companion galaxies, M32 and NGC 205, directly south and north, respectively. To the southeast lies NGC 752, an open cluster of stars, about 1300 light years away, best observed with binoculars. NGC 7662 (the Blue Snowball planetary nebula) needs a telescope to view it. Binary-star fans will enjoy Almach, which resolves in a small telescope to reveal an orange giant and a smaller, fainter blue companion. Alpheratz, the brightest star in Andromeda, is also informally part of Pegasus, as the northeast corner of that constellation's Great Square asterism.

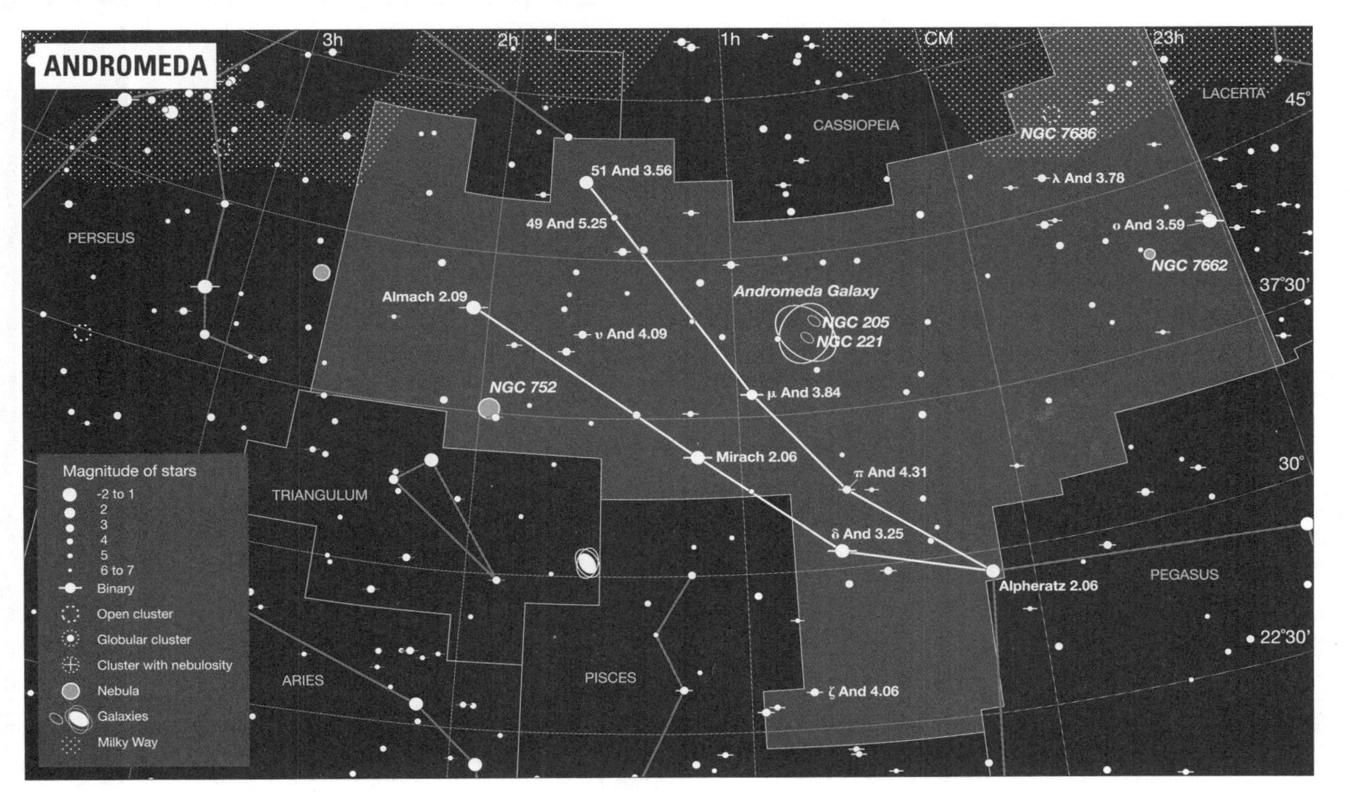

ANDROMEDA
3h
2h
1h
CM
23h
LACERTA
45°
CASSIOPEIA
NGC 7686
PERSEUS
51 And 3.56
λ And 3.78
49 And 5.25
ο And 3.59
NGC 7662
37°30'
Andromeda Galaxy
Almach 2.09
NGC 205
NGC 221
υ And 4.09
NGC 752
μ And 3.84
Mirach 2.06
π And 4.31
30°
TRIANGULUM
δ And 3.25
PEGASUS
Alpheratz 2.06
22°30'
ARIES
PISCES
ζ And 4.06
Magnitude of stars
-2 to 1
2
3
4
5
6 to 7
Binary
Open cluster
Globular cluster
Cluster with nebulosity
Nebula
Galaxies
Milky Way

Antlia & Pyxis

Coordinates of Brightest Stars Alpha Antliae (Mag 4.25): RA 10h 27m 9s, Dec -31° 04' 47"; Alpha Pyxidis (Mag 3.65): RA 8h 43m 36s, Dec -33° 11' 40"

Size of Constellation Boundaries 239 square degrees (Antlia); 221 square degrees (Pyxis)

Bounded by Hydra, Puppis, Vela, Centaurus

Constellation Legend The French astronomer Nicolas Louis de Lacaille, who in the 1750s was the first to map the southern skies systematically, is responsible for the naming of both these constellations. Antlia is meant to suggest an air pump (then recently invented by Denis Papin), while Pyxis represents a mariner's compass, and was originally part of the larger Argo Navis constellation (named after the ship the Greek hero Jason sailed in, in his quest for the Golden Fleece). Other former portions of Argo Navis are Carina, Puppis and Vela.

Constellation Sights Neither Antlia nor Pyxis offers much of interest to stargazers. Zeta Antliae is a multiple star that's easily separable in binoculars; in a telescope you'll see that the brighter of these stars is itself a double star. Pyxis offers up NGC 2627 and NGC 2818, both small, open clusters with a magnitude of about eight.

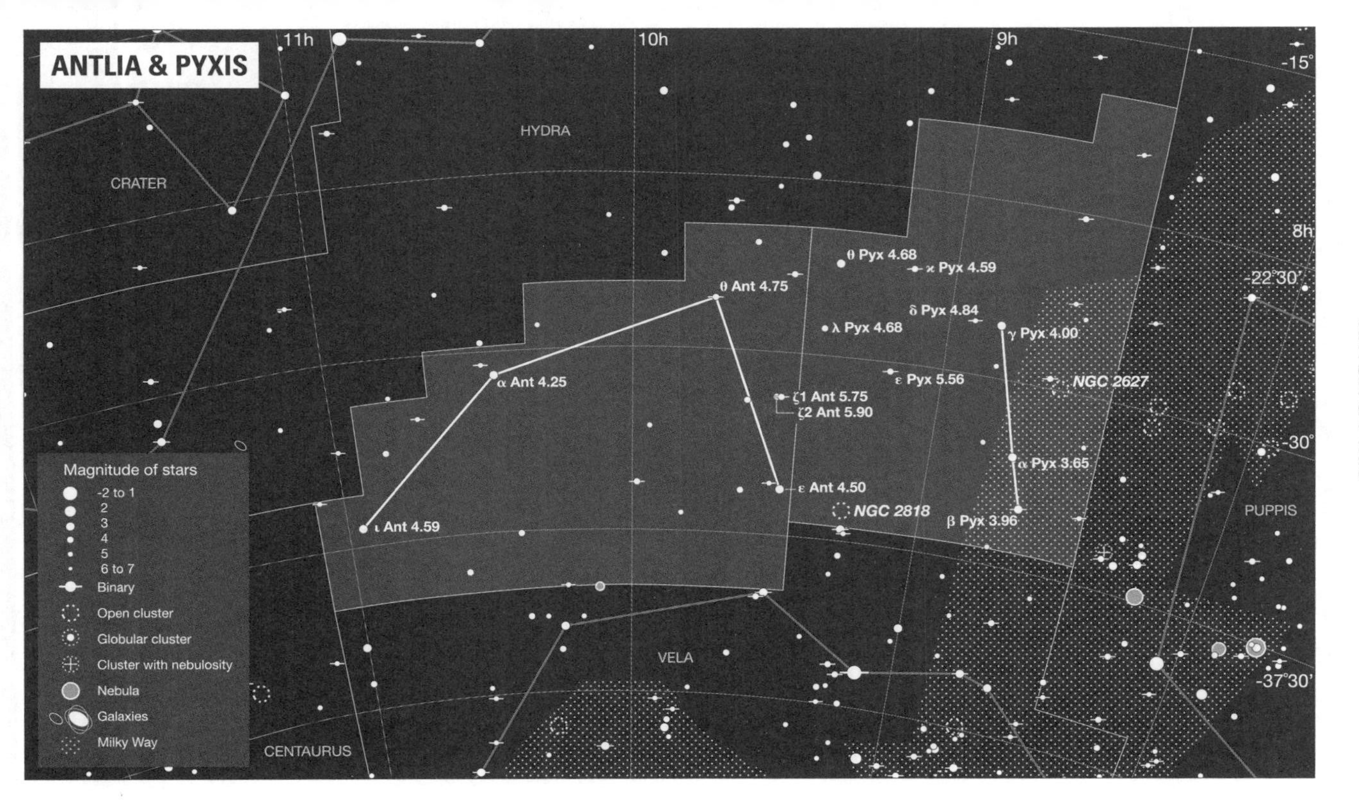
ANTLIA & PYXIS
11h
10h
9h
8h
-15°
-22°30'
-30°
-37°30'
HYDRA
CRATER
VELA
PUPPIS
CENTAURUS
θ Pyx 4.68
κ Pyx 4.59
θ Ant 4.75
δ Pyx 4.84
λ Pyx 4.68
γ Pyx 4.00
α Ant 4.25
ε Pyx 5.56
NGC 2627
ζ1 Ant 5.75
ζ2 Ant 5.90
α Pyx 3.65
ε Ant 4.50
NGC 2818
ι Ant 4.59
β Pyx 3.96
Magnitude of stars
-2 to 1
2
3
4
5
6 to 7
Binary
Open cluster
Globular cluster
Cluster with nebulosity
Nebula
Galaxies
Milky Way

Apus & Triangulum Australe

Coordinates of Brightest Stars Alpha Apodis (Mag 3.81): RA 14h 47m 52s, Dec -79° 03' 20"; Atria (Mag 1.90): RA 16h 48m 40s, Dec -69° 01' 58"

Size of Constellation Boundaries 206 square degrees (Apus); 110 square degrees (Triangulum Australe)

Bounded by Norma, Circinus, Musca, Chamaeleon, Octans, Pavo, Ara

Constellation Legend These two constellations were defined in the late sixteenth century by Dutch explorers Pieter Dirkszoon Keyser and Frederick de Houtman. Apus is generally thought to represent a bird of paradise (it was originally titled Avis Indica). Triangulum Australe is the Southern Triangle, named as such in order to differentiate it from the constellation Triangulum which resides north of the celestial equator.

Constellation Sights Apus's stars are dim and the constellation in general doesn't offer up much of note, save the wide, apparent double star Delta Apodis ("apparent" because the stars are a hundred light years from each other). Triangulum Australe, on the other hand, offers NGC 6025, an open cluster that is almost but not quite visible to the naked eye, but which is very easy to resolve in binoculars or a small telescope. Atria, also known as Alpha Trianguli Australis, is the constellation's brightest star. It is also worth looking at in binoculars, in particular for its rich orange colouration.

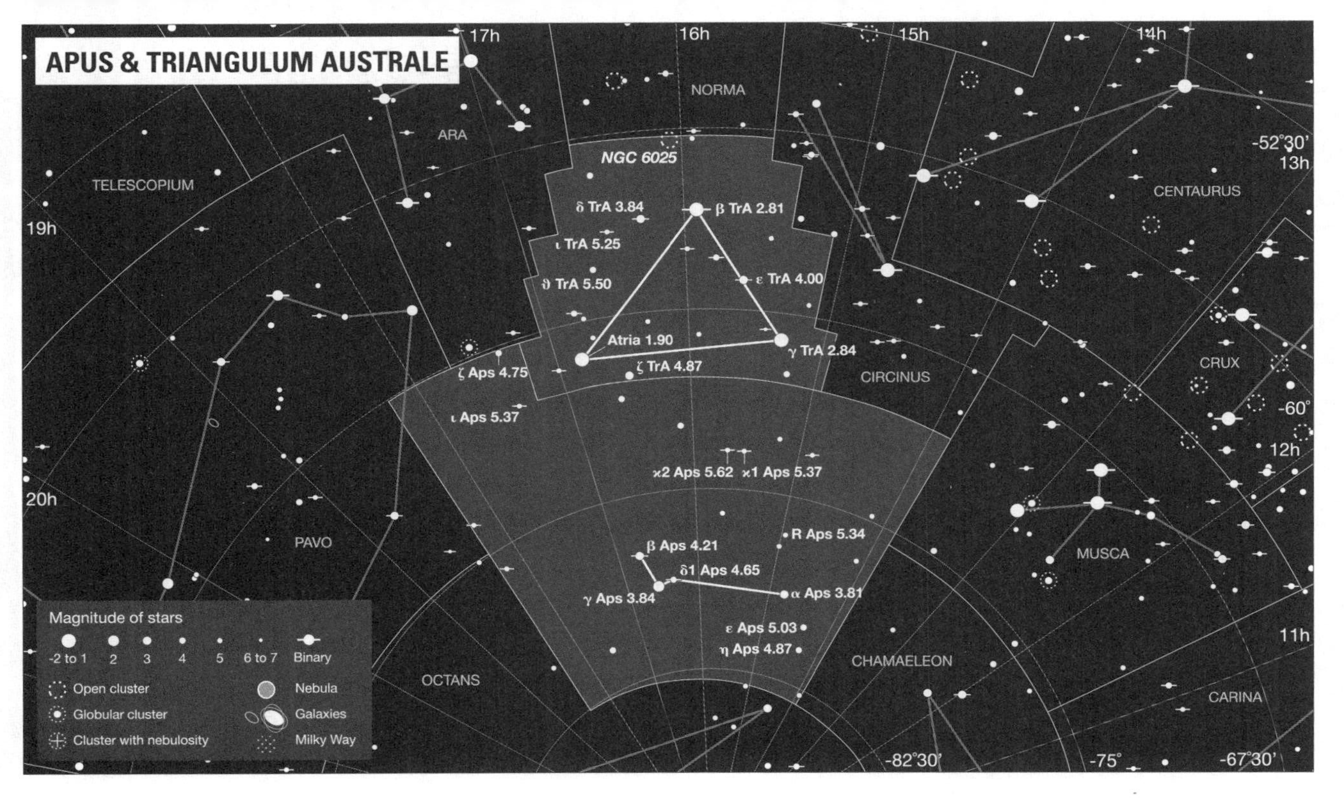
APUS & TRIANGULUM AUSTRALE
17h
16h
15h
14h
-52°30'
13h
19h
20h
-60°
12h
11h
-82°30'
-75°
-67°30'
NORMA
ARA
TELESCOPIUM
CENTAURUS
CRUX
CIRCINUS
MUSCA
PAVO
OCTANS
CHAMAELEON
CARINA
NGC 6025
δ TrA 3.84
β TrA 2.81
ι TrA 5.25
ϑ TrA 5.50
ε TrA 4.00
Atria 1.90
γ TrA 2.84
ζ TrA 4.87
ζ Aps 4.75
ι Aps 5.37
κ2 Aps 5.62
κ1 Aps 5.37
R Aps 5.34
β Aps 4.21
δ1 Aps 4.65
γ Aps 3.84
α Aps 3.81
ε Aps 5.03
η Aps 4.87
Magnitude of stars
-2 to 1
2
3
4
5
6 to 7
Binary
Open cluster
Globular cluster
Cluster with nebulosity
Nebula
Galaxies
Milky Way

Aquarius

Coordinates of Brightest Star Sadal Suud (Mag 2.87): RA 21h 31m 41s, Dec -5° 33' 39"

Size of Constellation Boundaries 980 square degrees

Bounded by Pegasus, Equuleus, Delphinus, Aquila, Capricornus, Piscis Austrinus, Sculptor, Cetus, Pisces

Constellation Legend Aquarius (from the latin "aqua" meaning water) is one of the signs of the Zodiac and is identified with the figure of the "water bearer". For the ancient Greeks, Aquarius is most closely connected with Ganymede, the young shepherd whom Zeus abducted to be his cupbearer, but many other cultures link the constellation to water and fertility. In ancient Egypt it is associated with Khnum, who was the guardian of the source of the Nile, and in India with the Hindu deity Varuna. For former hippies who are wondering when the "Age of Aquarius" actually begins, it's in about the year 2660 AD, which is when precession takes the vernal equinox into the constellation's current boundaries.

Constellation Sights The stars in Aquarius are fairly faint; the brightest, Sadal Suud and Sadal Melik, are only slightly brighter than third magnitude (Sadal Melik, incidentally, is almost exactly on the celestial equator). Zeta Aquarii is a binary that can be resolved with a small telescope. However, Aquarius is home to some notable nebulae, including the Helix nebula, which, at just three hundred light years, is one of the closest planetary nebulae to our planet. Binoculars or telescopes show this nebula as a patch of grey. Those of you with larger scopes (aperture of eight inches or more) will also be able to catch the Saturn nebula, so named because of its jutting bands of gas that look like planetary rings. If you're interested in globular clusters, M2 is just about visible with the naked eye but is seen best when using binoculars and telescopes.

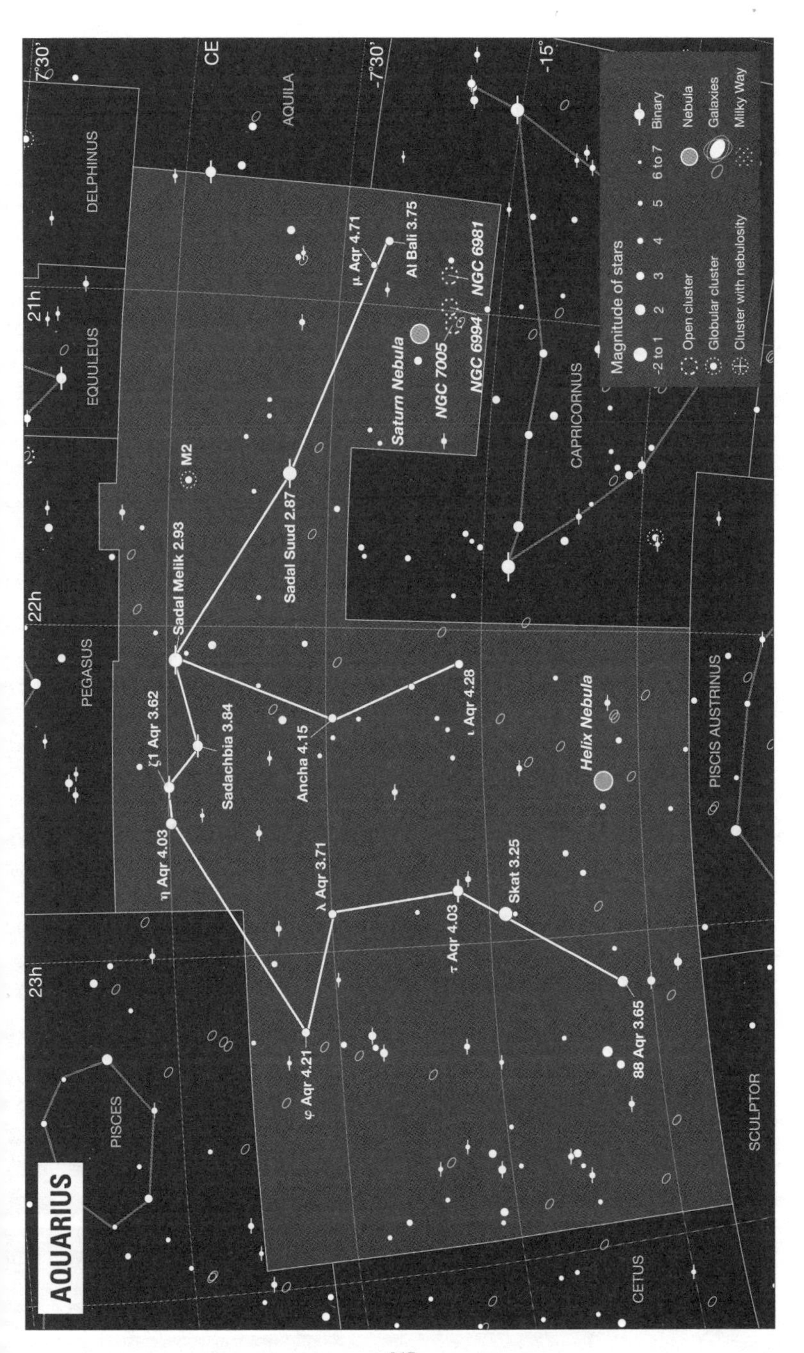
AQUARIUS
DELPHINUS
EQUULEUS
PEGASUS
PISCES
AQUILA
CAPRICORNUS
PISCIS AUSTRINUS
SCULPTOR
CETUS
21h
22h
23h
7°30'
-7°30'
-15°
μ Aqr 4.71
Al Bali 3.75
NGC 6981
NGC 6994
NGC 7005
Saturn Nebula
M2
Sadal Suud 2.87
Sadal Melik 2.93
ζ1 Aqr 3.62
Sadachbia 3.84
Ancha 4.15
ι Aqr 4.28
Helix Nebula
η Aqr 4.03
λ Aqr 3.71
τ Aqr 4.03
Skat 3.25
88 Aqr 3.65
φ Aqr 4.21
Magnitude of stars
-2 to 1
2
3
4
5
6 to 7
Binary
Open cluster
Globular cluster
Cluster with nebulosity
Nebula
Galaxies
Milky Way

Aquila

Coordinates of Brightest Stars Altair (Mag 0.75): RA 19h 50m 54s, Dec 8° 52' 26"

Size of Constellation Boundaries 652 square degrees

Bounded by Sagitta, Ophiuchus, Serpens Cauda, Scutum, Sagittarius, Capricornus, Aquarius, Delphinus

Constellation Legend Aquila means eagle, and this constellation is associated with the eagle servant of Zeus, who carried the god's thunderbolts and occasionally absconded a mortal or two at Zeus's request – most notably Ganymede (see Aquarius). In the Middle East, the constellation was called Al Nasr al-tair ("the flying eagle"), and Altair remains the name of its brightest star.

Constellation Sights At magnitude 0.75, Altair is one of the twenty brightest stars in the sky, and one of the three bright stars that create the Summer Triangle asterism (the other two being Deneb to the northeast in Cygnus, and Vega to the northwest in Lyra). Patient stargazers will also enjoy Eta Aquilae, one of the brightest Cepheid variables. From a top magnitude of about 3.6, it dims by almost a full order of magnitude, and its entire cycle takes just over a week to complete. A star cluster worth noting is NGC 6709, a small open cluster of about eighth magnitude which is only viewable with a telescope. A scope is also required for NGC 6755, an open cluster of about one hundred stars. Aquila lies in a dense area of the Milky Way, and its star fields reward observation.

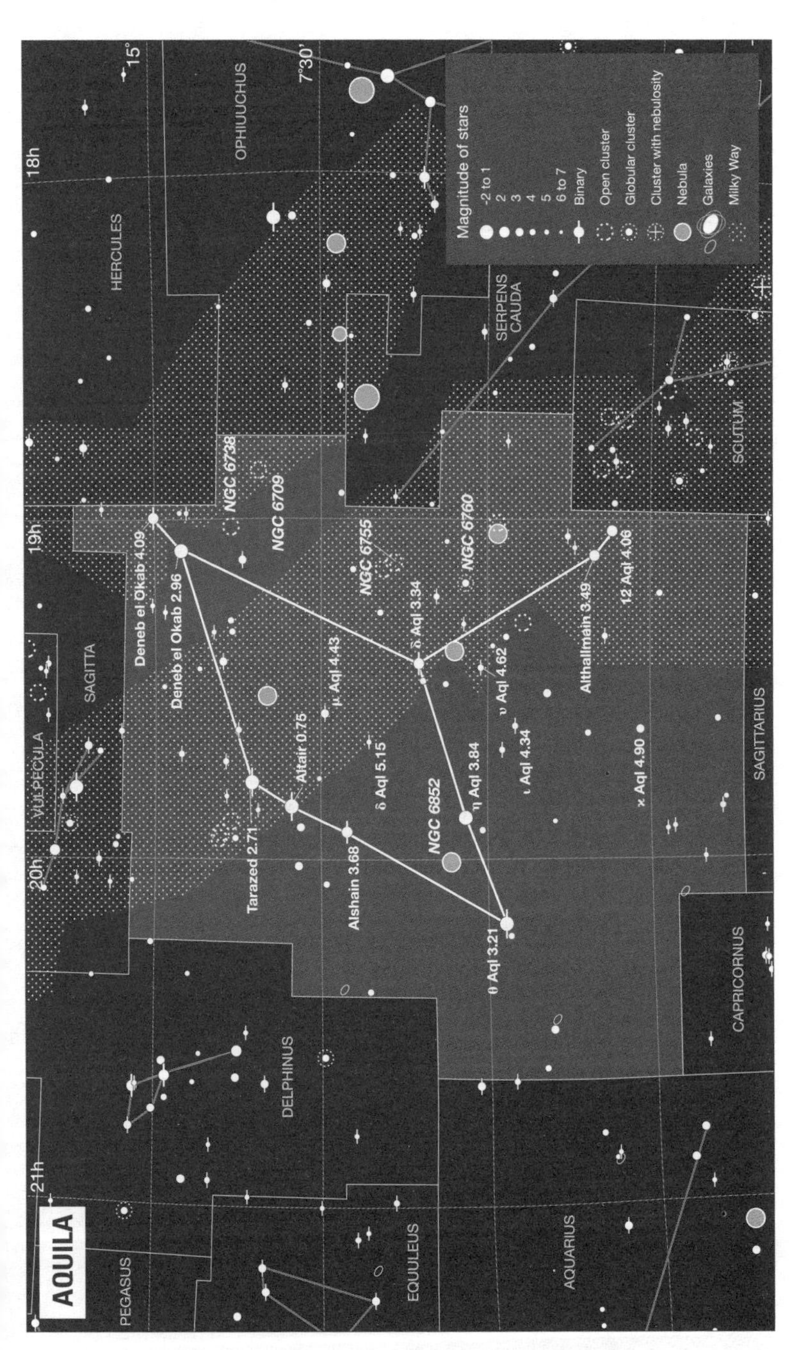

AQUILA
PEGASUS
EQUULEUS
DELPHINUS
AQUARIUS
CAPRICORNUS
SAGITTARIUS
SCUTUM
SERPENS CAUDA
OPHIUUCHUS
HERCULES
SAGITTA
VULPECULA
21h
20h
19h
18h
15°
7°30'
Deneb el Okab 4.09
Deneb el Okab 2.96
NGC 6738
NGC 6709
NGC 6755
NGC 6760
μ Aql 4.43
δ Aql 3.34
ν Aql 4.62
Althallmain 3.49
12 Aql 4.06
Altair 0.76
Tarazed 2.71
δ Aql 5.15
η Aql 3.84
ι Aql 4.34
κ Aql 4.90
Alshain 3.68
NGC 6852
θ Aql 3.21
Magnitude of stars
-2 to 1
2
3
4
5
6 to 7
Binary
Open cluster
Globular cluster
Cluster with nebulosity
Nebula
Galaxies
Milky Way

Ara

Coordinates of Brightest Star Alpha Arae (Mag 2.81): RA 17h 31m 51s, Dec -49° 52' 43"

Size of Constellation Boundaries 237 square degrees

Bounded by Scorpius, Norma, Triangulum Australe, Apus, Pavo, Telescopium, Corona Australis

Constellation Legend The ancient Greeks saw this constellation as a heavenly altar, upon which the gods pledged their allegiance to one another before their battle with the Titans. To the Romans it was Ara Centauri, the altar that Chiron the centaur used to sacrifice the wolf, Lupus. Indeed the stars within Ara were part of Centaurus and Lupus, before de Lacaille created the constellation in the 1750s. De Lacaille also invented the constellation Norma which now sits between Lupus and Ara.

Constellation Sights Ara houses NGC 6397, a large globular cluster that is about 7200 light years away – pretty close to us in relative terms and easily viewable in binoculars. NGC 6397 has a very compact core of stars, which have contracted in a process known as a "core collapse". It also appears to have main sequence stars in it, which is very unusual for globular clusters, which typically consist of older stars. One explanation is that the cluster is so densely packed that stars actually collide, forming new stars in the aftermath. Ara also features an open cluster, NGC 6193, also easy to view with binoculars.

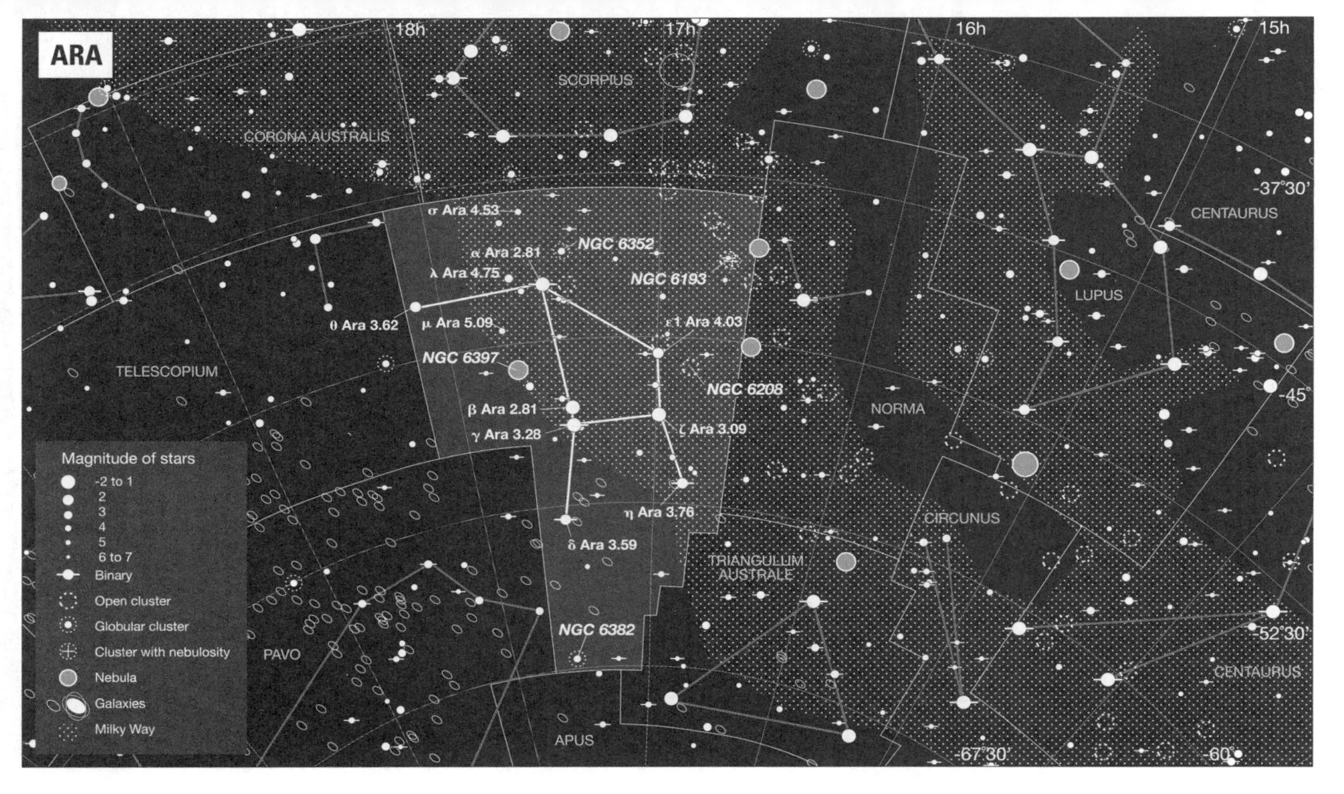
ARA
18h
17h
16h
15h
SCORPIUS
CORONA AUSTRALIS
-37°30'
CENTAURUS
σ Ara 4.53
α Ara 2.81
NGC 6352
λ Ara 4.75
NGC 6193
LUPUS
θ Ara 3.62
μ Ara 5.09
ε1 Ara 4.03
NGC 6397
TELESCOPIUM
NGC 6208
-45°
NORMA
β Ara 2.81
γ Ara 3.28
ζ Ara 3.09
η Ara 3.76
CIRCINUS
δ Ara 3.59
TRIANGULUM AUSTRALE
NGC 6382
PAVO
-52°30'
CENTAURUS
APUS
-67°30'
-60°
Magnitude of stars
-2 to 1
2
3
4
5
6 to 7
Binary
Open cluster
Globular cluster
Cluster with nebulosity
Nebula
Galaxies
Milky Way

Aries

Coordinates of Brightest Stars Hamal (Mag 2.00): RA 2h 07m 18s, Dec 23° 28' 27"

Size of Constellation Boundaries 441 square degrees

Bounded by Perseus, Triangulum, Pisces, Cetus, Taurus

Constellation Legend In Greek mythology, Aries was the flying ram with the Golden Fleece, sent by the god Hermes (Mercury) to rescue the royal siblings, Helle and Phryxus, from their murderous stepmother Ino. In the course of their escape, Helle fell into a bit of water known to this day as the Hellespont. Aries is historically the constellation that leads the Zodiac since, in the time of the ancient Greeks, the Sun was in Aries at the vernal equinox. Because of precession, however, the Sun is now in Pisces on the first day of spring. Hamal, the name of the constellation's brightest star, is the Arabic word for lamb.

Constellation Sights There's not all that much going on in this constellation. In terms of stars, the most interesting sight is Mesarthim, a double star whose components are actually both very similar blue-white stars, easily resolved in a small telescope. Mesarthim was one of the first double stars to be spotted, accidentally discovered in 1664 by English physicist Robert Hooke while he was watching a comet. Hooke's other contributions to astronomy include detailed early drawings of Mars, and the idea that Jupiter rotated on its axis.

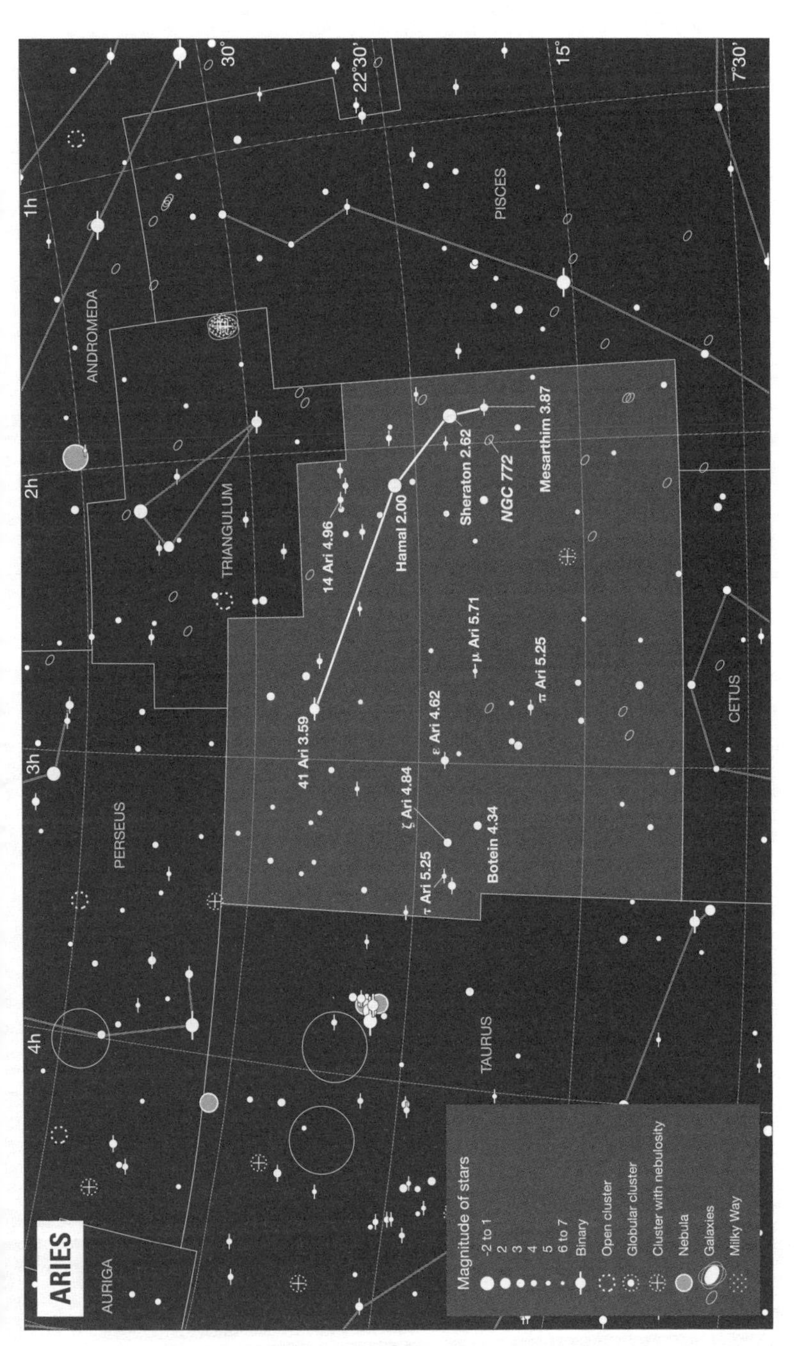
ARIES
AURIGA
PERSEUS
TAURUS
ANDROMEDA
TRIANGULUM
PISCES
CETUS
1h
2h
3h
4h
30°
22°30'
15°
7°30'
41 Ari 3.59
14 Ari 4.96
Hamal 2.00
Sheraton 2.62
NGC 772
Mesarthim 3.87
μ Ari 5.71
π Ari 5.25
ε Ari 4.62
ζ Ari 4.84
τ Ari 5.25
Botein 4.34
Magnitude of stars
-2 to 1
2
3
4
5
6 to 7
Binary
Open cluster
Globular cluster
Cluster with nebulosity
Nebula
Galaxies
Milky Way

Auriga

Coordinates of Brightest Star Capella (Mag 0.06): RA 5h 16m 52s, Dec 46° 00' 04"

Size of Constellation Boundaries 657 square degrees

Bounded by Camelopardalis, Perseus, Taurus, Gemini, Lynx

Constellation Legend This ancient constellation is associated with Erichthonius, the son of Hephestus (Vulcan), who as fourth king of Athens is credited with inventing the four-horse chariot. Auriga is often symbolized by a charioteer carrying a goat. The provenance of the goat in the whole story is a bit obscure (it may be the goat that suckled the Greek god Zeus as a baby), but it's fairly significant for the constellation, since the name of the brightest star in the constellation is Capella, which means "little she-goat" in Latin.

Constellation Sights Auriga has some truly dazzling stars. Capella is the sixth brightest star in the sky (it's actually a spectroscopic binary star) and shines with a clear yellow tint. Almaaz (which is Arabic for "he-goat") is a very interesting variable star, whose brightness dims by half when it is eclipsed by its companion star. This happens every 27 years (the next eclipse is due in 2009). Astronomers suspect Almaaz's companion star is accompanied by a large dark cloud made up of gas and dust. Below Almaaz is Hoedus I (who with Hoedus II represents the goat kids of Capella), another eclipsing binary with a period of 2.7 years. Auriga also offers up three interesting star clusters, visible through binoculars and telescopes: the Pinwheel cluster (M36), Auriga Salt-and-Pepper (M37) and M38. Each contains between 60 and 150 stars.

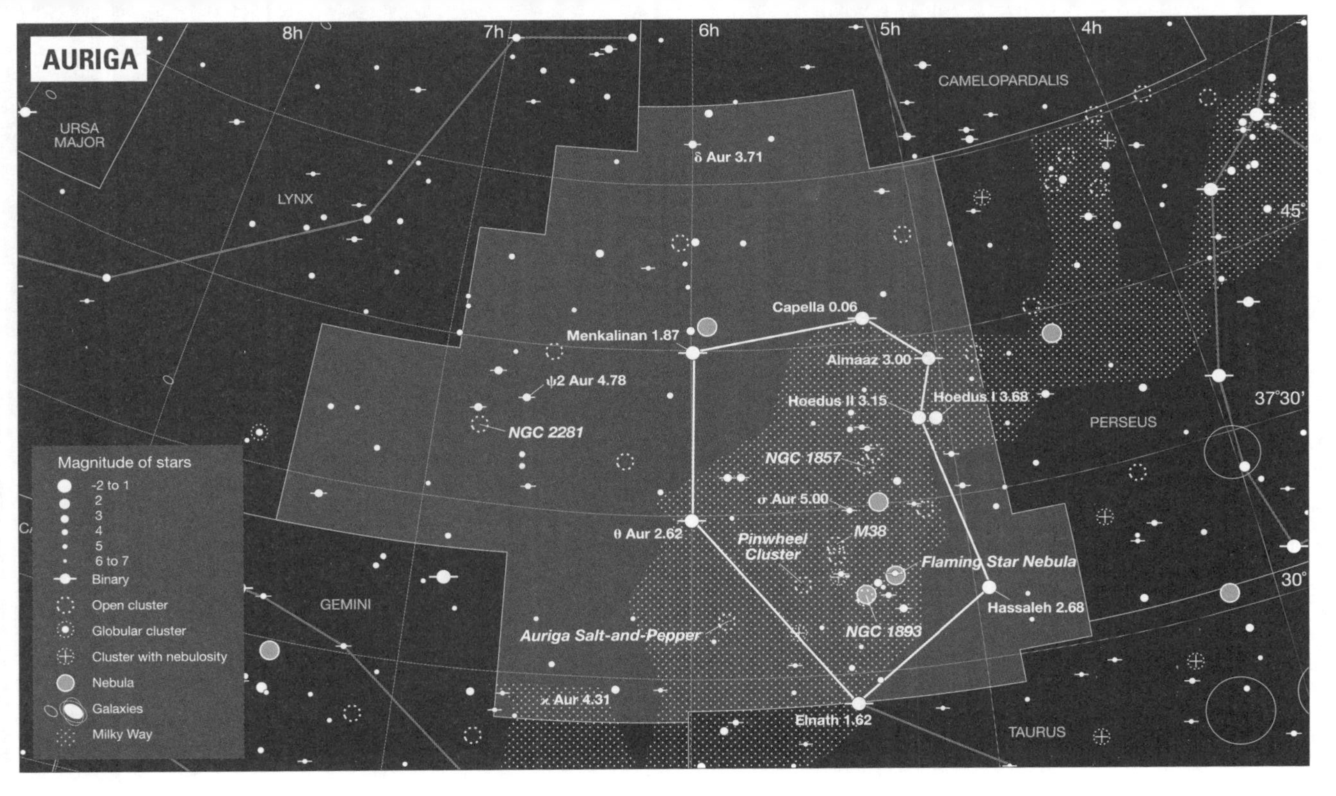
AURIGA
8h
7h
6h
5h
4h
URSA MAJOR
LYNX
CAMELOPARDALIS
δ Aur 3.71
45°
Capella 0.06
Menkalinan 1.87
Almaaz 3.00
ψ2 Aur 4.78
Hoedus II 3.15
Hoedus I 3.68
37°30'
PERSEUS
NGC 2281
NGC 1857
σ Aur 5.00
θ Aur 2.62
Pinwheel Cluster
M38
Flaming Star Nebula
30°
GEMINI
Hassaleh 2.68
Auriga Salt-and-Pepper
NGC 1893
χ Aur 4.31
Elnath 1.62
TAURUS
Magnitude of stars
-2 to 1
2
3
4
5
6 to 7
Binary
Open cluster
Globular cluster
Cluster with nebulosity
Nebula
Galaxies
Milky Way

Boötes

Coordinates of Brightest Star Arcturus (Mag -0.04): RA 14h 15m 46s, Dec 19° 10' 08"

Size of Constellation Boundaries 907 square degrees

Bounded by Draco, Ursa Major, Canes Venatici, Coma Berenices, Virgo, Serpens Caput, Corona Borealis, Hercules

Constellation Legend As a constellation, Boötes is so old that apparently even the ancient Greeks were unsure of its history. The precise origin of the word is obscure, but it is usually taken to mean "herdsman". The most common story with which it is linked is that of the nymph Callisto. Having been seduced by the god Zeus, Callisto was turned into a bear either by Hera (Zeus's jealous wife) or by Artemis (Diana), the goddess of chastity. Arcas, her son by Zeus, was out hunting and was about to kill her (unaware of who she was) when Zeus intervened and set them both in the sky. There Arcas acts as herdsman, keeping the Great Bear (Ursa Major) and other animal constellations in line. Another story connects Boötes to the son of Ceres (Demeter), the goddess of agriculture, and credits him with the invention of the plough.

Constellation Sights The big draw in Boötes is Arcturus (whose name means "guardian of the bear"), the brightest star in the northern celestial hemisphere, and the fourth brightest in the sky. Arcturus is visibly yellow-orange, and binoculars bring out the colour even more strongly. For sky watchers with telescopes, Boötes offers several challenges. There is Izar, a binary with a red primary and a blue partner, difficult to parse; Xi Boötis, another binary star which is resolvable with a small scope; and Alkalurops, a multiple star which appears as a double in a small telescope, but whose fainter star can itself be seen as two with the use of a larger telescope.

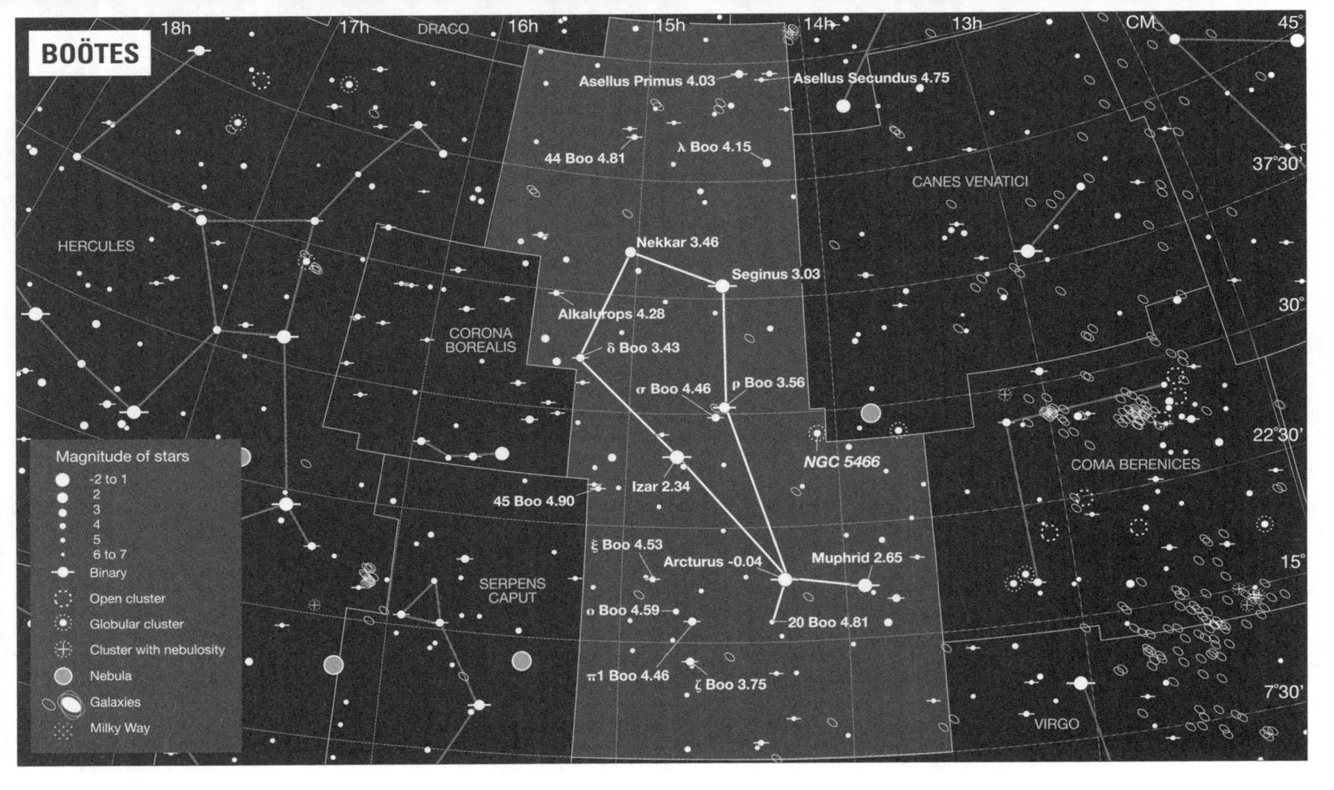
BOÖTES
18h
17h
DRACO
16h
15h
14h
13h
CM
45°
37°30'
30°
22°30'
15°
7°30'
Asellus Primus 4.03
Asellus Secundus 4.75
44 Boo 4.81
λ Boo 4.15
CANES VENATICI
HERCULES
Nekkar 3.46
Seginus 3.03
Alkalurops 4.28
CORONA BOREALIS
δ Boo 3.43
σ Boo 4.46
ρ Boo 3.56
NGC 5466
COMA BERENICES
Izar 2.34
45 Boo 4.90
ξ Boo 4.53
Arcturus -0.04
Muphrid 2.65
SERPENS CAPUT
ο Boo 4.59
20 Boo 4.81
π1 Boo 4.46
ζ Boo 3.75
VIRGO
Magnitude of stars
-2 to 1
2
3
4
5
6 to 7
Binary
Open cluster
Globular cluster
Cluster with nebulosity
Nebula
Galaxies
Milky Way

Caelum & Columba

Coordinates of Brightest Star Alpha Caeli (Mag 4.43): RA 4h 40m 34s, Dec -41° 51' 29"; Alpha Columbae (Mag 2.62): RA 5h 39m 39s, Dec -34° 04' 19"

Size of Constellation Boundaries 125 square degrees (Caelum); 270 square degrees (Columba)

Bounded by Lepus, Eridanus, Horologium, Dorado, Pictor, Puppis, Canis Major

Constellation Legend Caelum is Latin for "chisel", and is another constellation introduced by Nicolas Louis de Lacaille when he was mapping the southern sky in the mid-eighteenth century. Neighbouring Columba was first described by Dutch astronomer Petrus Plancius in 1592, who formed it from stars that were already part of Canis Major. Columba means "dove", and refers to the dove Noah released to look for land after the Flood, and which returned with the olive branch. Indeed, an earlier name for the constellation was Columba Noachi – "Noah's Dove".

Constellation Sights To be honest, Caelum is pretty much a dead loss for stargazing, and there's not much happening (that can be seen) in Columba either. However, there is one star in the constellation that is of interest. This is Mu Columbae, which is known as a "runaway star" because of the high speed – more than sixty miles per second – at which it hurtles through the galaxy. If we trace the star back, we find it comes from the same point in space as another "runaway star" known as AE Aurigae. Both these stars had a run-in with yet another star system, Iota Orionis, 2.5 million years ago, and the theory is that a close encounter between two sets of binaries caused the runaway stars to shoot across the cosmos at high speeds.

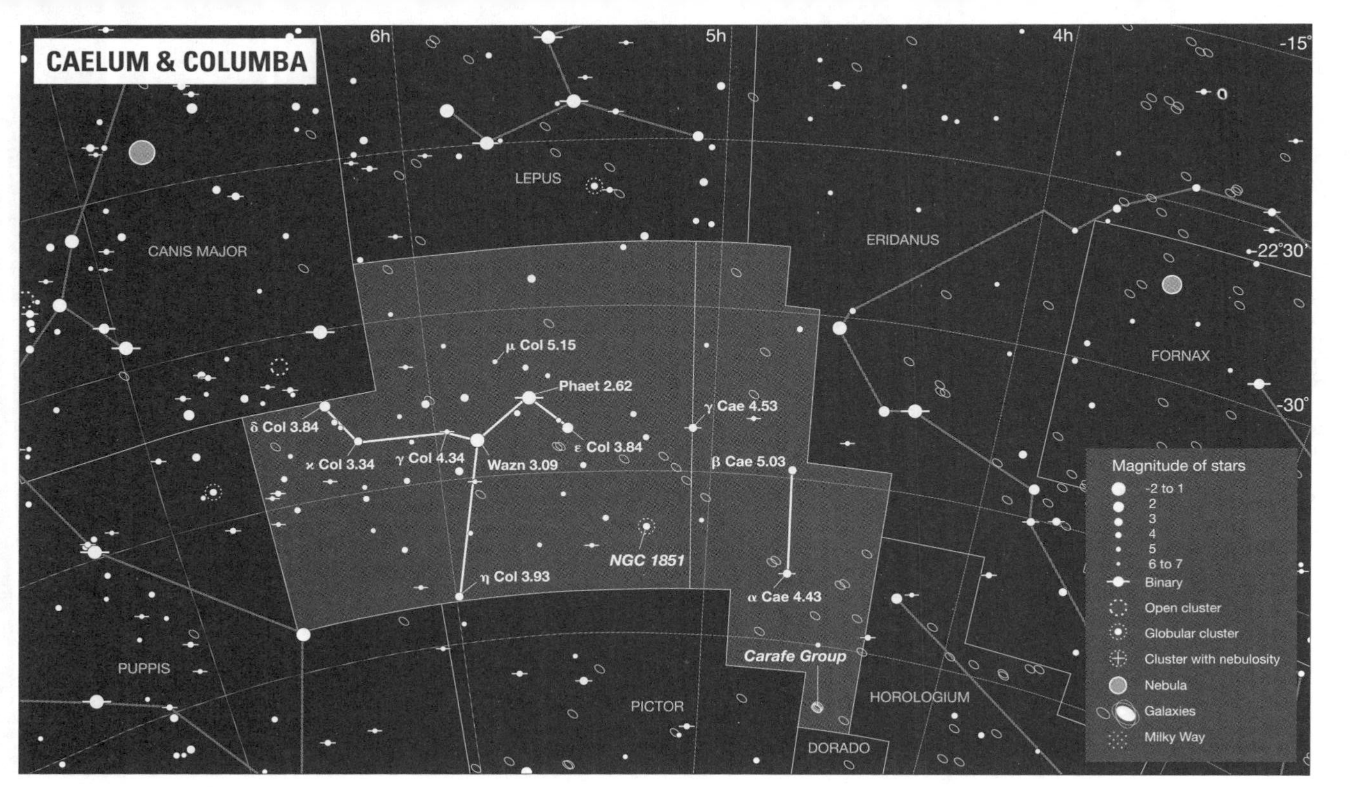
CAELUM & COLUMBA
6h
5h
4h
-15°
-22°30'
-30°
LEPUS
CANIS MAJOR
ERIDANUS
FORNAX
μ Col 5.15
Phaet 2.62
γ Cae 4.53
δ Col 3.84
ε Col 3.84
κ Col 3.34
γ Col 4.34
Wazn 3.09
β Cae 5.03
NGC 1851
η Col 3.93
α Cae 4.43
Carafe Group
PUPPIS
PICTOR
HOROLOGIUM
DORADO
Magnitude of stars
-2 to 1
2
3
4
5
6 to 7
Binary
Open cluster
Globular cluster
Cluster with nebulosity
Nebula
Galaxies
Milky Way

Camelopardalis

Coordinates of Brightest Star Beta Camelopardalis (Mag 4.00): RA 05h 03m 25s, Dec 60° 26' 47"

Size of Constellation Boundaries 757 square degrees

Bounded by Ursa Minor, Cepheus, Cassiopeia, Perseus, Auriga, Lynx, Ursa Major, Draco

Constellation Legend This patch of the far northern sky made it through several thousand years of human history without anyone bothering much about it. The Chinese saw several asterisms there (including an umbrella), but in the West it wasn't until the early seventeenth century that the German astronomer Jakob Bartsch decided to make a constellation out of it. It's supposed to be a giraffe (whose Greek name was "camel-leopard"), but it has also been connected to the camel that took Rebecca to Isaac in the biblical book of Genesis.

Constellation Sights Unfortunately, there aren't very many. No star in the constellation is brighter than fourth magnitude (meaning it's barely visible). There is also a small open cluster of fifteen stars (NGC 1502), the brightest of which is about fifth magnitude.

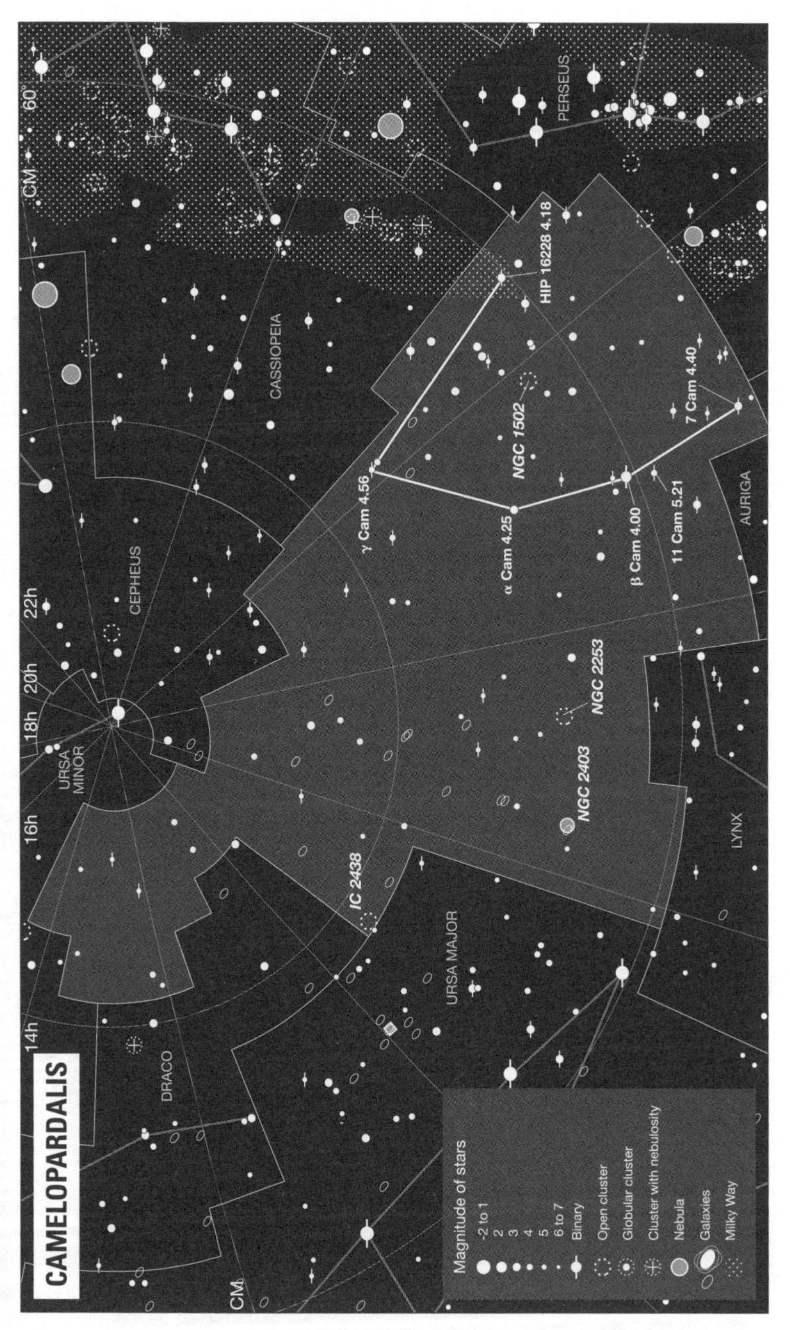
CAMELOPARDALIS
CM
60°
22h
20h
18h
16h
14h
CM
PERSEUS
CASSIOPEIA
CEPHEUS
URSA MINOR
DRACO
URSA MAJOR
AURIGA
LYNX
HIP 16228 4.18
NGC 1502
7 Cam 4.40
γ Cam 4.56
α Cam 4.25
β Cam 4.00
11 Cam 5.21
NGC 2253
NGC 2403
IC 2438
Magnitude of stars
-2 to 1
2
3
4
5
6 to 7
Binary
Open cluster
Globular cluster
Cluster with nebulosity
Nebula
Galaxies
Milky Way

Cancer

Coordinates of Brightest Star Altarf (Mag 3.5): RA 8h 16m 38s, Dec 09° 10' 43"

Size of Constellation Boundaries 506 square degrees

Bounded by Lynx, Gemini, Canis Minor, Hydra, Leo, Leo Minor

Constellation Legend One of the signs of the Zodiac, Cancer represents the crab sent by Hera (Juno) to harass Hercules while he battled with the multi-headed monster, the Hydra. Hercules took a few minutes to crush the crab before finishing off the Hydra itself. Historically, Cancer is known for being the constellation that the Sun is in at the summer solstice, which is why the line of latitude the Sun reaches on the solstice noon is known as the Tropic of Cancer. Currently, the Sun reaches the summer solstice in Gemini.

Constellation Sights Cancer has the rather dubious pleasure of being the only constellation in the sky in which its stars are dimmer than one of its deep-sky objects. The object in question is the Beehive Cluster (also known as the Praesepe Cluster or M44), an open cluster of 75 stars which, at third magnitude, is easily visible to the naked eye. In contrast, Cancer's brightest star, Altarf, has a magnitude of 3.5. Another open cluster, NGC 2682 (or M67), has more stars than the Beehive, but is further away and fainter, so needs viewing with binoculars or telescopes. Among Cancer's stars, Tegmine resolves as a double star in a small telescope and as a multiple star in larger ones, while Iota Cancri can be seen as a double star in small telescopes.

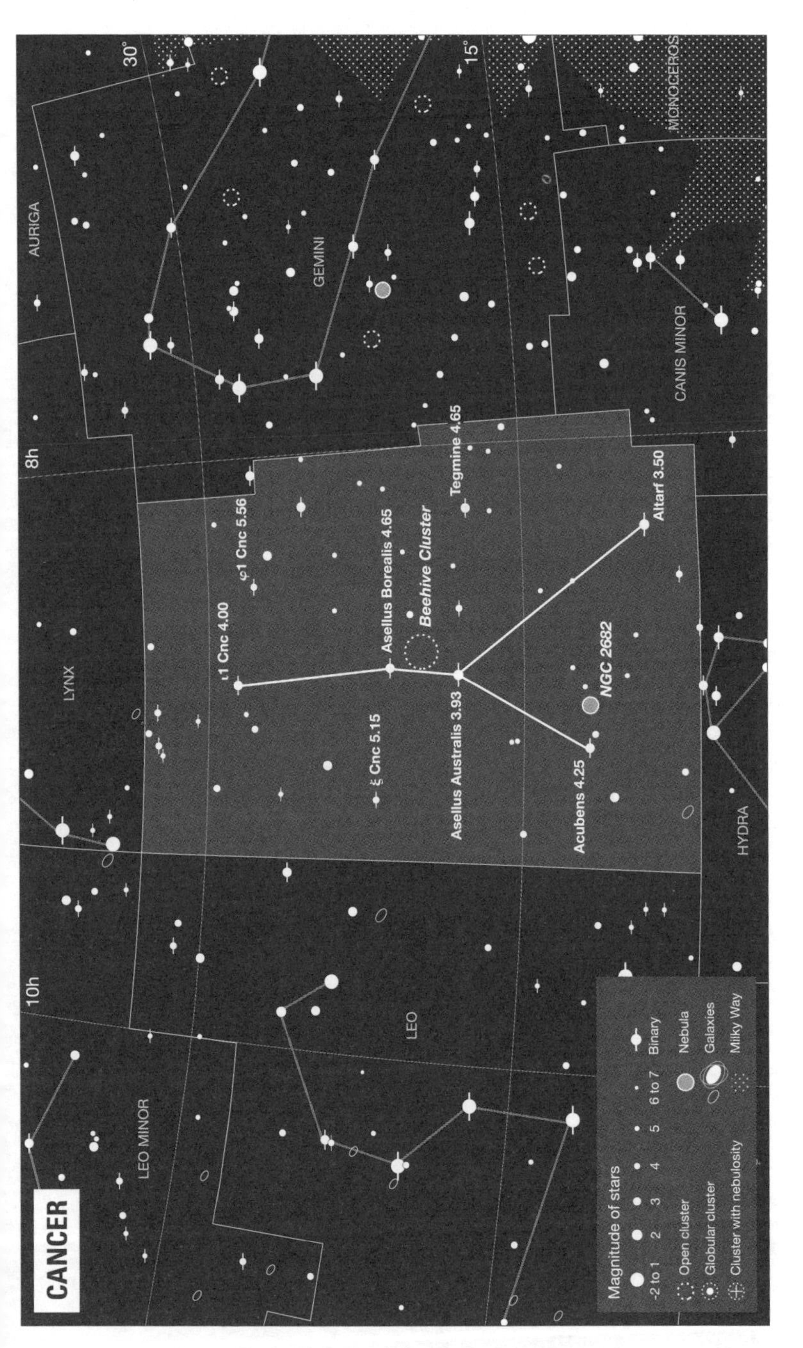
CANCER
AURIGA
LYNX
GEMINI
MONOCEROS
CANIS MINOR
HYDRA
LEO
LEO MINOR
30°
15°
8h
10h
ι1 Cnc 4.00
φ1 Cnc 5.56
Asellus Borealis 4.65
Beehive Cluster
Tegmine 4.65
Altarf 3.50
ξ Cnc 5.15
Asellus Australis 3.93
NGC 2682
Acubens 4.25
Magnitude of stars
-2 to 1
2
3
4
5
6 to 7
Binary
Nebula
Galaxies
Milky Way
Open cluster
Globular cluster
Cluster with nebulosity

Canes Venatici

Coordinates of Brightest Star Cor Caroli (Mag 2.87): RA 12h 56m 08s, Dec 38° 18' 17"

Size of Constellation Boundaries 465 square degrees

Bounded by Ursa Major, Coma Berenices, Boötes

Constellation Legend Canes Venatici represents the two hunting dogs of Arcas (Asterion and Chara), who is associated with the Boötes constellation. But while Boötes is one of the oldest of constellations, Canes Venatici is relatively recent, created in 1687 by the Polish astronomer Johannes Hevelius (prior to that point, the stars were grouped in Ursa Major). The constellation's primary star is known as Cor Caroli ("Charles's Heart"), named by Edmond Halley in 1725 as a tribute to British king Charles II, who founded the Royal Society (a forum for new scientific ideas) and ordered the construction of the Royal Observatory at Greenwich in 1675.

Constellation Sights Cor Caroli is in fact a double star, resolvable with a small telescope. But the big news in Canes Venatici is the Whirlpool galaxy (M51), one of the most famous galaxies around, thanks to its head-on spiral structure. You can spot the galaxy with binoculars and a small telescope will let you see the nucleus, but you'll need a telescope with an aperture of eight inches or more to see the arms. In the southern border of the constellation you'll also find NGC 5272 (M3), a really fine globular cluster that appears as a "fuzzy" star on the very edge of visibility. A four-inch scope or star should enable you to spot individual stars within the cluster.

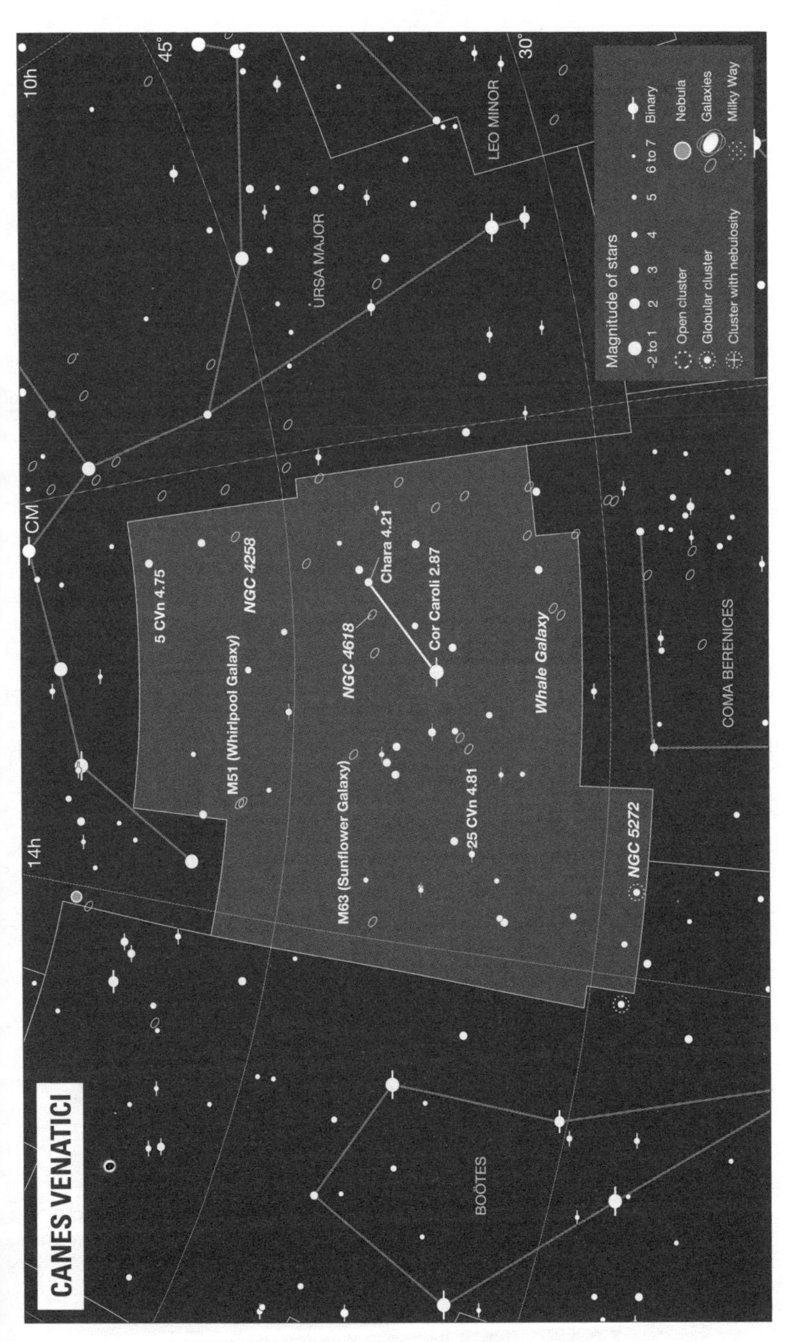
CANES VENATICI
10h
45°
30°
14h
CM
URSA MAJOR
LEO MINOR
COMA BERENICES
BOÖTES
5 CVn 4.75
NGC 4258
M51 (Whirlpool Galaxy)
NGC 4618
Chara 4.21
Cor Caroli 2.87
Whale Galaxy
M63 (Sunflower Galaxy)
25 CVn 4.81
NGC 5272
Magnitude of stars
-2 to 1
2
3
4
5
6 to 7
Binary
Open cluster
Globular cluster
Cluster with nebulosity
Nebula
Galaxies
Milky Way

Canis Major

Coordinates of Brightest Star Sirius (Mag -1.47): RA 6h 45m 15s, Dec -16° 43' 07"

Size of Constellation Boundaries 380 square degrees

Bounded by Monoceros, Lepus, Columba, Puppis

Constellation Legend Canis Major ("The Great Dog") is one of the hunting dogs associated with Orion (the other being Canis Minor). The constellation is famous for Sirius (also known as the Dog Star), which is the brightest star in the night sky. The expression "dog days" comes from the fact that the Romans noticed how Sirius rose before the Sun during the hottest part of the year. The Egyptians, who knew Sirius as Sothis, noticed how the same phenomenon occurred just as the Nile River flooded, and thought that it might actually be the cause. Sirius is one of the Sun's closest neighbours, at just over eight and a half light years distance.

Constellation Sights Sirius is actually a binary star; its companion is a small white dwarf, Sirius B (sometimes known as the "Pup"). Resolving Sirius B requires a large telescope, good eyesight, and some luck. Other sights include the whimsically named Mexican Jumping Star, or more prosaically NGC 2362, which is not a star at all but a dense, small, open cluster. It gets its strange name because many observers have noted that if they tap their telescope while it is centred on the cluster, the brightest star in the cluster (Tau Canis Majoris) appears to "jump" relative to its background (clearly, it's an illusion). M41 is another open cluster, with eighty or so stars, reputed to have been recorded by Aristotle around 325 BC. This would make it the faintest celestial object noted by ancient observers.

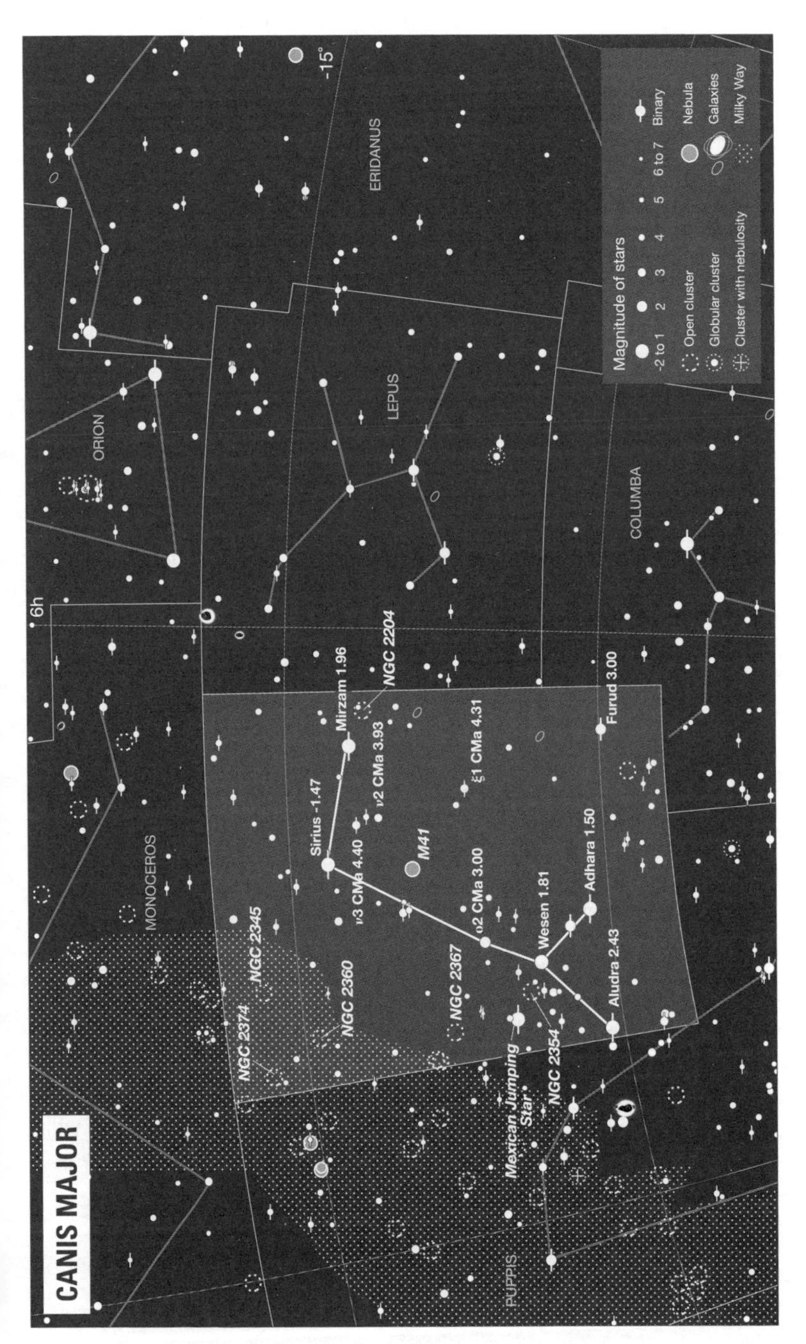
CANIS MAJOR
Magnitude of stars
-2 to 1
2
3
4
5
6 to 7
Binary
Open cluster
Globular cluster
Cluster with nebulosity
Nebula
Galaxies
Milky Way
-15°
6h
ERIDANUS
ORION
LEPUS
COLUMBA
MONOCEROS
PUPPIS
Sirius -1.47
Mirzam 1.96
NGC 2204
ν2 CMa 3.93
ξ1 CMa 4.31
ν3 CMa 4.40
M41
Furud 3.00
ο2 CMa 3.00
Adhara 1.50
Wesen 1.81
Aludra 2.43
NGC 2345
NGC 2374
NGC 2360
NGC 2367
NGC 2354
Mexican Jumping Star

Canis Minor & Monoceros

Coordinates of Brightest Stars Procyon (Mag 0.37): RA 7h 39m 25s, Dec 5° 13' 09"; Alpha Monocerotis (Mag 3.93): RA 7h 41m 15s, Dec 9° 33' 21"

Size of Constellation Boundaries 182 square degrees (Canis Minor); 482 square degrees (Monoceros)

Bounded by Gemini, Orion, Lepus, Canis Major, Puppis, Hydra, Cancer

Constellation Legend Canis Minor means "The Little Dog", and it is indeed one of the smaller constellations in the sky. Its primary star, Procyon (from the Greek phrase meaning "before the Dog") is so named because the star rises before Canis Major's primary star, Sirius. Monoceros is a modern constellation added by Petrus Plancius in the seventeenth century. It represents a unicorn.

Constellation Sights In Canis Minor, Procyon is interesting because of the number of characteristics it shares with Sirius. These include their brightness (Procyon is the eighth brightest star in the sky) and their binary nature: like Sirius, Procyon's primary star has a much smaller, much dimmer companion, which orbits its primary mate every four decades. However, Procyon's stars are separable in amateur instruments. Monoceros has a good number of clusters, in particular M50, an open cluster of stars sometimes known as the Heart-Shaped Cluster, which is visible with binoculars. NGC 2244 is another open cluster, which lies in the Rosette nebula. Unfortunately, the nebula itself is too faint to be seen satisfactorily in amateur telescopes or binoculars. The Christmas Tree Cluster looks triangular in telescopes (hence the name) and is home to the Cone nebula, also too faint for good viewing. Star enthusiasts will enjoy Beta Monocerotis, which resolves into a triple star in small scopes.

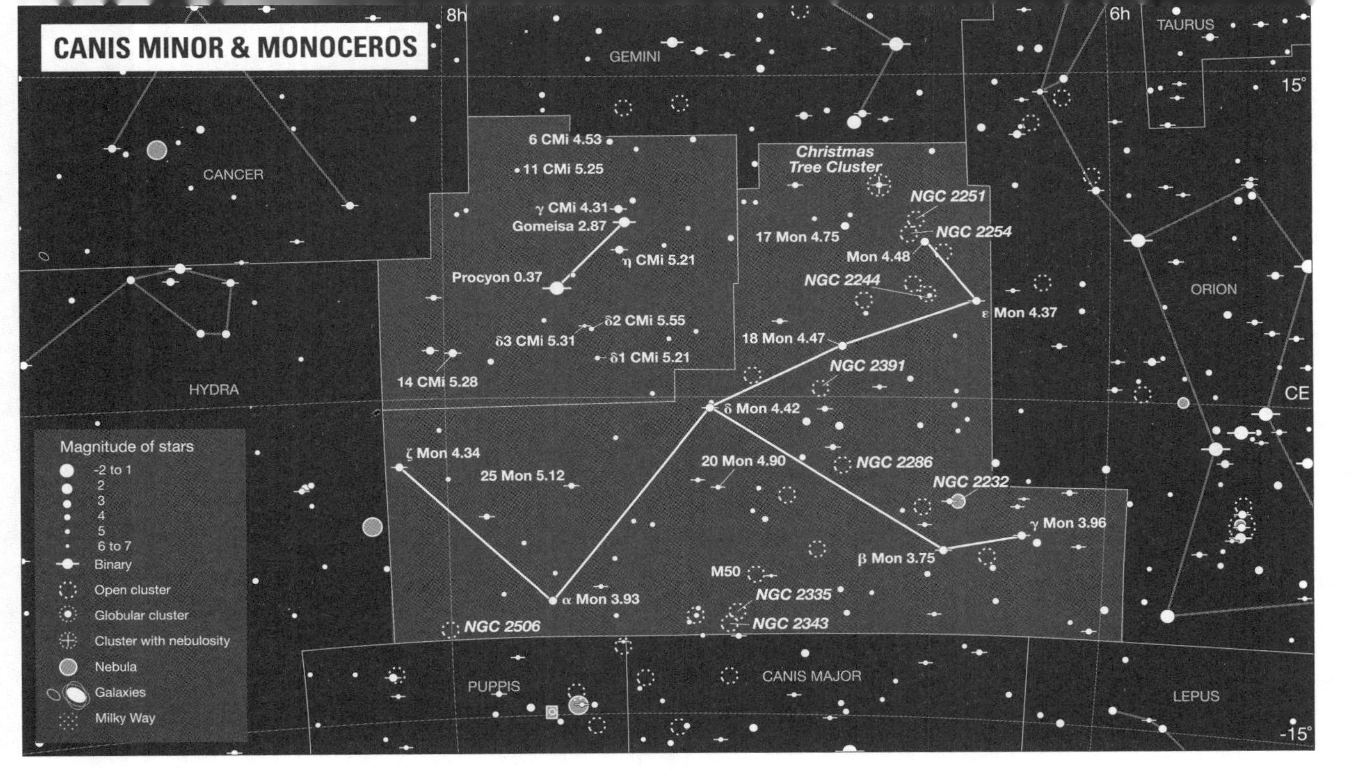
CANIS MINOR & MONOCEROS
8h
6h
15°
-15°
CE
GEMINI
TAURUS
CANCER
ORION
HYDRA
PUPPIS
CANIS MAJOR
LEPUS
6 CMi 4.53
11 CMi 5.25
γ CMi 4.31
Gomeisa 2.87
η CMi 5.21
Procyon 0.37
δ2 CMi 5.55
δ3 CMi 5.31
δ1 CMi 5.21
14 CMi 5.28
Christmas Tree Cluster
NGC 2251
NGC 2254
17 Mon 4.75
Mon 4.48
NGC 2244
ε Mon 4.37
18 Mon 4.47
NGC 2391
δ Mon 4.42
ζ Mon 4.34
25 Mon 5.12
20 Mon 4.90
NGC 2286
NGC 2232
γ Mon 3.96
β Mon 3.75
M50
NGC 2335
NGC 2343
α Mon 3.93
NGC 2506
Magnitude of stars
-2 to 1
2
3
4
5
6 to 7
Binary
Open cluster
Globular cluster
Cluster with nebulosity
Nebula
Galaxies
Milky Way

Capricornus

Coordinates of Brightest Star Deneb Algiedi (Mag 2.84): RA 21h 47m 10s, Dec 16° 06' 59"

Size of Constellation Boundaries 414 square degrees

Bounded by Aquarius, Aquila, Sagittarius, Microscopium, Pisces Austrinus

Constellation Legend Capricornus ("horned goat" in Latin), is one of the signs of the Zodiac, and one of the oldest constellations, having been identified as a group by the Sumerians. Mythologically it is associated with the god Pan, who turned himself into a half-goat, half-fish in a frantic attempt to escape the sea monster Typhon. The Chinese saw the constellation not as a goat (or a goat-fish), but as a bull, and the Egyptians associated it with the water god Chnum. Two and a half thousand years back, the Sun was in Capricornus during the winter solstice, and its southernmost latitude in the sky that day is still known as the Tropic of Capricorn. Today the Sun is in Sagittarius at the winter solstice.

Constellation Sights The star Alpha Capricorni is an interesting double star, and sharp eyes or binoculars can separate its two component stars, Geidi and Algedi. These two stars are actually separated by more than one thousand light years and do not have a direct relationship. Dabih is likewise a double star, although this time its two stars do orbit each other. Good binoculars or a telescope should resolve the second star. Capricornus is also home to NGC 7099 (also known as M30), a seventh-magnitude globular cluster.

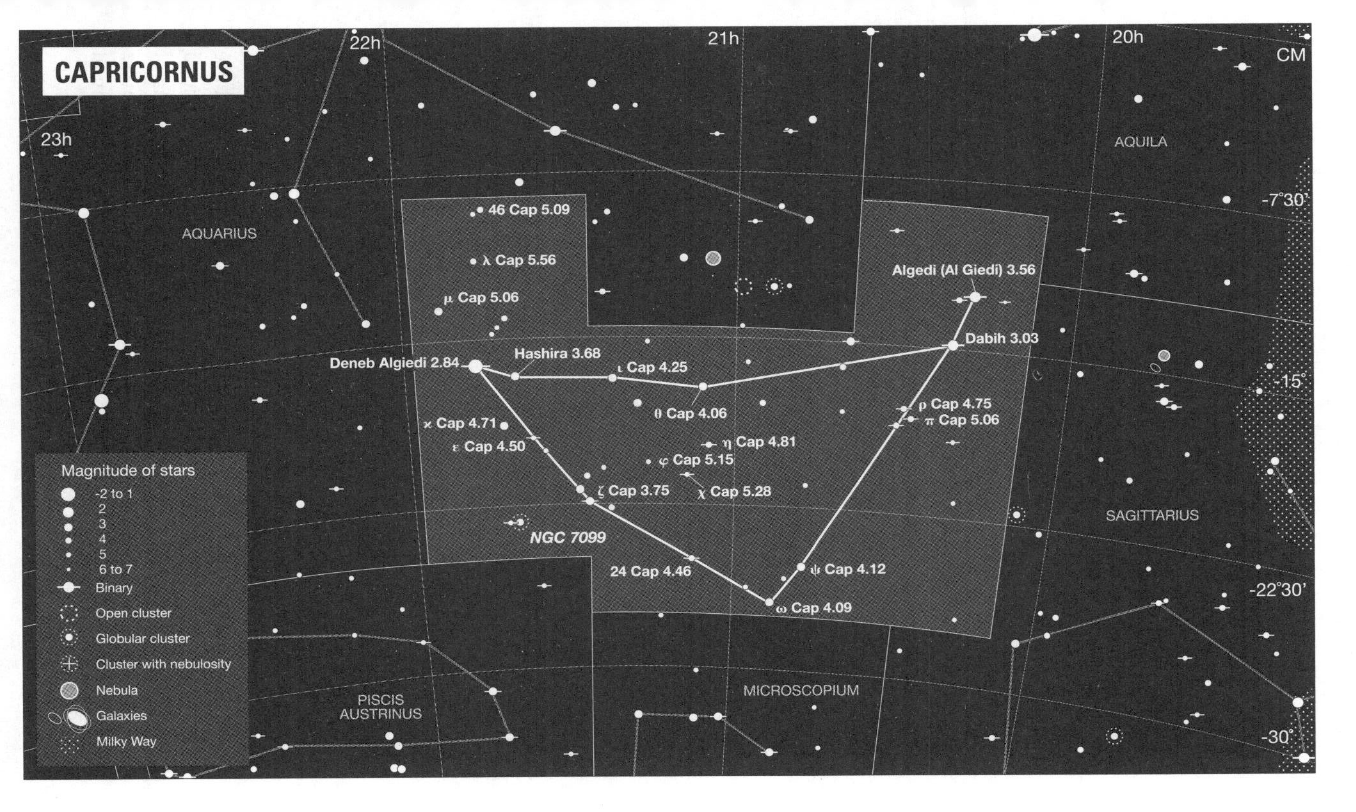
CAPRICORNUS
22h
21h
20h
23h
CM
AQUILA
-7°30'
-15°
-22°30'
-30°
AQUARIUS
SAGITTARIUS
MICROSCOPIUM
PISCIS AUSTRINUS
46 Cap 5.09
λ Cap 5.56
μ Cap 5.06
Algedi (Al Giedi) 3.56
Dabih 3.03
Deneb Algiedi 2.84
Hashira 3.68
ι Cap 4.25
θ Cap 4.06
ρ Cap 4.75
π Cap 5.06
κ Cap 4.71
ε Cap 4.50
η Cap 4.81
φ Cap 5.15
ζ Cap 3.75
χ Cap 5.28
NGC 7099
24 Cap 4.46
ψ Cap 4.12
ω Cap 4.09
Magnitude of stars
-2 to 1
2
3
4
5
6 to 7
Binary
Open cluster
Globular cluster
Cluster with nebulosity
Nebula
Galaxies
Milky Way

Carina & Volans

Coordinates of Brightest Stars Canopus (Mag -0.63): RA 6h 23m 57s, Dec -52° 41' 45"; Beta Volantis (Mag 3.75): RA 8h 25m 44s, Dec -66° 08' 39"

Size of Constellation Boundaries 494 square degrees (Carina); 141 square degrees (Volans)

Bounded by Vela, Puppis, Pictor, Dorado, Mensa, Chamaeleon, Musca, Centaurus

Constellation Legend Carina is the keel of what used to be the massive constellation of Argo Navis, the famed ship of the Greek hero Jason. Argo Navis was chopped up into four constellations (Carina, Vela, Puppis, Pyxis) by the International Astronomical Union in 1929, making these four effectively the "youngest" constellations. Volans, which represents a flying fish, was named Piscis Volans by Dutch explorers Keyser and de Houtman in the sixteenth century.

Constellation Sights In Carina, the object that sticks out is Canopus, the second brightest star in the sky, and actually hundreds of times more luminous than Sirius (whose brighter appearance is due to the fact that it's nearer to us). It's also the home of the variable Eta Carinae, now just below visual range, but with a history of flaring up and at times outshining even Canopus. Look at it in a telescope and you'll see that it's not perfectly round. What you are seeing is shells of gas shed by this massive star. Eta Carinae is housed within the Eta Carinae nebula, which you can see with binoculars; the Keyhole nebula (labelled here) lies in front of it. Carina is also home to the Southern Pleiades (also known as IC 2602), and eight stars in this open cluster are visible to the naked eye (one up on the original Pleiades). Other clusters that are visible unaided are NGC 2516, NGC 3114 and NGC 3532. After all the visual drama of Carina, Volans is positively bland, with only two double stars – Gamma and Epsilon Volantis – of any note. These are separable with a small telescope.

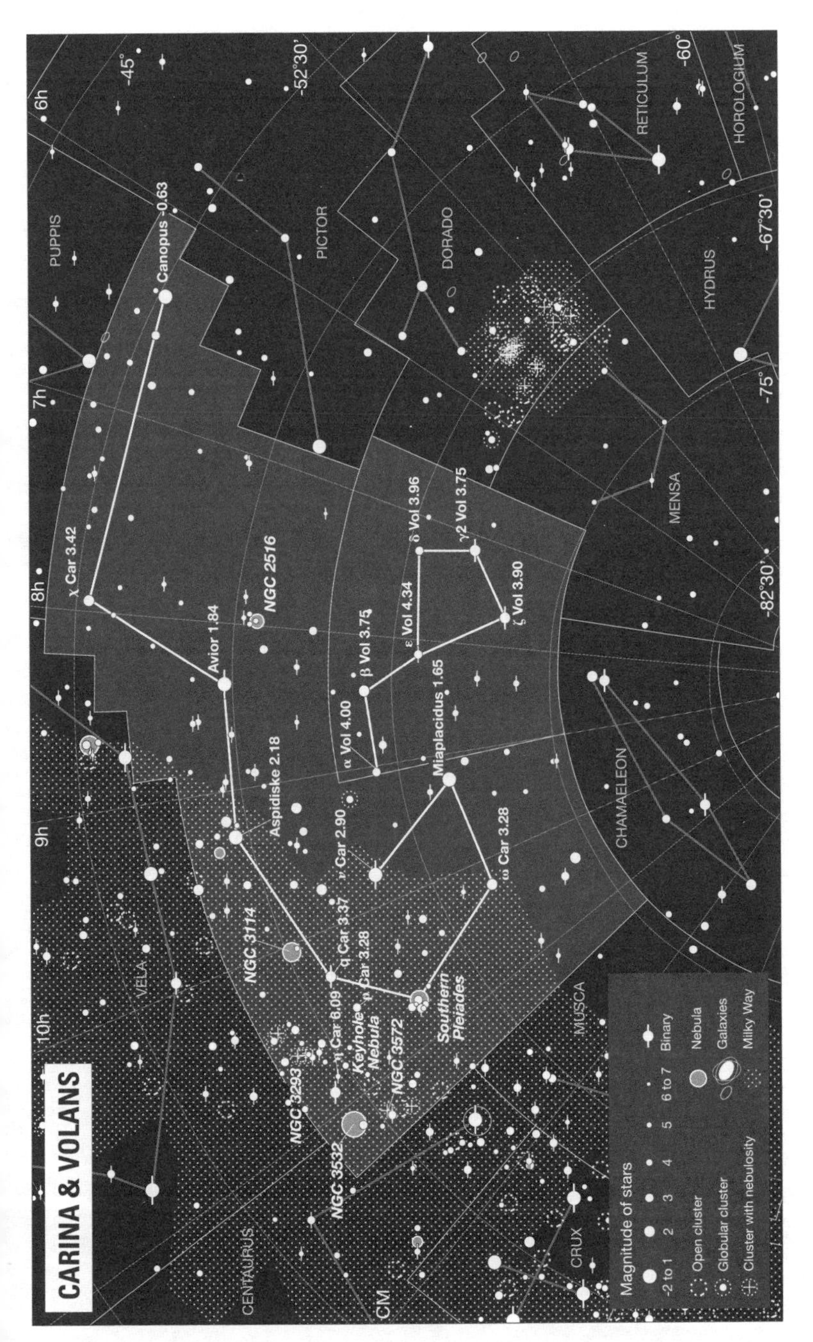
CARINA & VOLANS
6h
7h
8h
9h
10h
-45°
-52°30'
-60°
-67°30'
-75°
-82°30'
PUPPIS
PICTOR
DORADO
RETICULUM
HOROLOGIUM
HYDRUS
MENSA
CHAMAELEON
VELA
MUSCA
CENTAURUS
CRUX
CM
Canopus -0.63
χ Car 3.42
Avior 1.84
NGC 2516
Aspidiske 2.18
NGC 3114
q Car 3.37
p Car 3.28
η Car 6.09
Keyhole Nebula
NGC 3293
NGC 3532
NGC 3572
Southern Pleiades
υ Car 2.90
Miaplacidus 1.65
ω Car 3.28
α Vol 4.00
β Vol 3.75
ε Vol 4.34
δ Vol 3.96
γ2 Vol 3.75
ζ Vol 3.90
Magnitude of stars
-2 to 1
2
3
4
5
6 to 7
Binary
Open cluster
Globular cluster
Cluster with nebulosity
Nebula
Galaxies
Milky Way

Cassiopeia

Coordinates of Brightest Star Gamma Cassiopeiae (Mag 2.12): RA 0h 56m 51s, Dec 60° 43' 48"

Size of Constellation Boundaries 598 square degrees

Bounded by Cepheus, Lacerta, Andromeda, Perseus, Camelopardalis

Constellation Legend This one's named after the infamously vain Greek queen whose boast of her own personal beauty very nearly caused her daughter Andromeda (whose constellation lies directly south) to be consumed by a marauding sea monster sent by Neptune. Cassiopeia is typically seen as hanging upside down (southwards) as punishment for her vanity. Early Arabian cultures saw not a queen but a hand, with each of the major stars of the constellation representing a finger.

Constellation Sights Gamma Cassiopeiae is currently the constellation's brightest star but is wildly variable, swinging between magnitudes of between 1.6 and 3.3. Much of that variability comes from the star's rapid rotation, which causes it to fling off great, bright emissions of gas. Most of the other bright stars in the constellation are also to some extent variable. Two open clusters of note are Cassiopeia Salt-and-Pepper (M52), which has more than a hundred stars and requires a telescope to view individual stars, and the smaller cluster M103. Video-game enthusiasts will be amused at the "Pac Man" nebula, which looks (very vaguely) like the famous muncher caught mid-chew. You'll want a telescope for that one.

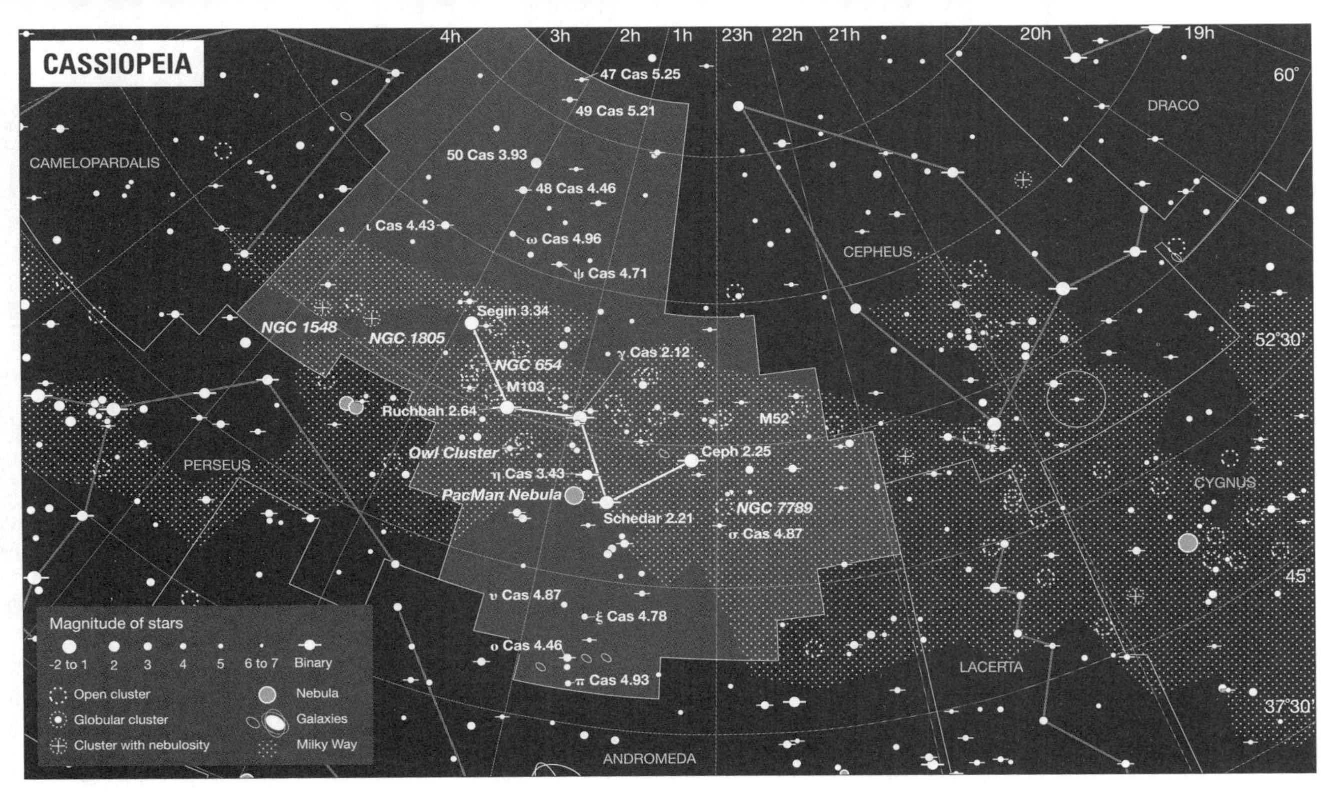
CASSIOPEIA
4h
3h
2h
1h
23h
22h
21h
20h
19h
60°
52°30'
45°
37°30'
DRACO
CAMELOPARDALIS
CEPHEUS
PERSEUS
CYGNUS
LACERTA
ANDROMEDA
47 Cas 5.25
49 Cas 5.21
50 Cas 3.93
48 Cas 4.46
ι Cas 4.43
ω Cas 4.96
ψ Cas 4.71
Segin 3.34
NGC 1548
NGC 1805
γ Cas 2.12
NGC 654
M103
Ruchbah 2.64
M52
Owl Cluster
Ceph 2.25
η Cas 3.43
PacMan Nebula
NGC 7789
Schedar 2.21
σ Cas 4.87
υ Cas 4.87
ξ Cas 4.78
ο Cas 4.46
π Cas 4.93
Magnitude of stars
-2 to 1
2
3
4
5
6 to 7
Binary
Open cluster
Globular cluster
Cluster with nebulosity
Nebula
Galaxies
Milky Way

Centaurus

Coordinates of Brightest Star Rigel Kentaurus (Mag -0.04): RA 14h 39m 36s, Dec -60° 50' 40"

Size of Constellation Boundaries 1060 square degrees

Bounded by Hydra, Antlia, Vela, Carina, Crux, Musca, Circinus, Lupus

Constellation Legend Centaurus is the centaur, a creature with the body of a horse, and the chest, arms and head of a man. Centaurus is generally regarded specifically as the centaur Chiron, who taught numerous Greek heroes including Jason (whose ship the Argo was previously represented in the sky very near where Centaurus lies). Chiron was accidentally slain by Hercules (who also has his own constellation), and was turned into a constellation by Zeus (Jupiter).

Constellation Sights Centaurus is loaded with interesting things to see. First of all is the multiple star Rigel Kentaurus, also known as Alpha Centauri, the third brightest star in the sky (after Sirius and Canopus) and our Sun's closest neighbour at just 4.4 light years away. Binoculars or a small telescope will separate this star into an orange and yellow pair, with a third member (dim, red Proxima Centauri, actually the closest to us by 0.2 light years) two degrees out. Omega Centauri is the largest, brightest globular cluster we can see. It's a hazy moon-sized patch of light to the naked eye, but individual stars resolve in telescopes. It's also worth looking out for Centaurus A (or NGC 5128), an elliptical galaxy visible with a small telescope, which shows a dust lane in larger scopes that is probably a result of a galactic collision. And for fans of planetary nebulae, there's the Blue Planetary (NGC 3918), which, as its name suggests, appears as a blue disc in small telescopes.

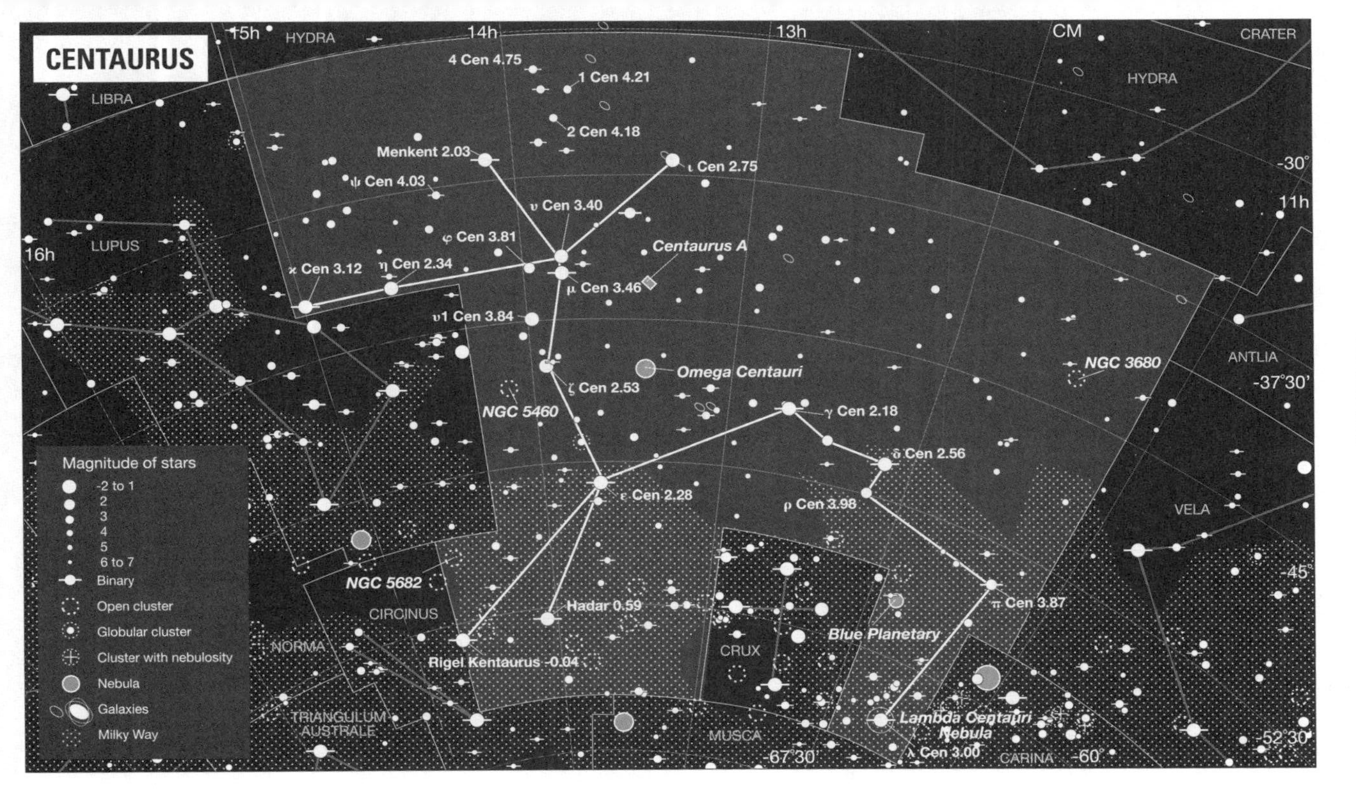
CENTAURUS
LIBRA
LUPUS
HYDRA
CRATER
CM
ANTLIA
VELA
CARINA
MUSCA
CRUX
CIRCINUS
NORMA
TRIANGULUM AUSTRALE
16h
15h
14h
13h
11h
-30°
-37°30'
-45°
-52°30'
-60°
-67°30'
4 Cen 4.75
1 Cen 4.21
2 Cen 4.18
Menkent 2.03
ψ Cen 4.03
ι Cen 2.75
υ Cen 3.40
φ Cen 3.81
Centaurus A
κ Cen 3.12
η Cen 2.34
μ Cen 3.46
υ1 Cen 3.84
Omega Centauri
ζ Cen 2.53
NGC 5460
γ Cen 2.18
δ Cen 2.56
ε Cen 2.28
ρ Cen 3.98
NGC 3680
NGC 5682
Hadar 0.59
π Cen 3.87
Blue Planetary
Rigel Kentaurus -0.04
Lambda Centauri Nebula
λ Cen 3.00
Magnitude of stars
-2 to 1
2
3
4
5
6 to 7
Binary
Open cluster
Globular cluster
Cluster with nebulosity
Nebula
Galaxies
Milky Way

Cepheus

Coordinates of Brightest Star Alderamin (Mag 2.43): RA 21h 18m 35s, Dec 62° 35' 43"

Size of Constellation Boundaries 588 square miles

Bounded by Ursa Minor, Draco, Cygnus, Lacerta, Cassiopeia, Camelopardalis

Constellation Legend This constellation represents the Greek king Cepheus, husband to Cassiopeia and father to Andromeda. In scientific circles, the constellation is rather more famous for giving its name to a specific type of variable star ("Cepheid variables"), which feature a relationship between size and brightness that is so standard that these variables are used as "standard candles" to help determine cosmic distances.

Constellation Sights The archetypal Cepheid variable is Delta Cephei, in the southeast corner of the constellation, which varies in brightness by nearly a full order of magnitude every five days and nine hours. A telescope will also spot its dimmer companion star. Alfirk (Arabic for "the flock") is a variable star of another type (called Beta Cephei stars); it is also a double star whose smaller companion will show up in a small telescope. Erakis (called the Garnet Star by William Herschel) shows reddish in binoculars and telescopes and is also variable in brightness because it pulsates in size over a period of two years or so. Far in the north of the constellation is the faint (ninth-magnitude) cluster NGC 188; this dim collection of stars is believed to be the oldest in our galaxy, with an age of more than five billion years.

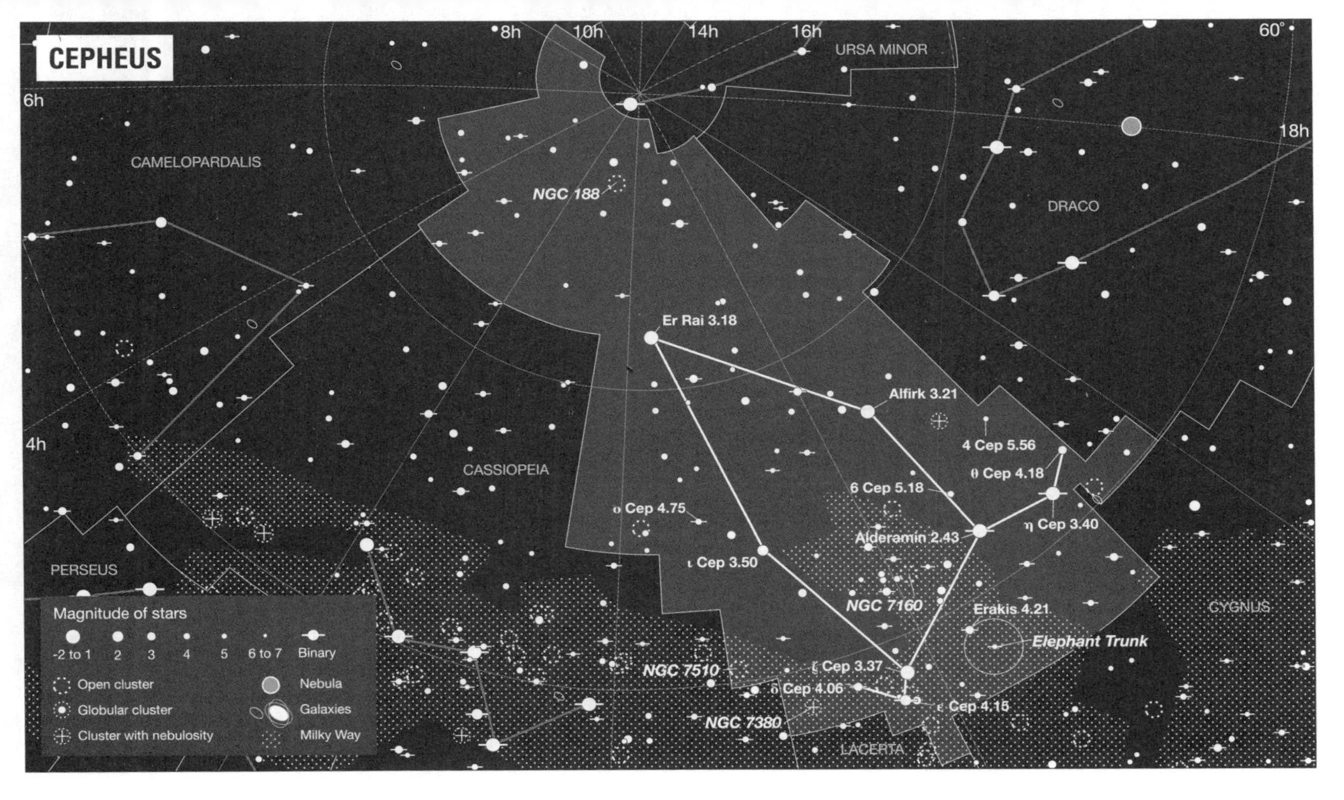
CEPHEUS
URSA MINOR
DRACO
CAMELOPARDALIS
CASSIOPEIA
PERSEUS
CYGNUS
LACERTA
8h
10h
14h
16h
60°
6h
18h
4h
NGC 188
Er Rai 3.18
Alfirk 3.21
4 Cep 5.56
θ Cep 4.18
6 Cep 5.18
η Cep 3.40
ο Cep 4.75
Alderamin 2.43
ι Cep 3.50
NGC 7160
Erakis 4.21
Elephant Trunk
ζ Cep 3.37
NGC 7510
δ Cep 4.06
ε Cep 4.15
NGC 7380
Magnitude of stars
-2 to 1
2
3
4
5
6 to 7
Binary
Open cluster
Nebula
Globular cluster
Galaxies
Cluster with nebulosity
Milky Way

Cetus

Coordinates of Brightest Star Deneb Kaitos (Mag 2.0): RA 0h 43m 42s, Dec -17° 58' 22"

Size of Constellation Boundaries 1231 square degrees

Bounded by Aries, Pisces, Aquarius, Sculptor, Fornax, Eridanus, Taurus

Constellation Legend Cetus has a couple of stories associated with it. The Greeks saw it as the sea monster that threatened Andromeda and was turned to stone by Perseus. In ancient Mesopotamia the Sumerians and the Babylonians thought it represented the fearsome god-monster Tiamat. The constellation is also associated with the biblical story of Jonah and the whale (cetus is the Latin for "whale").

Constellation Sights Mira is a red giant that swells and contracts on a roughly 330-day cycle, ranging in brightness from third magnitude down to ninth magnitude (and occasionally as bright as second magnitude). It was the first variable star to be discovered (in the late sixteenth century), and is the namesake for a type of long-period variable known as Mira variables. If you've ever wondered what the Sun might look like from another star system, take a look at Tau Ceti – at 11.9 light years away, it's close enough to our own Sun in size and luminosity that it's a prime candidate for SETI scans (SETI = Search for Extra-Terrestrial Intelligence). Galaxy-gazers should turn their telescopes to M77, which – though appearing small and fuzzy in a scope – is notable for being the brightest example of a Seyfert galaxy: a special, active class of galaxy with a notably bright centre.

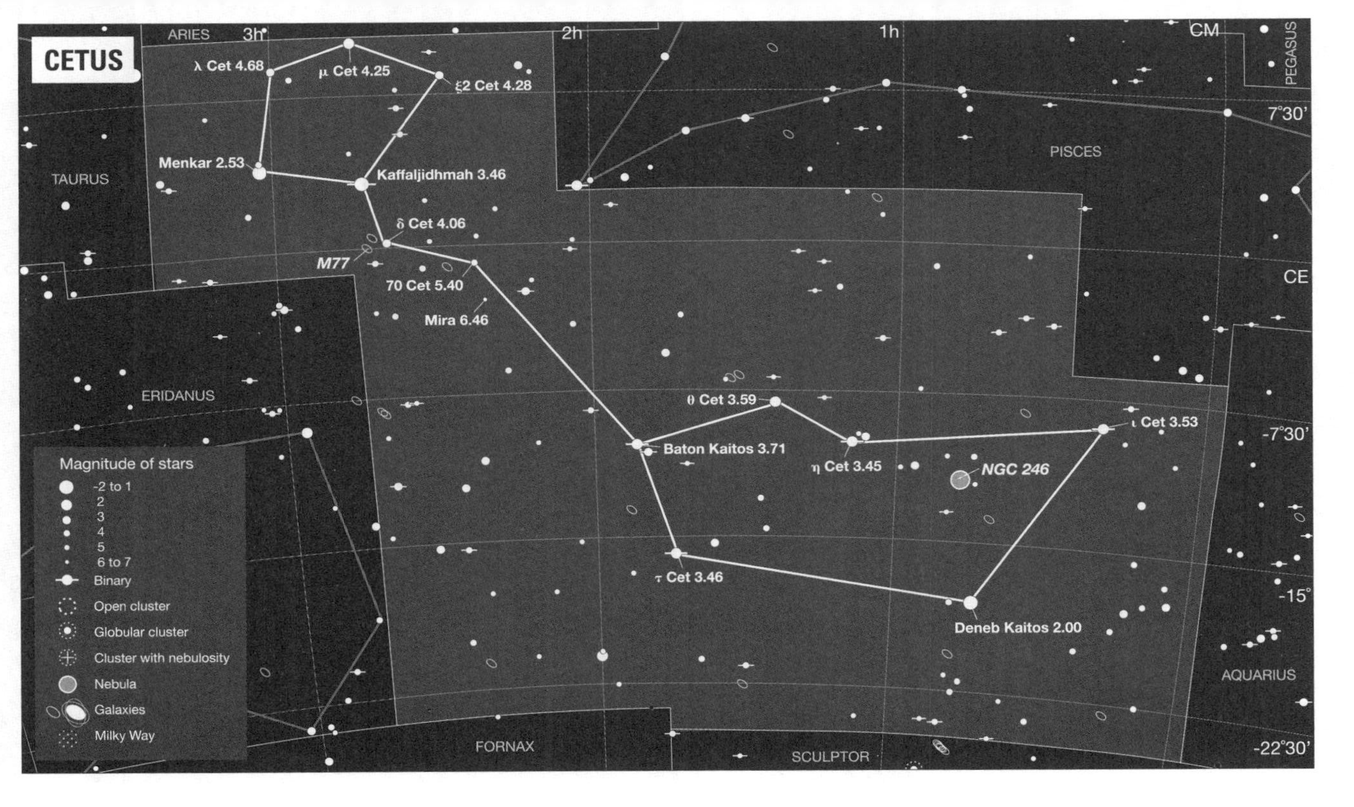
CETUS
ARIES
3h
2h
1h
CM
PEGASUS
λ Cet 4.68
μ Cet 4.25
ξ2 Cet 4.28
7°30'
PISCES
TAURUS
Menkar 2.53
Kaffaljidhmah 3.46
δ Cet 4.06
M77
70 Cet 5.40
Mira 6.46
CE
ERIDANUS
θ Cet 3.59
ι Cet 3.53
-7°30'
Baton Kaitos 3.71
η Cet 3.45
NGC 246
τ Cet 3.46
-15°
Deneb Kaitos 2.00
AQUARIUS
FORNAX
SCULPTOR
-22°30'
Magnitude of stars
-2 to 1
2
3
4
5
6 to 7
Binary
Open cluster
Globular cluster
Cluster with nebulosity
Nebula
Galaxies
Milky Way

Chamaeleon & Octans

Coordinates of Brightest Stars Alpha Chamaeleontis (Mag 4.03): RA 8h 18m 32s, Dec -76° 55' 35"; Nu Octanis (Mag 3.71): RA 21h 41m 29s, Dec -77° 22' 46"

Size of Constellation Boundaries 132 square degrees (Chamaeleon); 291 square degrees (Octans)

Bounded by Musca, Carina, Volans, Mensa, Hydra, Tucana, Indus, Pavo, Apus

Constellation Legend Chamaeleon was created in the late sixteenth century by Dutch explorers Keyser and de Houtman, and named after the famous camouflaging lizard by the astronomer Johann Bayer in 1603. Octans is another of the southern constellations devised by Nicolas Louis de Lacaille in the eighteenth century. As with all de Lacaille's constellations, it was named (by him) after a scientific instrument – this time the octant, which was used for navigational purposes and was a forerunner of the sextant.

Constellation Sights Chamaeleon is aptly named, since the constellation is hard to spot and doesn't present much of interest. Delta Chamaeleontis is a double star that can be resolved through binoculars, while NGC 3195 is a planetary nebula viewable with a good, medium-sized telescope. Octans' main attraction is Sigma Octanis, which with a declination of -88°, 56' 49", is the star currently closest to the celestial south pole. Unlike its northern counterpart Polaris, Sigma Octanis is hard to spot because its magnitude is just 5.43 – barely visible to the naked eye.

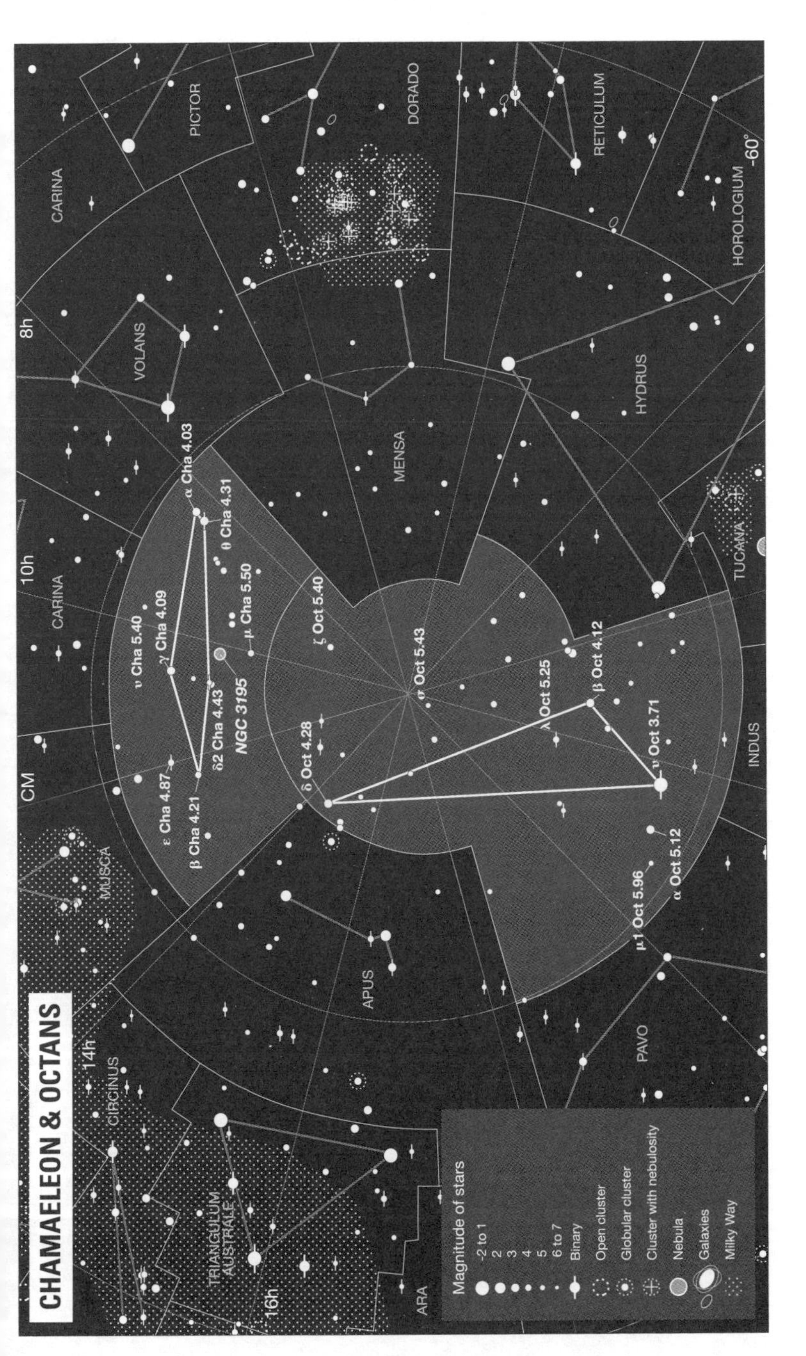
CHAMAELEON & OCTANS
CARINA
PICTOR
DORADO
RETICULUM
HOROLOGIUM
-60°
8h
VOLANS
MENSA
HYDRUS
TUCANA
10h
CARINA
α Cha 4.03
θ Cha 4.31
μ Cha 5.50
ν Cha 5.40
γ Cha 4.09
NGC 3195
δ2 Cha 4.43
ζ Oct 5.40
σ Oct 5.43
β Oct 4.12
λ Oct 5.25
ν Oct 3.71
INDUS
δ Oct 4.28
CM
ε Cha 4.87
β Cha 4.21
MUSCA
α Oct 5.12
μ1 Oct 5.96
APUS
PAVO
14h
CIRCINUS
TRIANGULUM AUSTRALE
16h
ARA
Magnitude of stars
-2 to 1
2
3
4
5
6 to 7
Binary
Open cluster
Globular cluster
Cluster with nebulosity
Nebula
Galaxies
Milky Way

Circinus

Coordinates of Brightest Star Alpha Circini (Mag 3.15): RA 14h 42m 30s, Dec -64° 59' 10"

Size of Constellation Boundaries 93 square degrees

Bounded by Lupus, Centaurus, Musca, Apus, Triangulum Australe, Norma

Constellation Legend This tiny constellation is yet another of Nicolas Louis de Lacaille's creations. It represents a pair of surveyor's compasses.

Constellation Sights Sadly, there's not much to look at in this constellation. Alpha Circinus, a double star resolvable in a small telescope, is mostly notable for its proximity to Rigel Kentaurus in the neighbouring constellation of Centaurus.

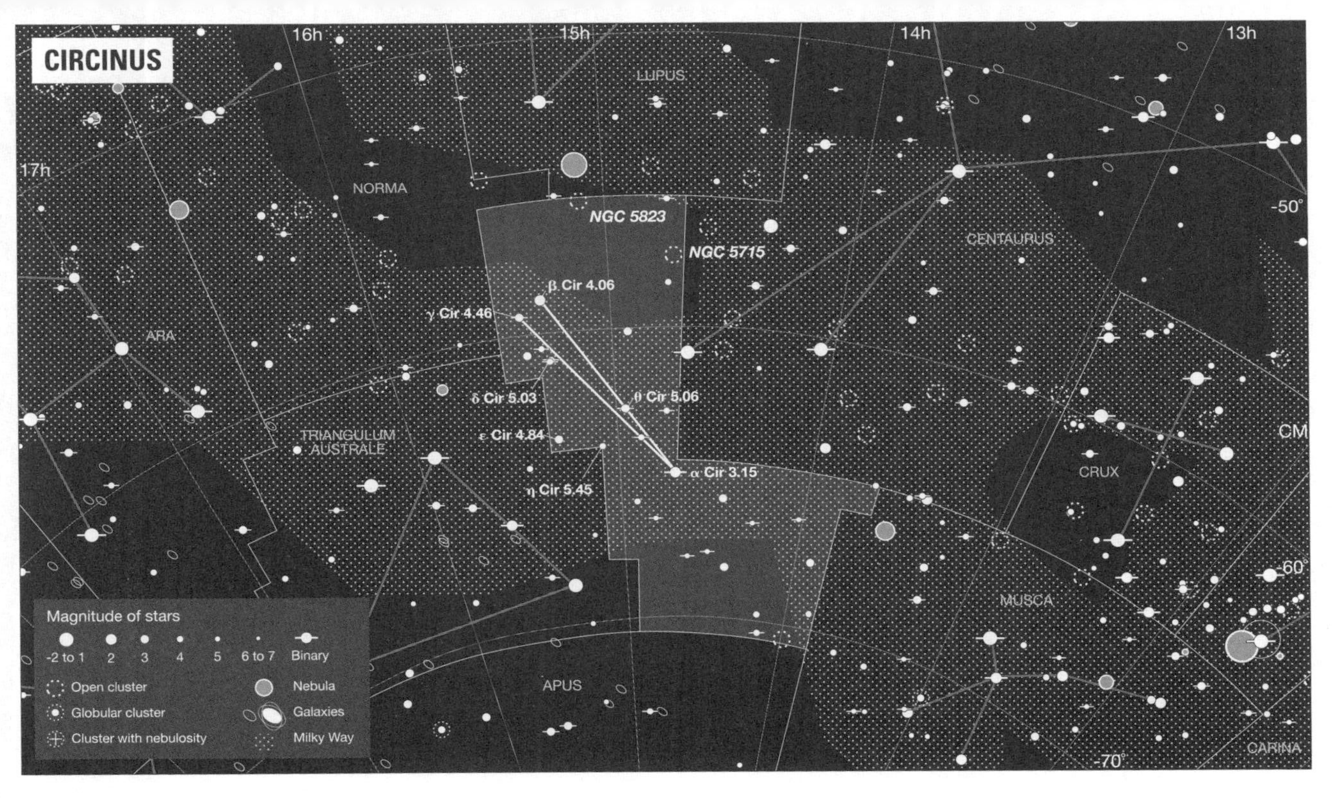
CIRCINUS
16h
15h
14h
13h
17h
-50°
-60°
-70°
LUPUS
NORMA
CENTAURUS
ARA
TRIANGULUM AUSTRALE
CRUX
MUSCA
CARINA
APUS
CM
NGC 5823
NGC 5715
β Cir 4.06
γ Cir 4.46
δ Cir 5.03
θ Cir 5.06
ε Cir 4.84
α Cir 3.15
η Cir 5.45
Magnitude of stars
-2 to 1
2
3
4
5
6 to 7
Binary
Open cluster
Globular cluster
Cluster with nebulosity
Nebula
Galaxies
Milky Way

Coma Berenices

Coordinates of Brightest Star Beta Comae Berenices (Mag 4.21): RA 13h 11m 58s, Dec 27° 51' 56"

Size of Constellation Boundaries 386 square degrees

Bounded by Canes Venatici, Ursa Major, Leo, Virgo, Boötes

Constellation Legend This group of faint stars (none brighter than fourth magnitude) was part of the constellation Leo until about 1590, when astronomer Tycho Brahe partitioned it off and made it an independent constellation. He named it in honour of Berenice, an Egyptian queen of the third century BC, who, according to legend, promised to trim her impressive tresses if her husband Ptolemy made it safely back from battle (Coma Berenices means "Berenice's Hair"). Coma Berenices lies in an area of space perpendicular to the plane of our own galaxy; when you're looking at this constellation, you're looking out beyond our own galaxy into the vast reaches of the cosmos.

Constellation Sights The southwest corner of Coma Berenices features a number of galaxies that are part of the huge Virgo cluster. A particularly interesting sight is the Black Eye galaxy (M64), which features an immense streak of dark dust arcing near the galaxy's centre; this "smudge" is visible in larger home telescopes. Moderate to large scopes can also get a good look at the Coma Pinwheel galaxy (M100), whose face-on position shows its spiral structure. NGC 4565 is another spiral galaxy, which can be viewed as a bright, narrow streak in small to moderate telescopes.

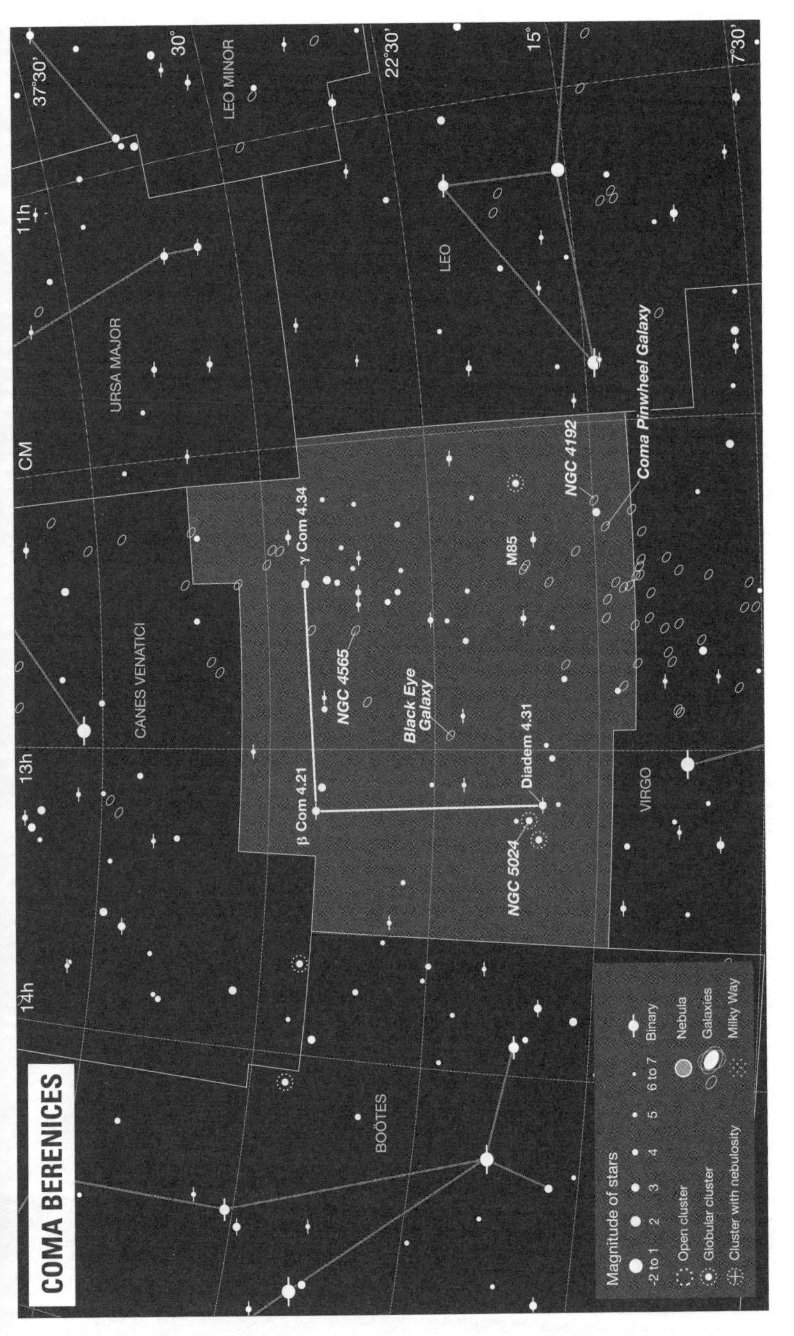

COMA BERENICES
37°30'
30°
22°30'
15°
7°30'
11h
13h
14h
CM
URSA MAJOR
LEO MINOR
LEO
CANES VENATICI
BOÖTES
VIRGO
γ Com 4.34
β Com 4.21
Diadem 4.31
NGC 4565
Black Eye Galaxy
NGC 4192
Coma Pinwheel Galaxy
M85
NGC 5024
Magnitude of stars
-2 to 1
2
3
4
5
6 to 7
Binary
Open cluster
Globular cluster
Cluster with nebulosity
Nebula
Galaxies
Milky Way

Corona Australis

Coordinates of Brightest Star Alpha Coronae Australis (Mag 4.09): RA 19h 9m 28s, Dec -37° 54' 05"

Size of Constellation Boundaries 179 square degree

Bounded by Sagittarius, Scorpius, Ara, Telescopium

Constellation Legend The "Southern Crown" has been known since ancient times and was one of the original 48 constellations listed by the astronomer Ptolemy. The Greeks seem to have associated it with Dionysus (Bacchus), the god of wine and misrule, who placed a crown or wreath in the sky in honour of his mother Semele. But the constellation is also connected with Sagittarius, the constellation directly to the north, and is sometimes depicted as arrows projecting from the hand of the Centaur (an alternative name is Corona Sagittarii, "Crown of Sagittarius"). The Chinese saw the constellation, not as a crown, but as a tortoise; as did the Arabs, who also likened it to a dish or a tent.

Constellation Sights While relatively dim (there are no stars above fourth magnitude), the greatest attraction of the constellation is its delicate and distinctive arc of stars – which is why the constellation was well known to ancient cultures. For telescope owners, NGC 6729 is a faint but interesting nebula, whose brightness varies in concert with the brightness of R Coronae Australis, an irregular variable star located within the nebula. The constellation also offers NGC 6541, a globular cluster visible with binoculars or a small telescope.

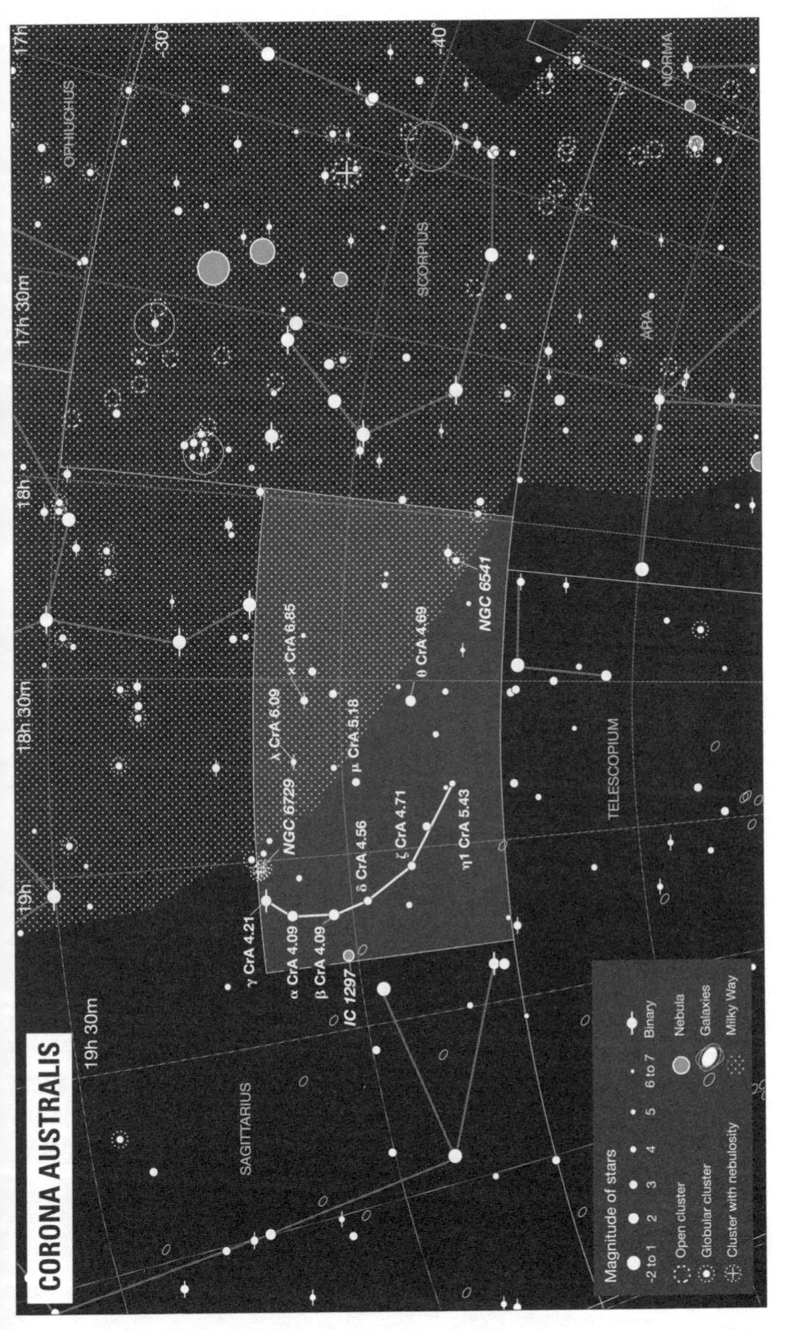
CORONA AUSTRALIS
17h
17h 30m
18h
18h 30m
19h
19h 30m
-30°
-40°
OPHIUCHUS
SCORPIUS
NORMA
ARA
TELESCOPIUM
SAGITTARIUS
NGC 6541
θ CrA 4.69
κ CrA 6.85
λ CrA 6.09
μ CrA 5.18
NGC 6729
δ CrA 4.56
ζ CrA 4.71
η1 CrA 5.43
γ CrA 4.21
α CrA 4.09
β CrA 4.09
IC 1297
Magnitude of stars
-2 to 1
2
3
4
5
6 to 7
Binary
Open cluster
Globular cluster
Cluster with nebulosity
Nebula
Galaxies
Milky Way

Corona Borealis

Coordinates of Brightest Star Gemma (Mag 2.21): RA 15h 34m 47s, Dec 26° 42' 20"

Size of Constellation Boundaries 179 square degrees

Bounded by Hercules, Boötes, Serpens Caput

Constellation Legend The "Northern Crown" represents the crown worn by the Cretan princess Ariadne, who eloped with the Athenian hero Theseus after helping him to slay the Minotaur. Theseus later abandoned her on the island of Naxos, but she was spotted by the wine god Dionysus (Bacchus) who fell in love with her. On the day of their wedding, Dionysus was so overjoyed that he threw Ariadne's crown into the sky to commemorate the occasion. For the Shawnee the constellation represented dancing maidens, one of whom became the wife of a mortal man – her absence is why the constellation doesn't form a complete circle.

Constellation Sights Gemma, the constellation's brightest star, is an eclipsing binary whose brightness dims slightly at each eclipse. Other double stars include Zeta Coronae Borealis, which is divisible in a small telescope, and Nu Coronae Borealis, a visual double star whose two components are not in orbit around each other. A variable star worth looking out for is R Coronae Borealis, which usually appears around sixth magnitude but can dim down as far as fifteenth magnitude for months before moving back to the realm of visibility. Corona Borealis is the home to a cluster of more than four hundred galaxies, which reside in the southwest corner of the constellation. However, these galaxies are very faint and very distant (more than a billion light years away), and only observers with the largest home telescopes will be able to find them.

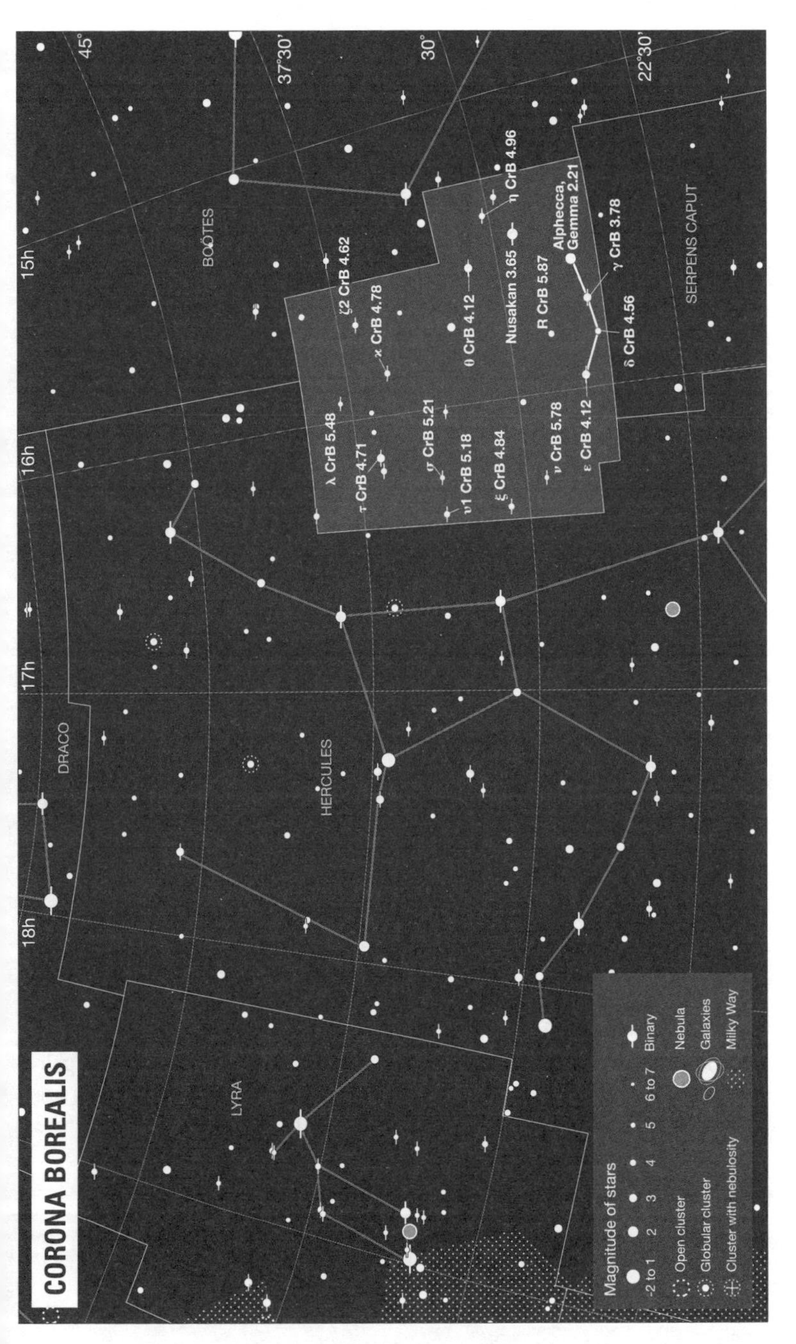
CORONA BOREALIS
45°
37°30'
30°
22°30'
15h
16h
17h
18h
BOÖTES
DRACO
HERCULES
LYRA
SERPENS CAPUT
η CrB 4.96
Alphecca, Gemma 2.21
γ CrB 3.78
ζ2 CrB 4.62
κ CrB 4.78
θ CrB 4.12
Nusakan 3.65
R CrB 5.87
δ CrB 4.56
λ CrB 5.48
τ CrB 4.71
σ CrB 5.21
υ1 CrB 5.18
ξ CrB 4.84
ν CrB 5.78
ε CrB 4.12
Magnitude of stars
-2 to 1
2
3
4
5
6 to 7
Binary
Open cluster
Globular cluster
Cluster with nebulosity
Nebula
Galaxies
Milky Way

Corvus & Crater

Coordinates of Brightest Stars Gienah (Mag 2.56): RA 12h 15m 48s, Dec -17° 33' 20"; Delta Crateris (Mag 3.53): RA 11h 19m 20s, Dec -14° 47' 29"

Size of Constellation Boundaries 184 square degrees (Corvus); 282 square degrees (Crater)

Bounded by Leo, Sextans, Hydra, Virgo

Constellation Legend The two small constellations of Corvus (the Crow) and Crater (the Cup) share the same pair of stories, both associated with the god Apollo. In the first, the crow was ordered by Apollo to bring him some water in a cup but, seeing some figs, he waited for them to ripen and was delayed. The crow's excuse was that a water snake had blocked the spring. Apollo punished the bird by setting him in the sky within view of the cup, but prevented from drinking by the closeness of the water snake (the constellation Hydra). The second story has the crow reporting to Apollo that the god's lover Coronis has been unfaithful to him. As a result of being the bearer of bad news, the bird's feathers were changed from white to black.

Constellation Sights These constellations don't offer much for the casual astronomer. Corvus's Algorab is a double star whose companion is very faint (ninth magnitude) and you'll need a telescope to see it. Corvus is also home to the Antennae galaxies, a pair of galaxies which are interacting gravitationally: images of the two galaxies show long streams of gas and stars arcing between them. Though you won't be able to see the streams in a home telescope, you should be able to make out the pair themselves with a moderately sized scope. Crater has even less going for it; its several galaxies are all less than eleventh magnitude and therefore not easily seen by home telescopes.

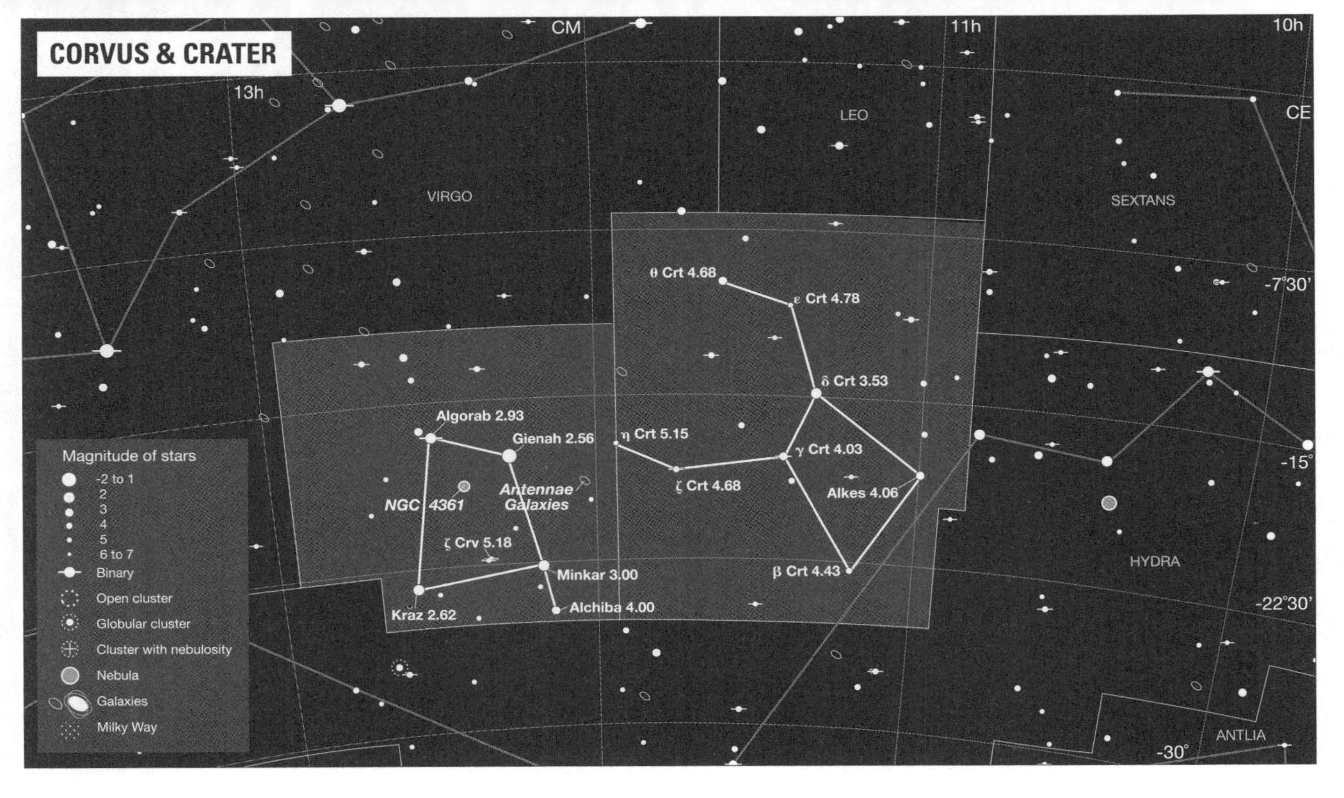
CORVUS & CRATER
CM
13h
11h
10h
CE
LEO
VIRGO
SEXTANS
HYDRA
ANTLIA
-7°30'
-15°
-22°30'
-30°
θ Crt 4.68
ε Crt 4.78
δ Crt 3.53
γ Crt 4.03
Alkes 4.06
β Crt 4.43
ζ Crt 4.68
η Crt 5.15
Algorab 2.93
Gienah 2.56
Antennae Galaxies
NGC 4361
ζ Crv 5.18
Minkar 3.00
Alchiba 4.00
Kraz 2.62
Magnitude of stars
-2 to 1
2
3
4
5
6 to 7
Binary
Open cluster
Globular cluster
Cluster with nebulosity
Nebula
Galaxies
Milky Way

Crux

Coordinates of Brightest Star Acrux (Mag 0.75): RA 12h 26m 36s, Dec -63° 06' 46"

Size of Constellation Boundaries 68 square degrees

Bounded by Centaurus, Musca

Constellation Legend Originally a part of the constellation Centaurus, where it formed the centaur's hind legs, Crux's distinctive cross shape led to its recognition as a constellation in its own right in the sixteenth century. Also known as the "Southern Cross", Crux is the smallest of all the officially recognized constellations, but nevertheless arguably the best known in the southern skies due to its brightness. In Australia, where it is visible all year round, it was adopted as an integral part of the country's national flag in 1901.

Constellation Sights The shape of the constellation itself is impressive enough, not least because three of the four stars in the cross are first-degree stars (Delta Crucis, the fourth, has a magnitude of 2.78). The brightest of the stars, Acrux, is a double star in small telescopes, as is Gacrux. Crux is also home to the Jewel Box (NGC 4755), the "jewel" of which is an eighth-magnitude red supergiant, surrounded by blue-white stars. Directly south of the Jewel Box lies the Coalsack, a large and dark nebula more than five hundred light years away, which blocks the stars lying behind it from view.

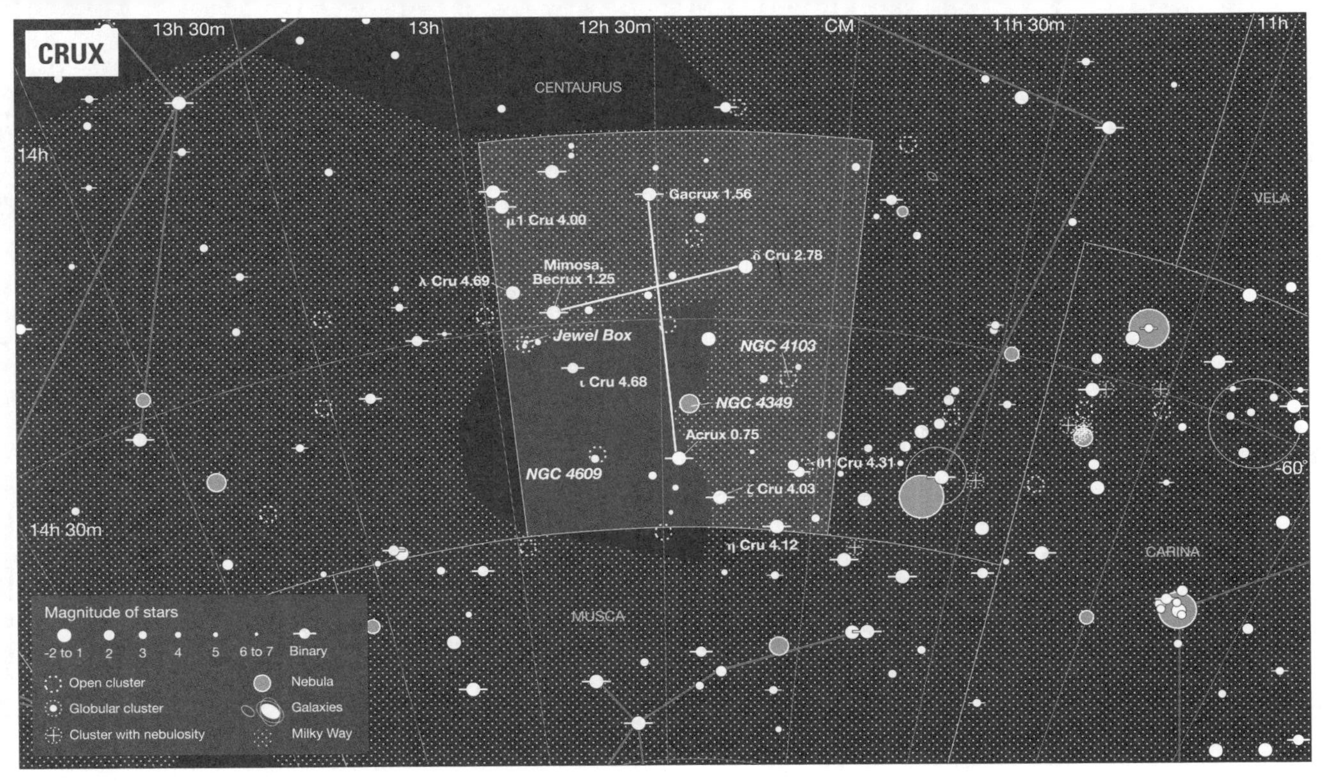
CRUX
13h 30m
13h
12h 30m
CM
11h 30m
11h
14h
14h 30m
-60°
CENTAURUS
VELA
CARINA
MUSCA
Gacrux 1.56
μ1 Cru 4.00
δ Cru 2.78
λ Cru 4.69
Mimosa,
Becrux 1.25
Jewel Box
NGC 4103
ι Cru 4.68
NGC 4349
Acrux 0.75
θ1 Cru 4.31
NGC 4609
ζ Cru 4.03
η Cru 4.12
Magnitude of stars
-2 to 1
2
3
4
5
6 to 7
Binary
Open cluster
Nebula
Globular cluster
Galaxies
Cluster with nebulosity
Milky Way

Cygnus

Coordinates of Brightest Star Deneb (Mag 1.25): RA 20h 41m 26s, Dec 45° 17' 18"

Size of Constellation Boundaries 804 square degrees

Bounded by Cepheus, Draco, Lyra, Vulpecula, Pegasus, Lacerta

Constellation Legend Cygnus (the Swan) is commonly associated with a fairly sordid adventure of Zeus, the principal Greek god, who transformed himself into a swan in order to have his way with Leda, queen of Sparta. Leda managed to accommodate her husband the same day and the result was the twins, Castor and Pollux: the first mortal, the second immortal. Cygnus is also known as the "Northern Cross" because of its shape.

Constellation Sights Cygnus houses some prominent nebulae visible in binoculars and telescopes, among them the Veil nebula (the remnant of a supernova that blew up fifty thousand years ago) and the North American nebula, whose shape is reminiscent of the North American continent. There's also the Blinking Planetary, a planetary nebula that gets its name because of the way it appears to "blink" off if you look at it and then glance quickly away! M39 is an open cluster of about 25 stars, which is visible to the naked eye and opens up in binoculars. Among its stars, Deneb (from the Arabic word for "tail") is interesting because it is the most distant first-magnitude star in the sky, some thirty-two hundred light years away. Albireo, which represents the constellation's head, resolves in telescopes into a lovely double star with highly constrasting colours: the primary star is yellow-orange while the smaller is a striking blue. Not included in this chart (because home scopes can't see it) but interesting to note is Cygnus X-1, an extremely strong X-ray source which astronomers believe is a black hole.

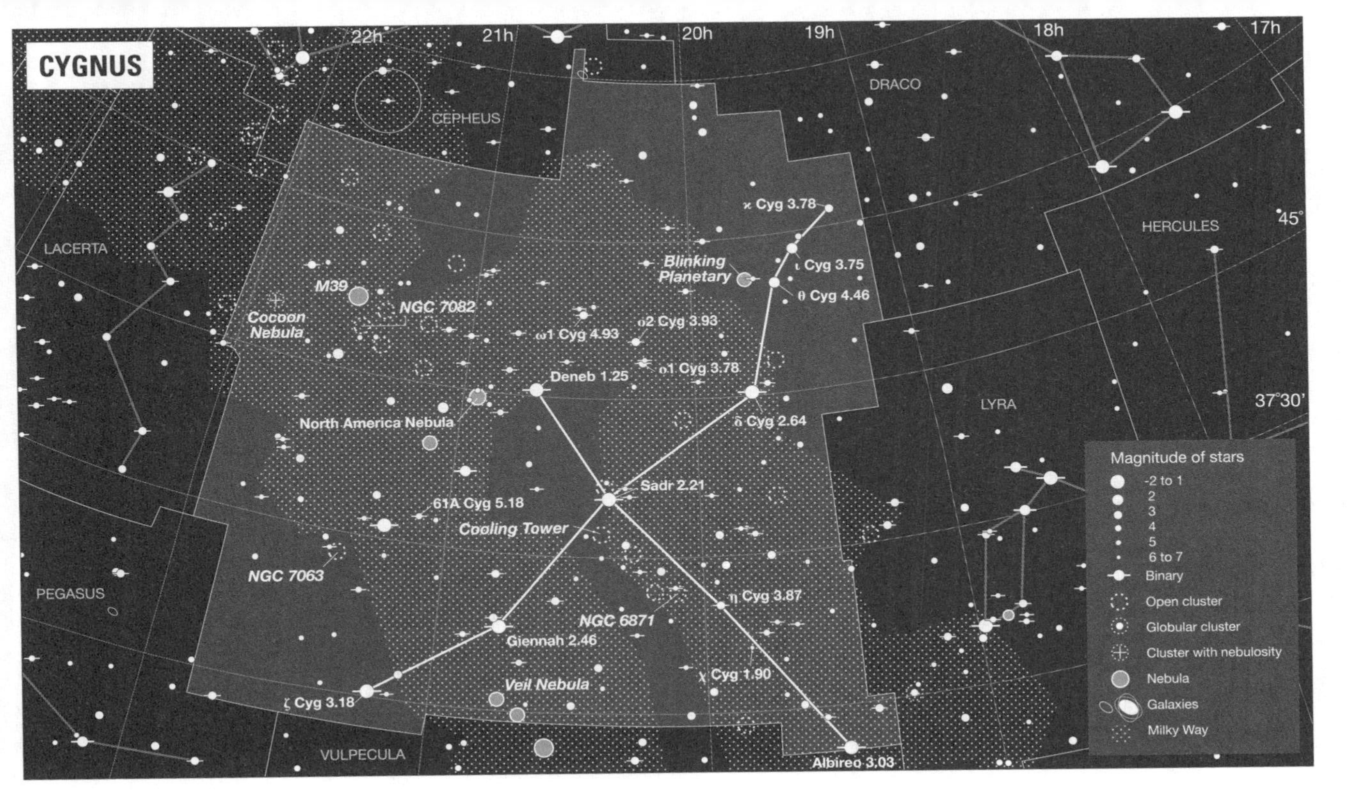
CYGNUS
22h
21h
20h
19h
18h
17h
45°
37°30'
DRACO
CEPHEUS
HERCULES
LACERTA
LYRA
PEGASUS
VULPECULA
κ Cyg 3.78
ι Cyg 3.75
θ Cyg 4.46
Blinking Planetary
M39
NGC 7082
Cocoon Nebula
ω1 Cyg 4.93
o2 Cyg 3.93
o1 Cyg 3.78
Deneb 1.25
δ Cyg 2.64
North America Nebula
Sadr 2.21
61A Cyg 5.18
Cooling Tower
NGC 7063
η Cyg 3.87
NGC 6871
Giennah 2.46
χ Cyg 1.90
Veil Nebula
ζ Cyg 3.18
Albireo 3.03
Magnitude of stars
-2 to 1
2
3
4
5
6 to 7
Binary
Open cluster
Globular cluster
Cluster with nebulosity
Nebula
Galaxies
Milky Way

Delphinus & Equuleus

Coordinates of Brightest Stars Rotanev (Mag 3.62): RA 20h 37m 33s, Dec 14° 36' 10"; Kitalpha (Mag 3.90): RA 21h 15m 49s, Dec 5° 15' 26"

Size of Constellation Boundaries 189 square degrees (Delphinus); 72 square degrees (Equuleus)

Bounded by Vulpecula, Sagitta, Aquila, Aquarius, Pegasus

Constellation Legend Both Delphinus (the Dolphin) and Equuleus (the Horse) were known to the ancient Greeks. Delphinus is said to represent the dolphin that rescued Arion, a famed singer and poet, who flung himself from a ship to escape the crew who were plotting to kill him for his riches. Ancient Arab astronomers, on the other hand, saw the constellation as having the shape of a camel. Equuleus, though one of the oldest constellations, has no particular legend attached to it, save that it was known as the "little horse" because it rises in the sky before Pegasus, the larger and more famous equine constellation directly to the east.

Constellation Sights Delphinus's main attraction is Job's Coffin, the diamond-shaped asterism formed by Sualocin, Rotanev, Delta Delphini and Gamma Delphini. The last of these is also a double star, separable in a small telescope. Delphinus's two brightest stars, incidentally, were given their names by Niccolo Cacciatore, a nineteenth-century Italian astronomer, by reversing the spelling of his name in its Latinized form – Nicolaus Venator. Over in Equuleus, Gamma Equulei is another double star, resolvable by using binoculars or a small telescope.

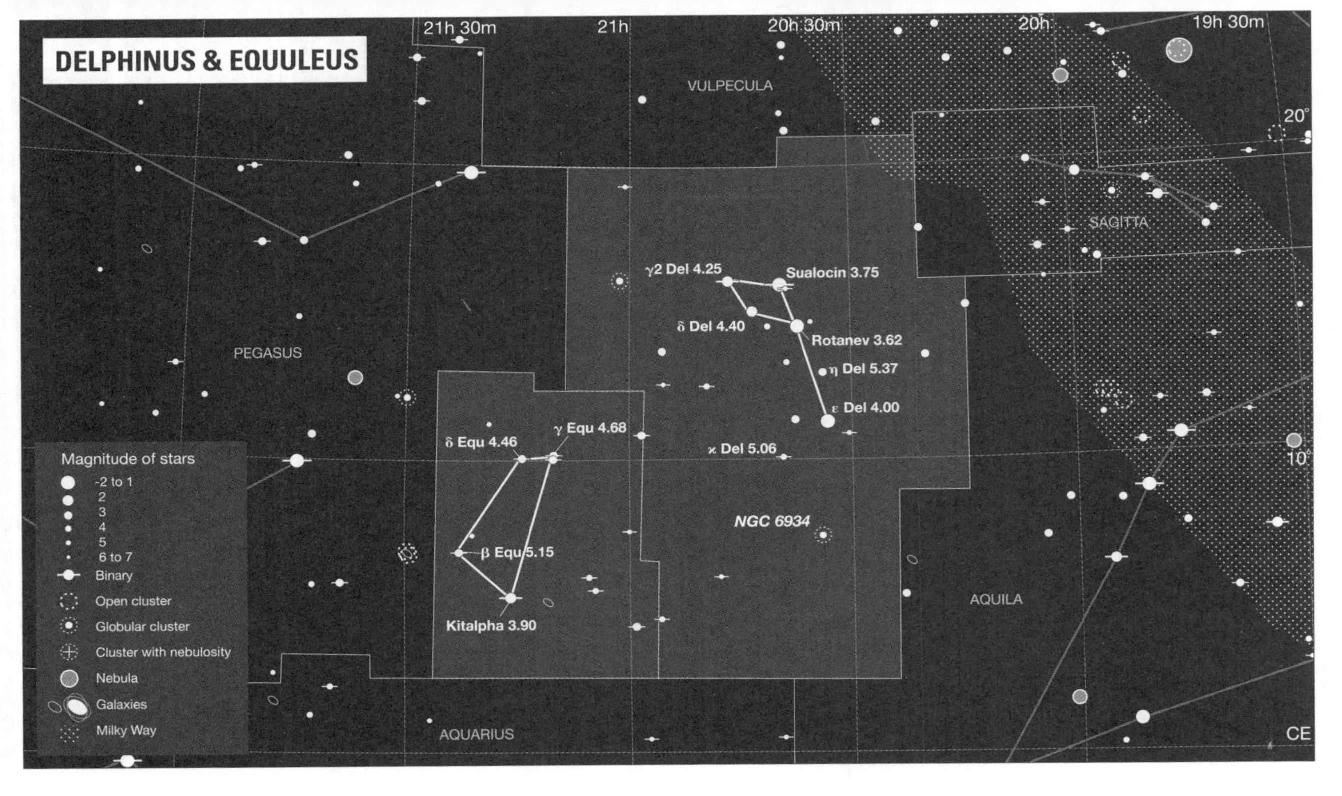
DELPHINUS & EQUULEUS
21h 30m
21h
20h 30m
20h
19h 30m
20°
10°
VULPECULA
SAGITTA
PEGASUS
AQUILA
AQUARIUS
CE
γ2 Del 4.25
Sualocin 3.75
δ Del 4.40
Rotanev 3.62
η Del 5.37
ε Del 4.00
κ Del 5.06
NGC 6934
δ Equ 4.46
γ Equ 4.68
β Equ 5.15
Kitalpha 3.90
Magnitude of stars
-2 to 1
2
3
4
5
6 to 7
Binary
Open cluster
Globular cluster
Cluster with nebulosity
Nebula
Galaxies
Milky Way

Dorado & Mensa

Coordinates of Brightest Stars Alpha Doradus (Mag 3.28): RA 4h 33m 60s, Dec -55° 02' 20"; Alpha Mensae (Mag 5.06): RA 6h 10m 14s, Dec -74° 45' 10"

Size of Constellation Boundaries 179 square degrees (Dorado); 153 square degrees (Mensa)

Bounded by Pictor, Caelum, Horologium, Reticulum, Hydrus, Octans, Chamaeleon, Volans

Constellation Legend Dorado is often called the "golden fish", named not after the popular pet, but after the tropical dolphinfish (also known as *mahi mahi*), whose yellow and green scales appear gold underwater. The constellation was credited to Keyser and de Houtman, two navigators who were part of the first Dutch expedition to the East Indies in 1595. Mensa means "table" in Latin, and this constellation was originally named "Mons Mensae", after Table Mountain in South Africa where Nicolas Louis de Lacaille mapped it from his nearby observatory in the 1750s.

Constellation Sights The most impressive sight of these two constellations is the Large Magellanic Cloud, a satellite galaxy of our own Milky Way which straddles the constellations' borders. Floating in space some 170,000 light years away, it is bright enough to be visible to the naked eye even in the glare of a full moon. Within this vast star cloud lies the Tarantula nebula (NGC 2070), an immense diffuse nebula (the largest known nebula of this type) also visible to the naked eye. Apart from the Magellanic Cloud, neither Dorado nor Mensa offers much for the stargazer (Mensa's brightest star is fifth magnitude). Those interested in variable stars are advised to check out Beta Doradus, a Cepheid variable that ranges in magnitude from 3.7 to just under 4.0 in a cycle that takes 9.8 days.

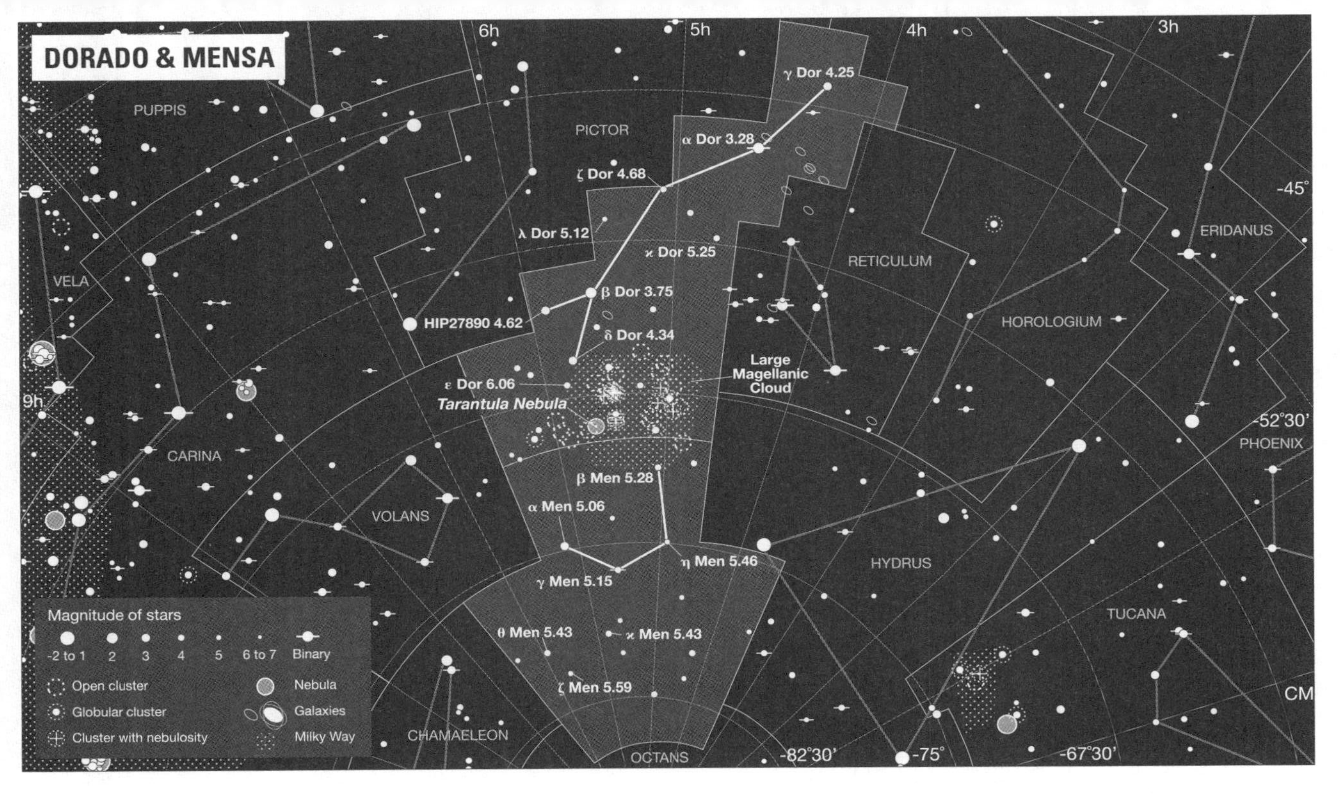
DORADO & MENSA
6h
5h
4h
3h
9h
PUPPIS
PICTOR
VELA
CARINA
VOLANS
CHAMAELEON
OCTANS
RETICULUM
HOROLOGIUM
ERIDANUS
PHOENIX
HYDRUS
TUCANA
CM
-45°
-52°30'
-67°30'
-75°
-82°30'
γ Dor 4.25
α Dor 3.28
ζ Dor 4.68
λ Dor 5.12
κ Dor 5.25
β Dor 3.75
HIP27890 4.62
δ Dor 4.34
ε Dor 6.06
Tarantula Nebula
Large Magellanic Cloud
β Men 5.28
α Men 5.06
η Men 5.46
γ Men 5.15
θ Men 5.43
κ Men 5.43
ζ Men 5.59
Magnitude of stars
-2 to 1
2
3
4
5
6 to 7
Binary
Open cluster
Globular cluster
Cluster with nebulosity
Nebula
Galaxies
Milky Way

Draco

Coordinates of Brightest Star Eltanin (Mag 2.21): RA 17h 56m 36s, Dec 51° 29' 15"

Size of Constellation Boundaries 1083 square miles

Bounded by Polaris, Ursa Minor, Camelopardalis, Ursa Major, Boötes, Hercules, Lyra, Cygnus, Cepheus

Constellation Legend Draco has always been seen as a dragon, and there are a number of dragon myths associated with it. The Sumerians and Babylonians saw it as Tiamat, the fierce dragon deity who battled the other gods for control of the universe. In Greek mythology, its corresponding dragon fought the goddess Athena during the battle between the gods and the Titans – the dragon lost and was hurled into the sky. Draco is also associated with the dragon slain by Cadmus, and more particularly with Ladon, the dragon guarding Hera's golden apple tree, slain by Hercules as one of his twelve labours. As it happens, the constellation Hercules is due south of where Draco's head lies, with Hercules' foot apparently lying on it.

Constellation Sights Nu Draconis is a very easy double star to view, with nearly identical stellar components which you can separate in binoculars or with a small telescope. Psi Draconis is a similarly easy double. Along the dragon's tail you'll find Thuban, which (though not terribly interesting in itself) was the pole star at the time the Great Pyramids were built in Egypt (c.2700 BC) and will be again, thanks to precession, in about 21,000 years. Those interested in nebulae will want to examine the Cat's Eye nebula, visible through a good telescope as a blue disc.

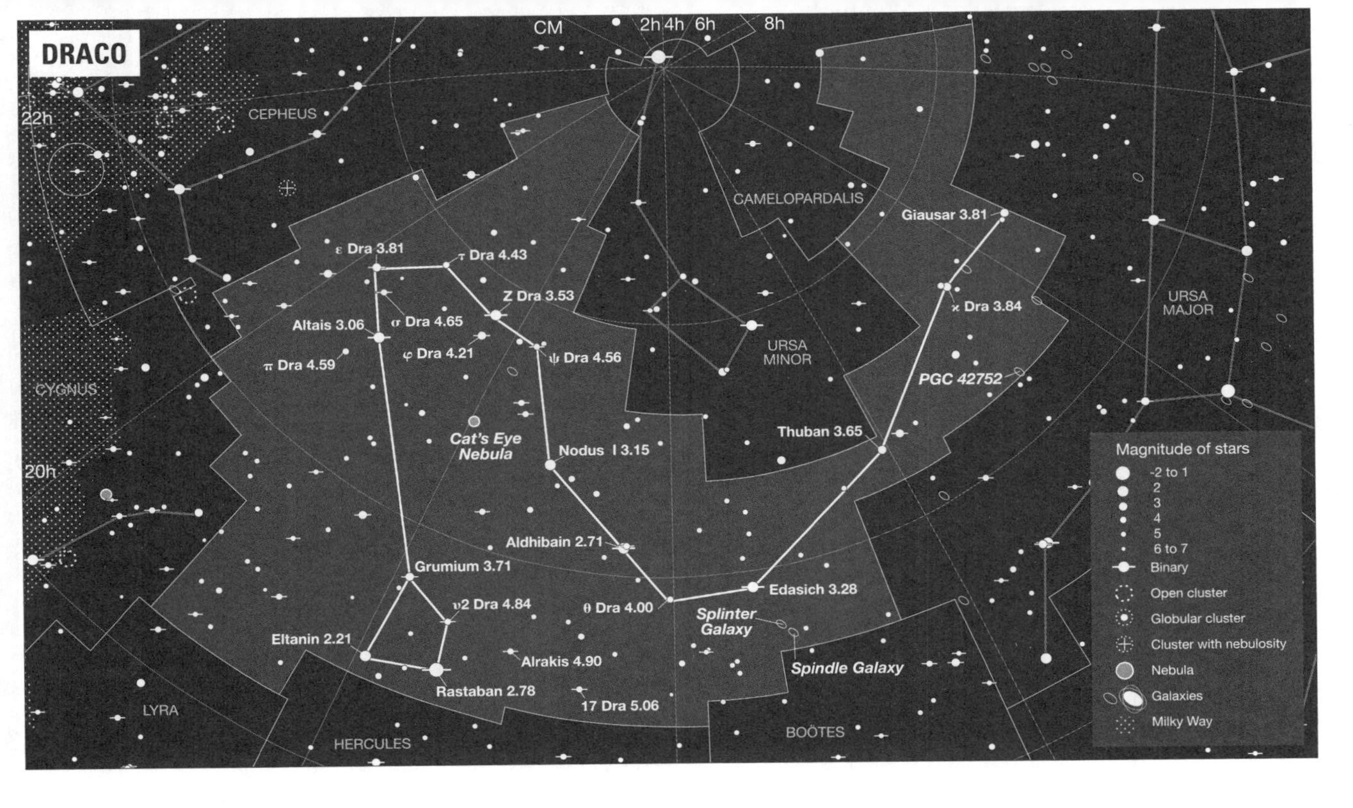
DRACO
CM
2h
4h
6h
8h
22h
20h
CEPHEUS
CAMELOPARDALIS
Giausar 3.81
ε Dra 3.81
τ Dra 4.43
Z Dra 3.53
κ Dra 3.84
URSA MAJOR
Altais 3.06
σ Dra 4.65
φ Dra 4.21
ψ Dra 4.56
URSA MINOR
PGC 42752
π Dra 4.59
CYGNUS
Cat's Eye Nebula
Nodus I 3.15
Thuban 3.65
Magnitude of stars
-2 to 1
2
3
4
5
6 to 7
Binary
Open cluster
Globular cluster
Cluster with nebulosity
Nebula
Galaxies
Milky Way
Aldhibain 2.71
Grumium 3.71
ν2 Dra 4.84
θ Dra 4.00
Edasich 3.28
Splinter Galaxy
Eltanin 2.21
Alrakis 4.90
Spindle Galaxy
Rastaban 2.78
17 Dra 5.06
LYRA
HERCULES
BOÖTES

Eridanus & Fornax

Coordinates of Brightest Stars Achernar (Mag 0.43): RA 1h 24m 43s, Dec -57° 13' 26"; Alpha Fornacis (Mag 3.78): RA 3h 12m 4s, Dec -28° 58' 38"

Size of Constellation Boundaries 1138 square degrees (Eridanus); 398 square degrees (Fornax)

Bounded by Taurus, Cetus, Sculptor, Phoenix, Hydrus, Horologium, Caelum, Lepus, Orion

Constellation Legend Eridanus was the name of a mythical river, and may well have corresponded to any number of important rivers of ancient civilizations, from the Tigris or Euphrates to the Nile or the Po. According to Greek mythology, Eridanus was the river into which Phaethon was hurled by Zeus after he lost control of the chariot of the sun god Apollo and started burning up the Earth. Fornax has rather more prosaic origins. Originally called Fornax Chemica ("chemical furnace"), it is one of the many constellations discovered by the French astronomer Nicolas Louis de Lacaille, and was named in honour of fellow countryman Antoine Lavoisier, one of the leading chemists of his day.

Constellation Sights Eridanus features Achernar, the ninth brightest star of all. Its name means "river's end" and is appropriate as it's the southernmost star in a constellation that stretches sixty degrees north to south – the greatest north–south range of any constellation in the night sky. Also notable are Beid, which resolves into a three-component multiple star (a larger-aperture telescope may be needed to see all three), and Acamar, another double star you can see separate in a small telescope. Fornax is home to the Fornax cluster of galaxies, which includes the unusual elliptical NGC 1316, also known as Fornax A, a strong source of radio waves.

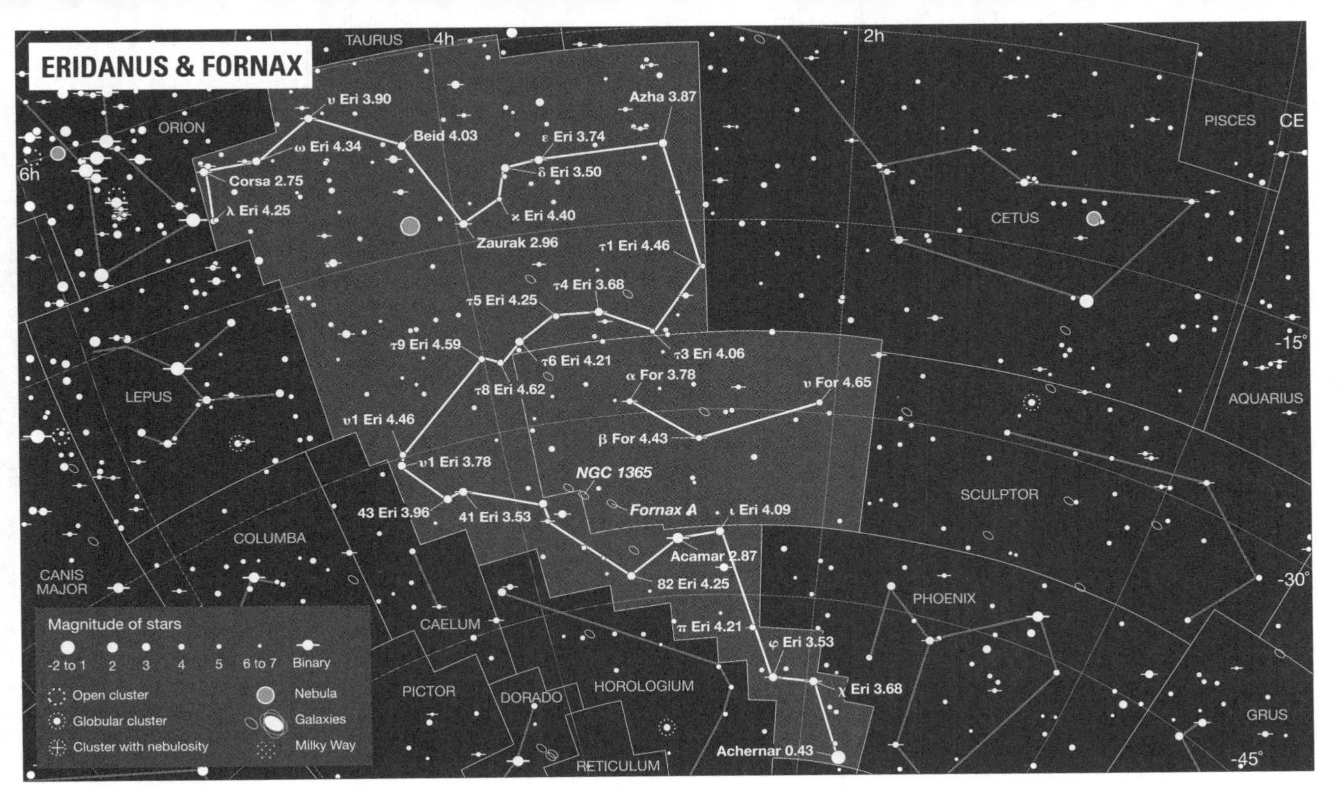
ERIDANUS & FORNAX
TAURUS
4h
2h
6h
ORION
PISCES
CE
CETUS
-15°
AQUARIUS
SCULPTOR
-30°
PHOENIX
GRUS
-45°
LEPUS
COLUMBA
CANIS MAJOR
CAELUM
PICTOR
DORADO
HOROLOGIUM
RETICULUM
υ Eri 3.90
Beid 4.03
ω Eri 4.34
Corsa 2.75
λ Eri 4.25
Azha 3.87
ε Eri 3.74
δ Eri 3.50
κ Eri 4.40
Zaurak 2.96
τ1 Eri 4.46
τ4 Eri 3.68
τ5 Eri 4.25
τ9 Eri 4.59
τ6 Eri 4.21
τ3 Eri 4.06
τ8 Eri 4.62
α For 3.78
ν For 4.65
υ1 Eri 4.46
β For 4.43
υ1 Eri 3.78
NGC 1365
43 Eri 3.96
41 Eri 3.53
Fornax A
ι Eri 4.09
Acamar 2.87
82 Eri 4.25
π Eri 4.21
φ Eri 3.53
χ Eri 3.68
Achernar 0.43
Magnitude of stars
-2 to 1
2
3
4
5
6 to 7
Binary
Open cluster
Globular cluster
Cluster with nebulosity
Nebula
Galaxies
Milky Way

Gemini

Coordinates of Brightest Star Pollux (Mag 1.15): RA 7h 45m 19s, Dec 28° 01' 15"

Size of Constellation Boundaries 514 square degrees

Bounded by Auriga, Taurus, Orion, Monoceros, Canis Minor, Cancer, Lynx

Constellation Legend There are several variants of the Castor and Pollux myth, but the gist of it is that they were the twin sons of Leda, who was seduced by Zeus disguised as a swan (see the Cygnus constellation). Leda also managed to sleep with her husband around the same time, so (in one version of the story) Castor is mortal while Pollux is immortal. The two boys were so close, however, that when Castor was slain, Pollux refused to be parted from him and Zeus's solution was to place them together in the sky – hence the name Gemini, or twins. Gemini's two brightest stars are named Castor and Pollux; fittingly for the immortal twin, Pollux is slightly brighter.

Constellation Sights Castor is a multiple star that, when viewed through a small telescope, resolves into a double star, with a smaller companion a short distance away from it. All three of these stars are binaries as well, making Castor a six-star system. M35 is an open cluster that's visible to the naked eye (fifth magnitude) and its stars are visible with binoculars. The Clown Face nebula (NGC 2392, also known as the Eskimo nebula) shows up as a blue disc in smaller telescopes, while larger scopes should allow you to make out the face of the clown itself. Disappointingly, no celestial red nose is visible.

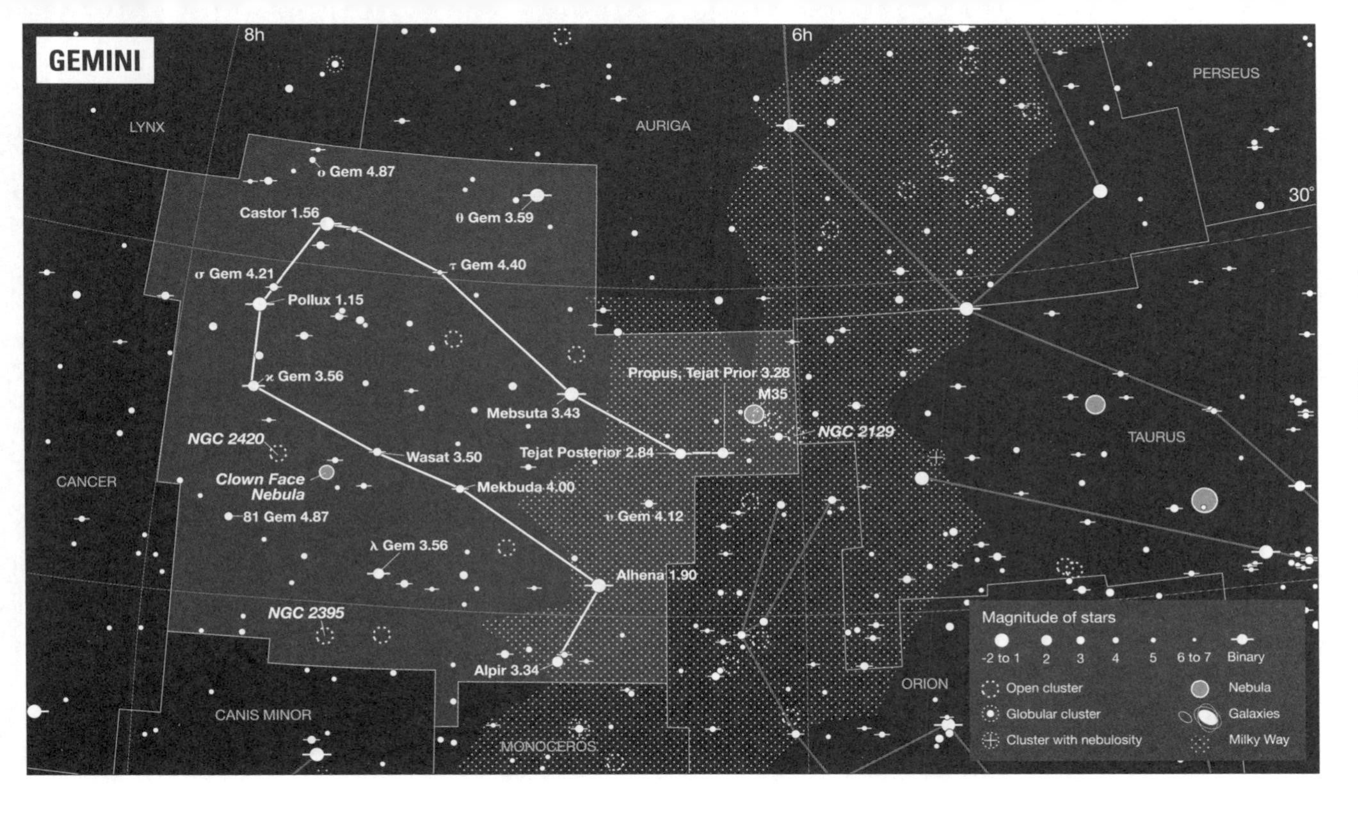
GEMINI
8h
6h
30°
LYNX
AURIGA
PERSEUS
TAURUS
CANCER
CANIS MINOR
MONOCEROS
ORION
ο Gem 4.87
Castor 1.56
θ Gem 3.59
τ Gem 4.40
σ Gem 4.21
Pollux 1.15
ϰ Gem 3.56
Mebsuta 3.43
Propus, Tejat Prior 3.28
M35
NGC 2129
NGC 2420
Wasat 3.50
Tejat Posterior 2.84
Clown Face Nebula
Mekbuda 4.00
81 Gem 4.87
υ Gem 4.12
λ Gem 3.56
Alhena 1.90
NGC 2395
Alpir 3.34
Magnitude of stars
-2 to 1
2
3
4
5
6 to 7
Binary
Open cluster
Globular cluster
Cluster with nebulosity
Nebula
Galaxies
Milky Way

Grus & Piscis Austrinus

Coordinates of Brightest Stars Al Nair (Mag 1.71): RA 22h 8m 14s, Dec -46° 56' 58"; Fomalhaut (Mag 1.15): RA 22h 57m 39s, Dec -29° 36' 34"

Size of Constellation Boundaries 366 square degrees (Grus); 245 square degrees (Piscis Austrinis)

Bounded by Aquarius, Capricornus, Microscopium, Indus, Tucana, Phoenix, Sculptor

Constellation Legend These two constellations used to be just one. Piscis Austrinus, which has been seen as a fish since the days of the Assyrians, is supposedly drinking or swimming in the water poured by the water-bearer Aquarius, which is directly north of it (the constellation's brightest star, Fomalhaut, has a name that means "fish's mouth"). Grus, in the shape of a crane, has no mythological history to speak of, although it was also known by another name, Phoenicopterus (Flamingo) in the seventeenth century. It was removed from the Piscis Austrinus constellation in 1603 upon publication of Johannes Bayer's *Uranometria*, which included twelve new southern constellations.

Constellation Sights Grus has a couple of nice binaries: both of the stars of Delta Gruis can be seen with the naked eye, as can those of Mu Gruis. Grus is also home to the Grus Quartet, four spiral galaxies that interact with each other, the brightest of which is tenth magnitude. Piscis Austrinus has Fomalhaut, one of the twenty brightest stars in the sky. Many astronomers believe that there is a planetary system hidden in the disc of dust circling it. Beta and Gamma Piscis Austrinis are both double stars resolvable in telescopes, although Gamma is more difficult to resolve in smaller telescopes, and the secondary star is quite faint (ninth magnitude).

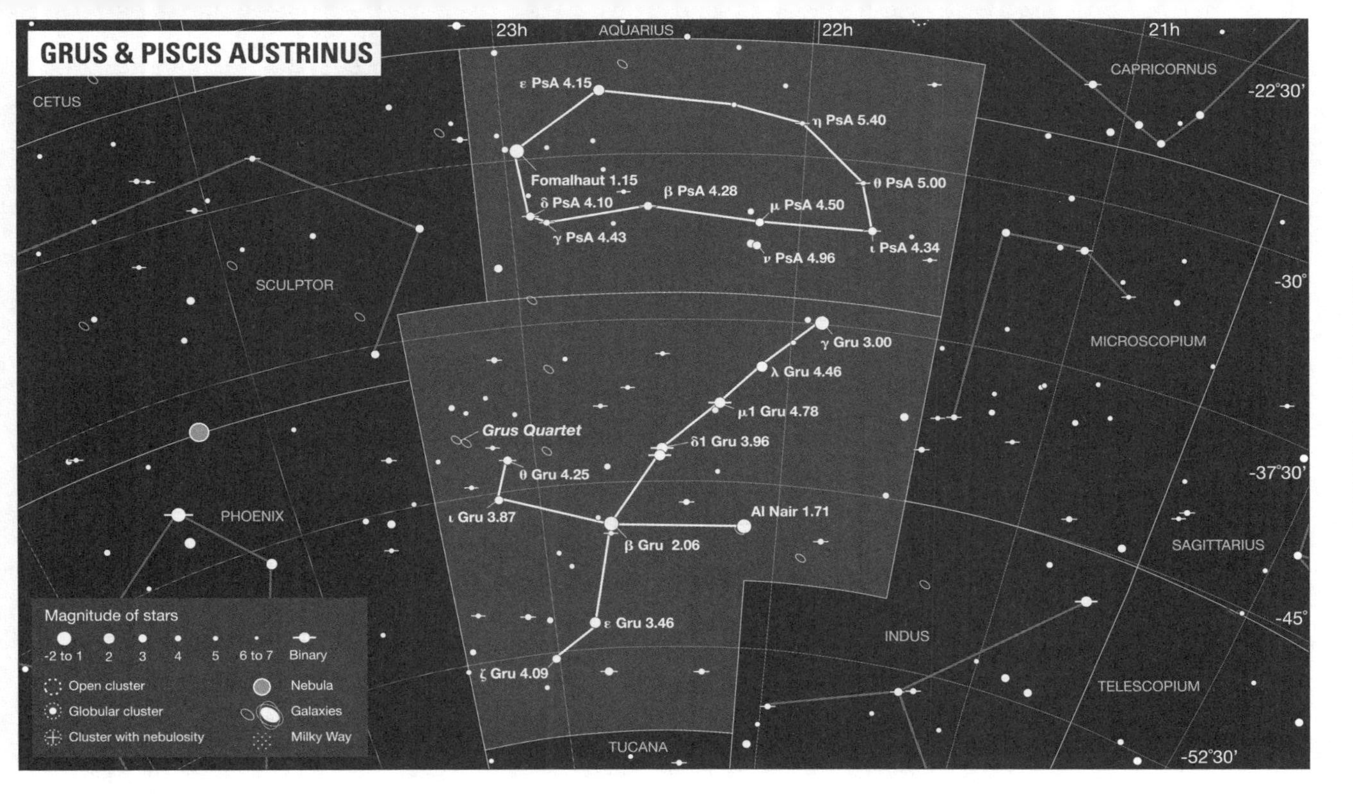
GRUS & PISCIS AUSTRINUS
23h
22h
21h
-22°30'
-30°
-37°30'
-45°
-52°30'
CETUS
AQUARIUS
CAPRICORNUS
SCULPTOR
MICROSCOPIUM
PHOENIX
SAGITTARIUS
INDUS
TELESCOPIUM
TUCANA
ε PsA 4.15
η PsA 5.40
Fomalhaut 1.15
β PsA 4.28
θ PsA 5.00
δ PsA 4.10
μ PsA 4.50
γ PsA 4.43
ι PsA 4.34
ν PsA 4.96
γ Gru 3.00
λ Gru 4.46
μ1 Gru 4.78
Grus Quartet
δ1 Gru 3.96
θ Gru 4.25
ι Gru 3.87
Al Nair 1.71
β Gru 2.06
ε Gru 3.46
ζ Gru 4.09
Magnitude of stars
-2 to 1
2
3
4
5
6 to 7
Binary
Open cluster
Nebula
Globular cluster
Galaxies
Cluster with nebulosity
Milky Way

Hercules

Coordinates of Brightest Star Kornephoros (Mag 2.75): RA 16h 30m 13s, Dec 21° 29' 00"

Size of Constellation Boundaries 1225 square degrees

Bounded by Draco, Boötes, Corona Borealis, Serpens Caput, Ophiuchus, Aquila, Sagitta, Vulpecula, Lyra

Constellation Legend Hercules is the most famous of ancient Greek heroes. He is best known for the legends of the twelve, supposedly impossible, labours that he accomplished. A number of other constellations relate to him and his adventures, including Leo, Draco, Cancer, Hydra, and the obsolete constellation of Argo Navis (now four separate constellations), named after the ship on which Hercules accompanied Jason on his quest for the Golden Fleece. The Babylonians knew the constellation as Gilgamesh, a mythical hero, who, like Hercules, was fathered by a god and was famed for his great strength and battles against monsters. Hercules hangs upside down in the sky, with his head pointing south and his feet on the head of Draco, directly north.

Constellation Sights The most dazzling sight in Hercules is M13, which is the brightest globular cluster north of the celestial equator, and contains at least one hundred thousand stars, though some astronomers set the number much higher. The naked eye sees it as a fuzzy star-like object and a small telescope will make out some individual stars. NGC 6341 (also known as M92) is another worthwhile globular cluster. Sharp eyes can spot it under good conditions but generally binoculars are in order, and a larger telescope is needed to make out the stars within it. Among Hercules' stars, Rasalgethi is a variable red giant whose brightness fluctuates with no set period (but over roughly a hundred days), and which shows a companion in a small telescope. 95 Herculis is a pretty double discernible with a small telescope.

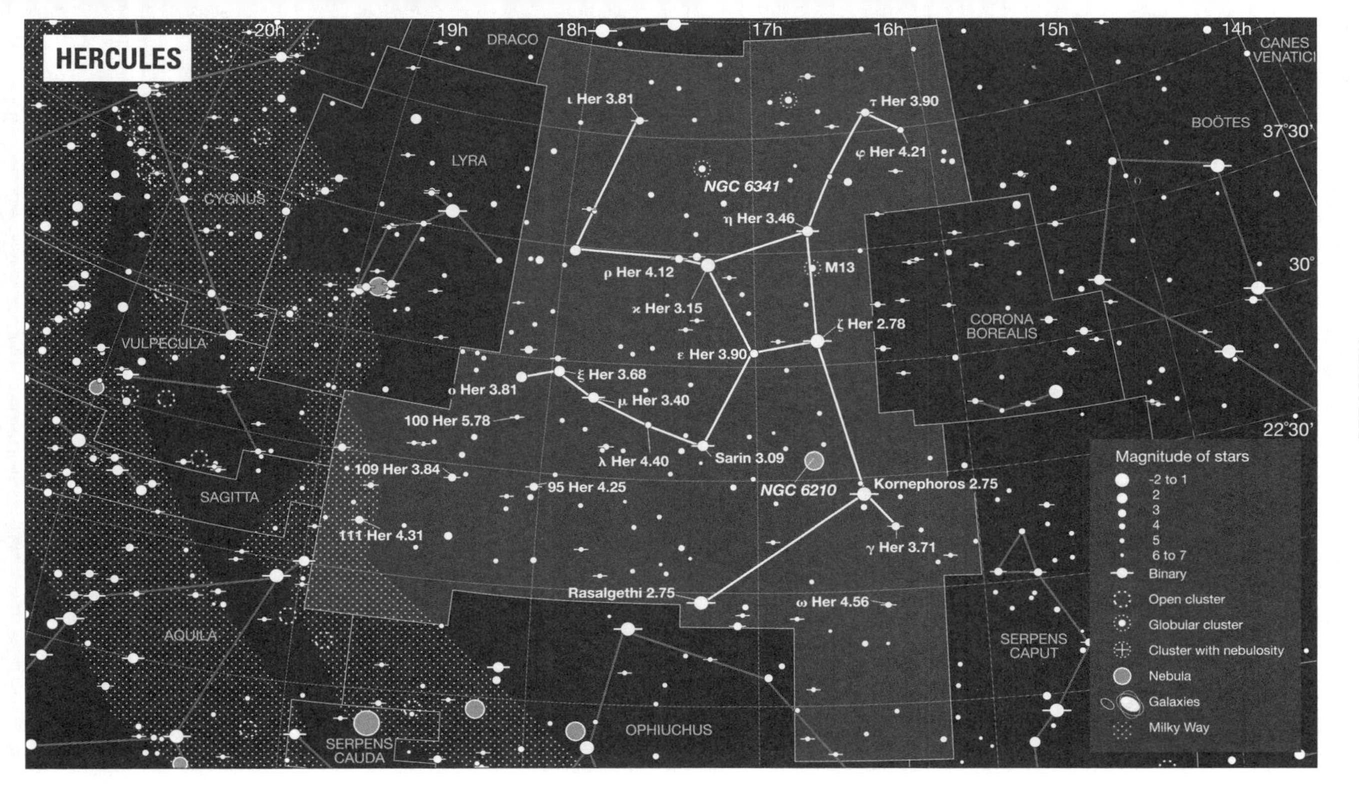
HERCULES
20h
19h
18h
17h
16h
15h
14h
37°30'
30°
22°30'
DRACO
CANES VENATICI
BOÖTES
LYRA
CYGNUS
VULPECULA
CORONA BOREALIS
SAGITTA
AQUILA
SERPENS CAUDA
OPHIUCHUS
SERPENS CAPUT
ι Her 3.81
τ Her 3.90
φ Her 4.21
NGC 6341
η Her 3.46
M13
ρ Her 4.12
κ Her 3.15
ζ Her 2.78
ε Her 3.90
ξ Her 3.68
ο Her 3.81
μ Her 3.40
100 Her 5.78
λ Her 4.40
Sarin 3.09
109 Her 3.84
95 Her 4.25
NGC 6210
Kornephoros 2.75
111 Her 4.31
γ Her 3.71
Rasalgethi 2.75
ω Her 4.56
Magnitude of stars
-2 to 1
2
3
4
5
6 to 7
Binary
Open cluster
Globular cluster
Cluster with nebulosity
Nebula
Galaxies
Milky Way

Horologium & Reticulum

Coordinates of Brightest Stars Alpha Horologii (Mag 3.84): RA 4h 14m 0s, Dec -42° 17' 15"; Alpha Reticuli (Mag 3.31): RA 4h 14m 25s, Dec -62° 28' 00"

Size of Constellation Boundaries 249 square degrees (Horologium); 114 square degrees (Reticulum)

Bounded by Eridanus, Hydrus, Dorado, Caelum

Constellation Legend Both of these constellations are credited to the astronomer Nicolas Louis de Lacaille. It sometimes seems as if de Lacaille was determined to commemorate every mechanical contraption ever invented in the names of his constellations. Horologium is named after the pendulum clock, while Reticulum honours the reticle, a tool used by astronomers to line up stars and measure their positions. It is also known as the Net, and sometimes as the Rhomboidal Net, a legacy of Isaak Habrecht of Strasbourg who first drew it, and who referred to it as the Rhombus.

Constellation Sights Neither constellation offers up much of interest. Horologium is home to globular cluster NGC 1261, which is faint but will resolve in a small telescope. Reticulum offers Zeta Reticuli, a double star with two sun-like components. You can separate the star if you have particularly sharp eyesight, otherwise a pair of binoculars is needed.

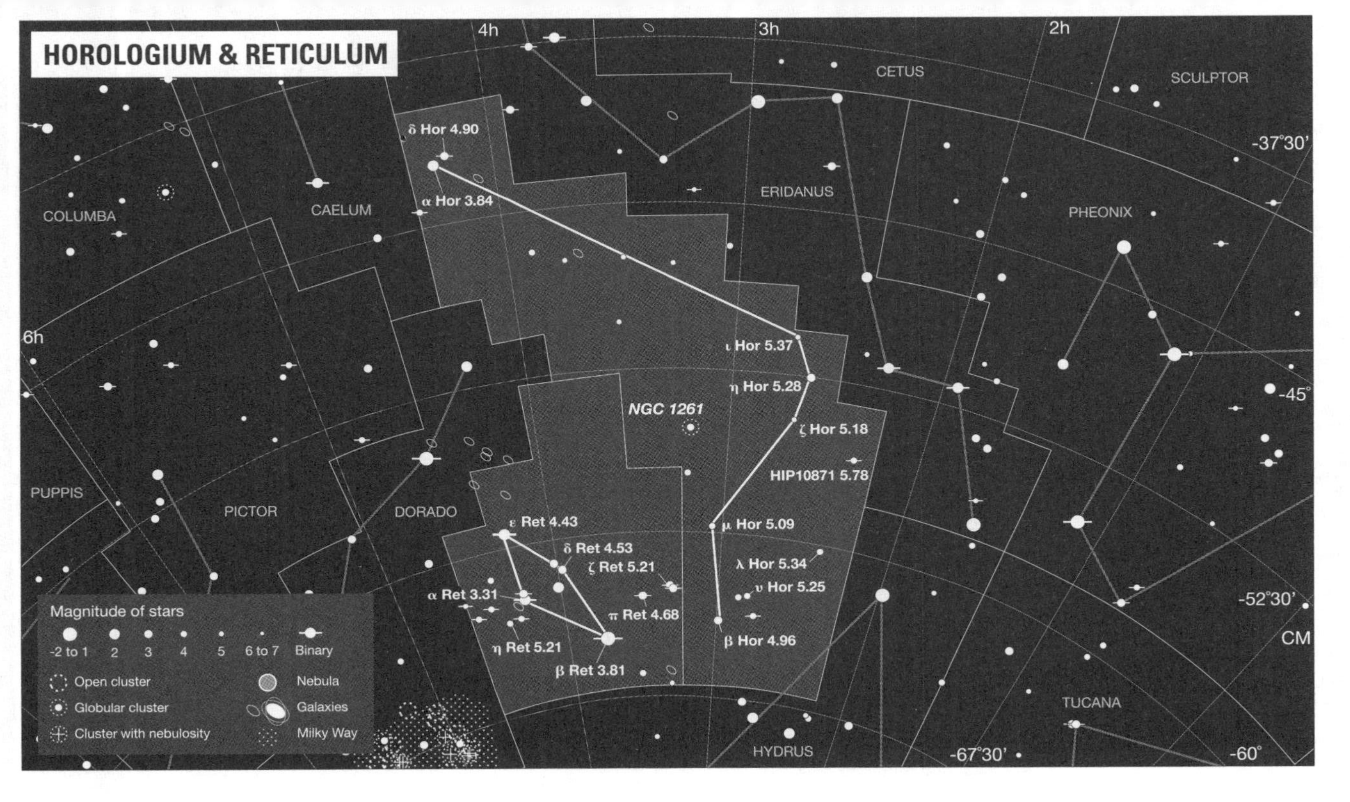
HOROLOGIUM & RETICULUM
4h
3h
2h
6h
CETUS
SCULPTOR
ERIDANUS
PHEONIX
COLUMBA
CAELUM
PUPPIS
PICTOR
DORADO
TUCANA
HYDRUS
CM
-37°30'
-45°
-52°30'
-60°
-67°30'
δ Hor 4.90
α Hor 3.84
ι Hor 5.37
η Hor 5.28
NGC 1261
ζ Hor 5.18
HIP10871 5.78
μ Hor 5.09
λ Hor 5.34
υ Hor 5.25
β Hor 4.96
ε Ret 4.43
δ Ret 4.53
ζ Ret 5.21
α Ret 3.31
π Ret 4.68
η Ret 5.21
β Ret 3.81
Magnitude of stars
-2 to 1
2
3
4
5
6 to 7
Binary
Open cluster
Globular cluster
Cluster with nebulosity
Nebula
Galaxies
Milky Way

Hydra

Coordinates of Brightest Star Alphard (Mag 1.96): RA 9h 27m 35s, Dec -8° 40' 07"

Size of Constellation Boundaries 1303 square degrees

Bounded by Cancer, Canis Minor, Monoceros, Puppis, Pyxis, Antlia, Centaurus, Libra, Virgo, Corvus, Crater, Sextans, Leo

Constellation Legend Hydra literally translates as "water snake", and is therefore connected to the story associated with the Corvus constellation. However, it is more usually taken to represent the many-headed monster of the same name that Hercules was sent to kill as one of his labours – a difficult task since whenever he lopped off a head another two grew back in its place. He solved the problem by applying a burning iron to the Hydra's wounds with each blow, cauterizing the wounds so as to stop new heads from growing. Hydra is the biggest of the constellations, stretching ninety degrees across the sky from head to tail, yet it is easy to miss because most of its stars are of magnitude 3.0 or dimmer.

Constellation Sights Alphard is the brightest star in the constellation, at magnitude 1.96, and is also the brightest in any direction for a considerable stretch of sky. Perhaps this is why it is called Alphard, Arabic for "the solitary one". Down near the tail, R Hydrae is an interesting Mira variable whose brightness drops dramatically from third magnitude to eleventh over a thirteen-month cycle. M48 is an open cluster of about eighty stars you can just about make out with the naked eye in good conditions, while the Southern Pinwheel galaxy presents a head-on view of a spiral galaxy. It's visible in small telescopes, but you'll need a larger scope to make out the arms. It was the first galaxy discovered outside our own Local Group. The Eye nebula (also known as the Ghost of Jupiter) will appear as a blue disc in a small telescope, with the eye visible in a larger one.

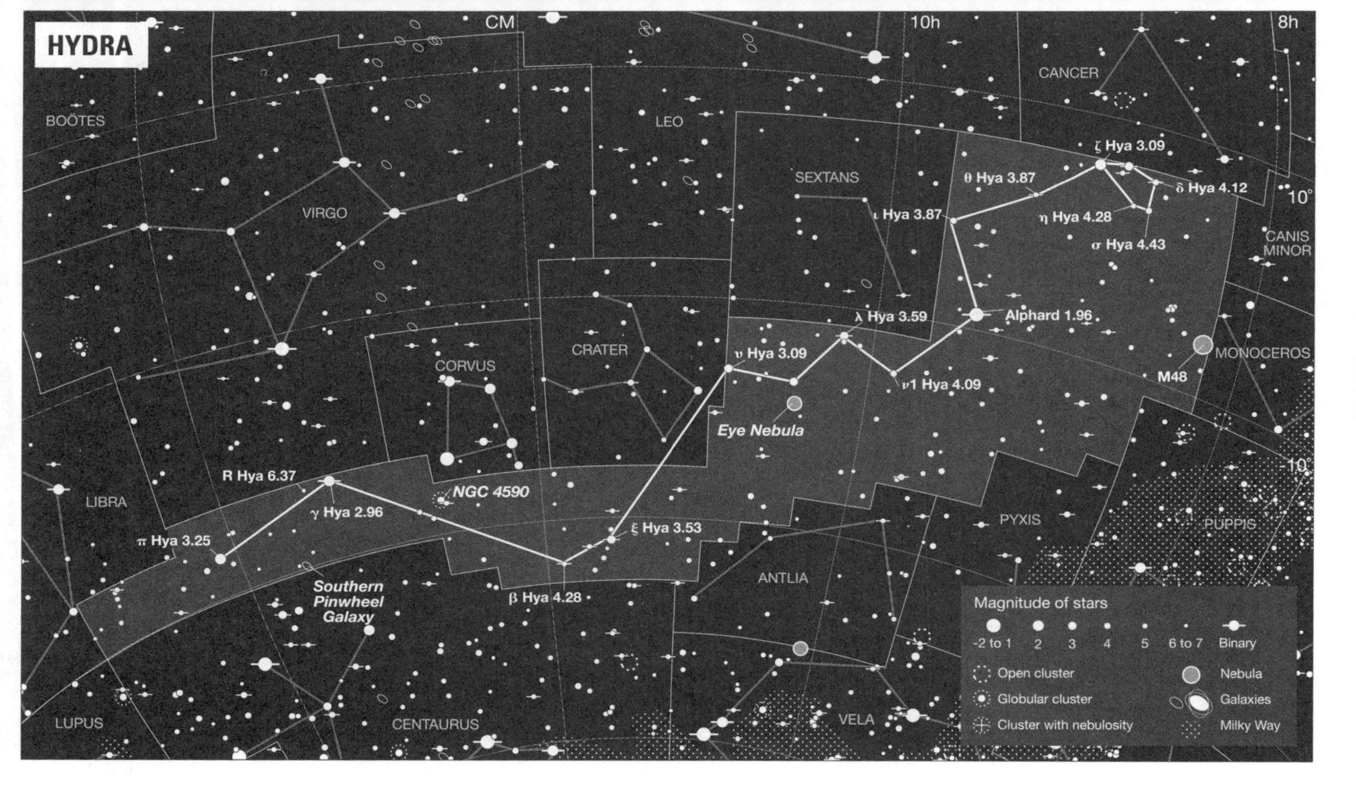

HYDRA
CM
10h
8h
10°
-10°
BOÖTES
LEO
CANCER
SEXTANS
VIRGO
CANIS MINOR
CRATER
CORVUS
MONOCEROS
LIBRA
PYXIS
PUPPIS
ANTLIA
LUPUS
CENTAURUS
VELA
ζ Hya 3.09
θ Hya 3.87
δ Hya 4.12
ι Hya 3.87
η Hya 4.28
σ Hya 4.43
λ Hya 3.59
Alphard 1.96
υ Hya 3.09
ν1 Hya 4.09
M48
Eye Nebula
R Hya 6.37
NGC 4590
γ Hya 2.96
ξ Hya 3.53
π Hya 3.25
Southern Pinwheel Galaxy
β Hya 4.28
Magnitude of stars
-2 to 1
2
3
4
5
6 to 7
Binary
Open cluster
Globular cluster
Cluster with nebulosity
Nebula
Galaxies
Milky Way

Hydrus & Tucana

Coordinates of Brightest Stars Beta Hydri (Mag 2.81): RA 0h 25m 45s, Dec -77° 14' 25"; Alpha Tucanae (Mag 2.84): RA 22h 18m 30s, Dec -60° 14' 51"

Size of Constellation Boundaries 243 square degrees (Hydrus); 295 square degrees (Tucana)

Bounded by Phoenix, Grus, Indus, Octans, Mensa, Dorado, Reticulum, Horologium, Eridanus

Constellation Legend Hydrus and Tucana were added to the creatures of the celestial bestiary by the Dutch navigators Pieter Dirkszoon Keyser and Frederick de Houtman, who charted the southern skies on a voyage to the East Indies in 1595. Hydrus represents a small water snake, a southern counterpart to the Hydra constellation (although unlike that of Hydra, this snake is drawn simply from nature rather than myth) whilst Tucana represents a toucan, or rather the shape of the bird's bill.

Constellation Sights The most prominent object in either constellation is the Small Magellanic Cloud, the smaller of the two easily visible satellite galaxies of the Milky Way. Without any magnification it registers as a large, bright patch several times wider than the Moon, but with binoculars or a telescope you'll see much more, including nebulae and clusters. Close by is another dazzling sight: 47 Tucanae (also known as NGC 104), the second largest and brightest globular cluster in the sky, also visible to the naked eye. NGC 362, another globular cluster, is visible in binoculars just north of the Small Magellanic Cloud. On the star front, Beta Tucanae is a multiple star that appears as a double in binoculars and a multiple in telescopes. Beta Hydri in Hydrus, a Sun-like star some 24 light years distant, is the brightest star near the southern pole.

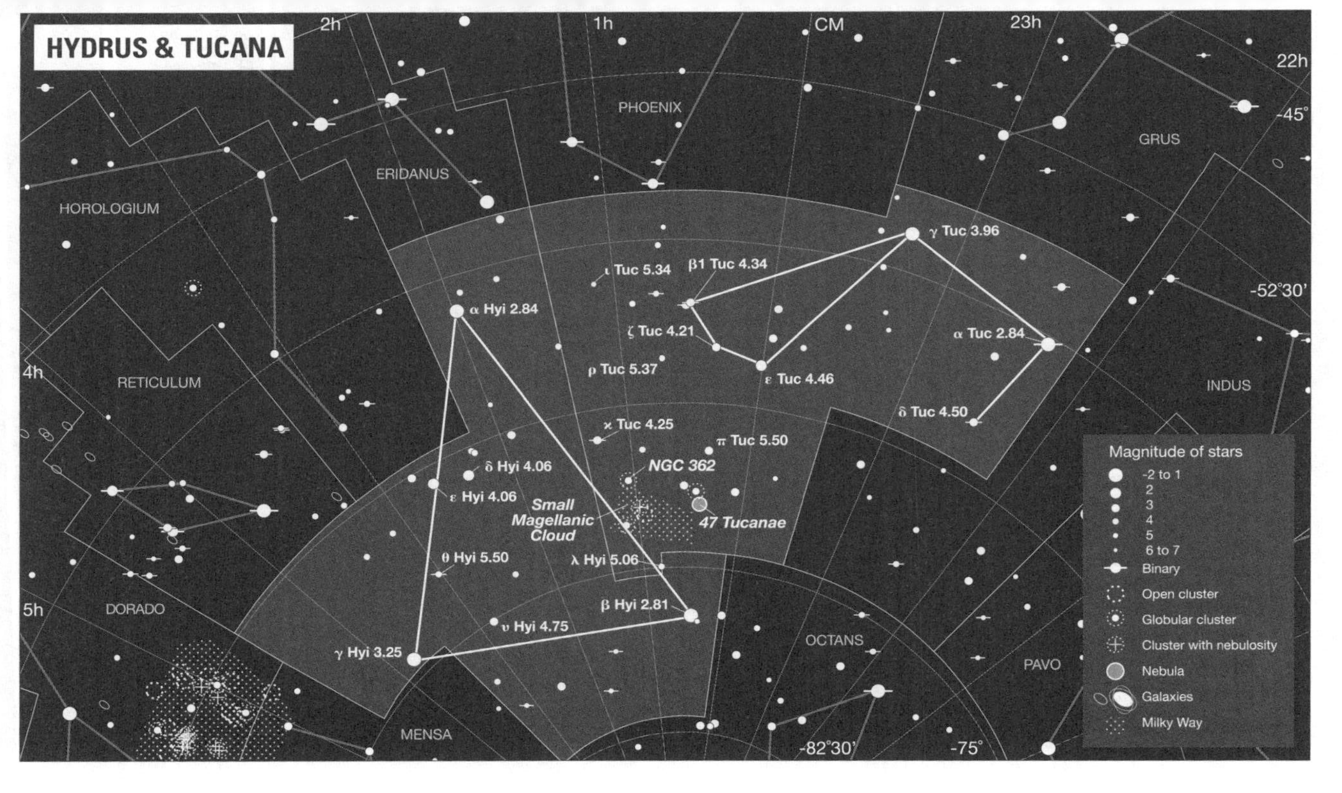

HYDRUS & TUCANA
2h
1h
CM
23h
22h
-45°
-52°30'
4h
5h
-82°30'
-75°
PHOENIX
GRUS
ERIDANUS
HOROLOGIUM
RETICULUM
INDUS
DORADO
MENSA
OCTANS
PAVO
γ Tuc 3.96
ι Tuc 5.34
β1 Tuc 4.34
α Hyi 2.84
ζ Tuc 4.21
α Tuc 2.84
ρ Tuc 5.37
ε Tuc 4.46
δ Tuc 4.50
κ Tuc 4.25
π Tuc 5.50
δ Hyi 4.06
NGC 362
ε Hyi 4.06
Small Magellanic Cloud
47 Tucanae
θ Hyi 5.50
λ Hyi 5.06
β Hyi 2.81
υ Hyi 4.75
γ Hyi 3.25
Magnitude of stars
-2 to 1
2
3
4
5
6 to 7
Binary
Open cluster
Globular cluster
Cluster with nebulosity
Nebula
Galaxies
Milky Way

Indus & Microscopium

Coordinates of Brightest Stars Alpha Indi (Mag 3.09): RA 20h 37m 34s, Dec -47° 17' 00"; Gamma Microscopii (Mag 4.65): RA 21h 01m 17s, Dec -32° 14' 55"

Size of Constellation Boundaries 294 square degrees (Indus); 210 square degrees (Microscopium)

Bounded by Capricornus, Sagittarius, Telescopium, Pavo, Octans, Tucana, Grus, Picis Austrinus

Constellation Legend Indus is "The Indian", although its title refers to Native Americans rather than those who live on the Indian subcontinent. Its position in the sky was charted by Pieter Dirkszoon Keyser and Frederick de Houtman in the late sixteenth century, a time of great exploration for Europeans. It's not known if they had a particular tribe or person in mind, but there is some conjecture that they may have been thinking of the native peoples of Tierra del Fuego and Patagonia. Microscopium, the microscope, is yet another piece of scientific equipment to be immortalized in the sky by de Lacaille.

Constellation Sights Neither Indus nor Microscopium holds much of interest for stargazers. Theta Indi is a double star whose components separate in a small telescope; Alpha Microscopium is also a double star, although the dimmer star is only of tenth magnitude. Of minor interest is Epsilon Indi, one of the closest stars to our own Sun – less than twelve light years away. It is less than fifteen percent as luminous as the Sun, making it the least luminous star that can be seen without the use of binoculars or telescope. It was thought to be a single star, but scientists recently located its companion, Epsilon Indi B, the nearest brown dwarf yet discovered.

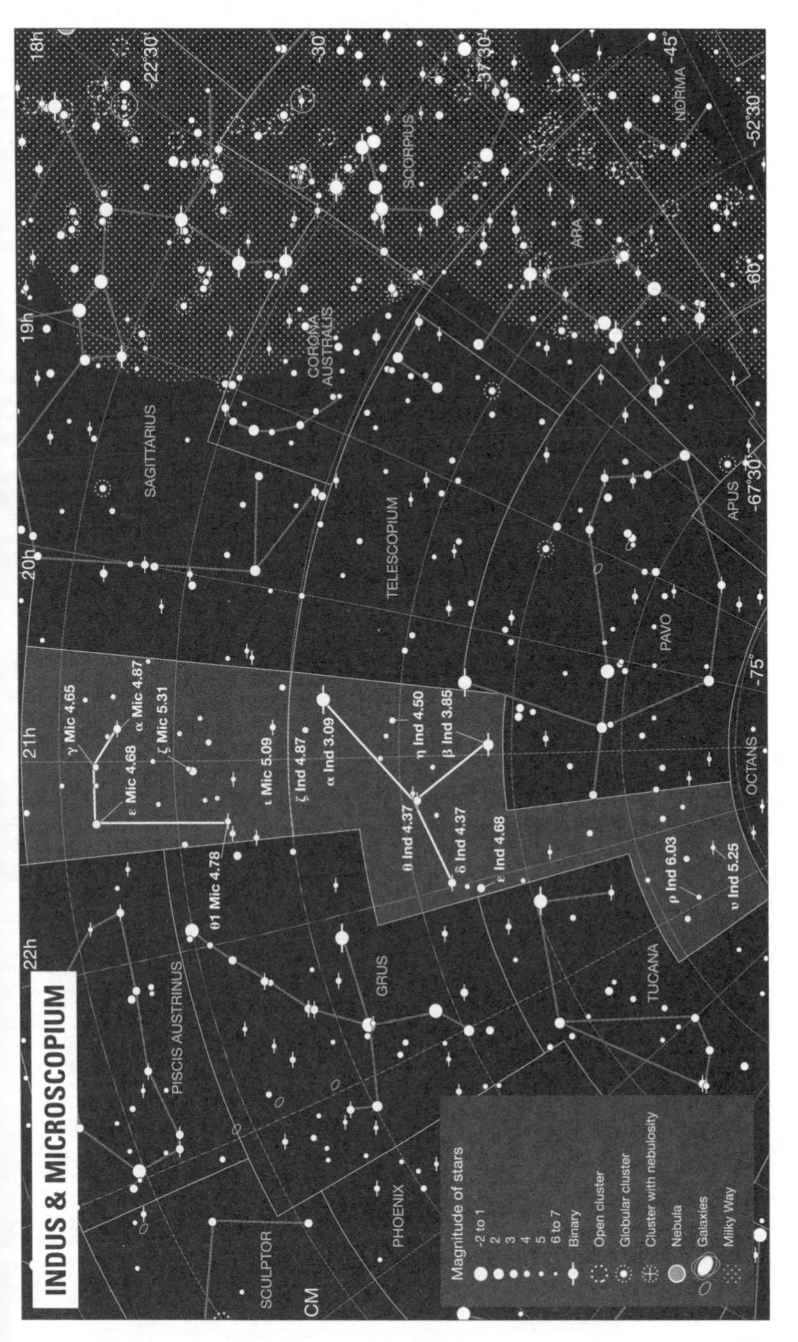

INDUS & MICROSCOPIUM
18h
19h
20h
21h
22h
-22°30'
-30°
-37°30'
-45°
-52°30'
-60°
-67°30'
-75°
SAGITTARIUS
CORONA AUSTRALIS
SCORPIUS
NORMA
ARA
TELESCOPIUM
PAVO
APUS
OCTANS
TUCANA
GRUS
PISCIS AUSTRINUS
PHOENIX
SCULPTOR
CM
γ Mic 4.65
α Mic 4.87
ζ Mic 5.31
ε Mic 4.68
ι Mic 5.09
θ1 Mic 4.78
ζ Ind 4.87
α Ind 3.09
η Ind 4.50
β Ind 3.85
θ Ind 4.37
δ Ind 4.37
ε Ind 4.68
ρ Ind 6.03
ν Ind 5.25
Magnitude of stars
-2 to 1
2
3
4
5
6 to 7
Binary
Open cluster
Globular cluster
Cluster with nebulosity
Nebula
Galaxies
Milky Way

Lacerta

Coordinates of Brightest Star Alpha Lacertae (Mag 3.75): RA 22h 31m 14s, Dec 50° 17' 41"

Size of Constellation Boundaries 201 square degrees

Bounded by Cepheus, Cygnus, Pegasus, Andromeda, Cassiopeia

Constellation Legend Lacerta, the Lizard, is one of the more obscure northern constellations, wedged between Andromeda and Cygnus. It is credited to the seventeenth-century Polish astronomer Johannes Hevelius and appears in John Flamsteed's *Catalog of Stars* of 1726. However, a similar constellation had already been drawn up by Augustin Royer, who named it "The Sceptre and Hand of Justice" in honour of King Louis XIV of France, and a later one was created by Johann Ellert Bode as a tribute to Frederick the Great, king of Prussia. Neither held their position, so Lacerta remained. It has since become well known to astronomers for its nova activity; several novae have been spotted in the constellation over the last century.

Constellation Sights Open cluster NGC 7243 is visible in binoculars and features about forty stars. Beyond that, the most interesting object in Lacerta is one amateurs are extremely unlikely to see: BL Lacertae, which was originally thought to be a variable star but which is now known to be a type of active galactic nucleus known as a blazar. BL Lacertae is fourteenth magnitude at best, putting it out of the reach of all but the largest home telescopes.

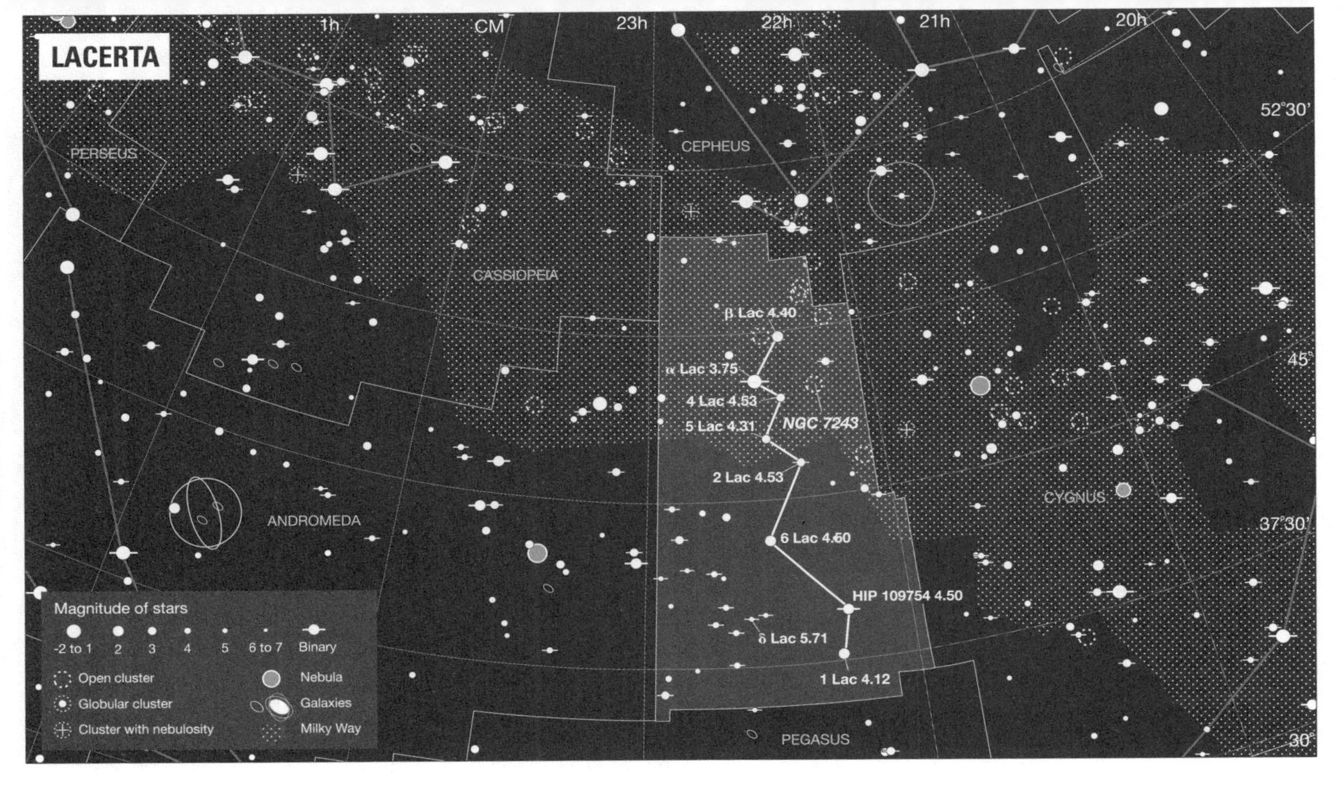
LACERTA
1h
CM
23h
22h
21h
20h
52°30'
45°
37°30'
30°
PERSEUS
CEPHEUS
CASSIOPEIA
ANDROMEDA
CYGNUS
PEGASUS
β Lac 4.40
α Lac 3.75
4 Lac 4.53
5 Lac 4.31
NGC 7243
2 Lac 4.53
6 Lac 4.60
HIP 109754 4.50
δ Lac 5.71
1 Lac 4.12
Magnitude of stars
-2 to 1
2
3
4
5
6 to 7
Binary
Open cluster
Globular cluster
Cluster with nebulosity
Nebula
Galaxies
Milky Way

Leo Minor & Lynx

Coordinates of Brightest Stars 46 Leonis Minoris (Mag 3.78): RA 10h 53m 19s, Dec 34° 12' 06"; Alpha Lyncis (Mag 3.12): RA 9h 21m 03s, Dec 34° 22' 57"

Size of Constellation Boundaries 232 square degrees (Leo Minor); 545 square degrees (Lynx)

Bounded by Camelopardalis, Auriga, Gemini, Cancer, Leo, Ursa Major

Constellation Legend Both Leo Minor (the Lesser Lion) and Lynx were charted by Johannes Hevelius in 1687. The lynx is a wild cat known to have excellent night vision, which is why Hevelius gave the constellation its name: nearly all its stars are faint, and Hevelius suggested that stargazers would need the eyes of a lynx to find them.

Constellation Sights Leo Minor has almost nothing in terms of quality stargazing. It has one notable double star, Beta Leonis Minoris, but its components are not separable in most amateur telescopes. For sky watchers with larger scopes Lynx offers the Intergalactic Wanderer, also known as NGC 2419. This globular cluster is almost 300,000 light years away, making it one of the most distant globular clusters associated with the Milky Way galaxy. Lynx also offers double stars, including 38 Lyncis, near Alpha Lyncis, whose individual stars are resolvable in a moderate-sized scope.

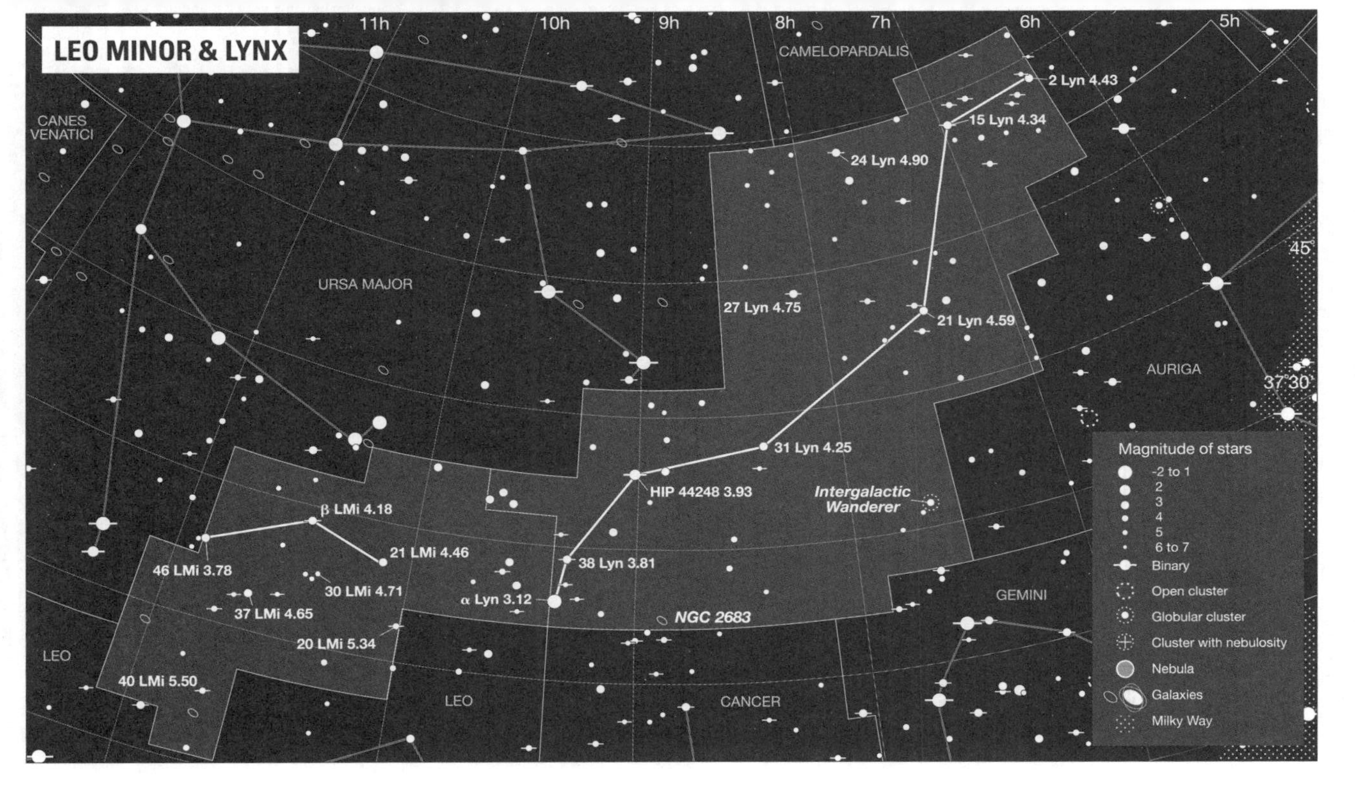
LEO MINOR & LYNX
11h
10h
9h
8h
7h
6h
5h
45°
37°30'
CAMELOPARDALIS
CANES VENATICI
URSA MAJOR
AURIGA
GEMINI
LEO
CANCER
2 Lyn 4.43
15 Lyn 4.34
24 Lyn 4.90
27 Lyn 4.75
21 Lyn 4.59
31 Lyn 4.25
HIP 44248 3.93
Intergalactic Wanderer
38 Lyn 3.81
α Lyn 3.12
NGC 2683
β LMi 4.18
21 LMi 4.46
46 LMi 3.78
30 LMi 4.71
37 LMi 4.65
20 LMi 5.34
40 LMi 5.50
Magnitude of stars
-2 to 1
2
3
4
5
6 to 7
Binary
Open cluster
Globular cluster
Cluster with nebulosity
Nebula
Galaxies
Milky Way

Leo & Sextans

Coordinates of Brightest Stars Regulus (Mag 1.34): RA 10h 08m 22s, Dec 11° 57' 20"; Alpha Sextantis (Mag 4.46): RA 10h 07m 00s, Dec -0°, 22' 59"

Size of Constellation Boundaries 947 square degrees (Leo); 314 square degrees (Sextans)

Bounded by Leo Minor, Cancer, Hydra, Crater, Virgo, Coma Berenices, Ursa Major

Constellation Legend The constellation Leo is a sign of the Zodiac, and was seen as a lion by several ancient civilizations including the Persians, the Babylonians and the Egyptians. The Greeks associated the constellation with the Nemean Lion, which Hercules had to slay as one of his twelve labours. This was no mean feat, as the lion had an impenetrable hide. Hercules eventually accomplished the task by throttling the lion using his bare hands. Sextans derives its name from the sextant, an instrument used for navigation by measuring the positions of stars.

Constellation Sights Leo's first sight is its asterism, the Sickle. This consists of six stars at the head of the constellation, and it is easily viewed with the naked eye. Regulus, the brightest star in Leo, shows itself as a double star in binoculars. Algieba, another double star which can be resolved by the eye alone, is an even better sight in a telescope since both of its stars are a blazing yellow-and-orange colour. Galaxy hunters will enjoy the Leo Triplet, a trio of ninth-magnitude galaxies, the most prominent of which is NGC 3627 (also known as M66). Sextans also contains the Spindle galaxy (NGC 311), also of the ninth magnitude.

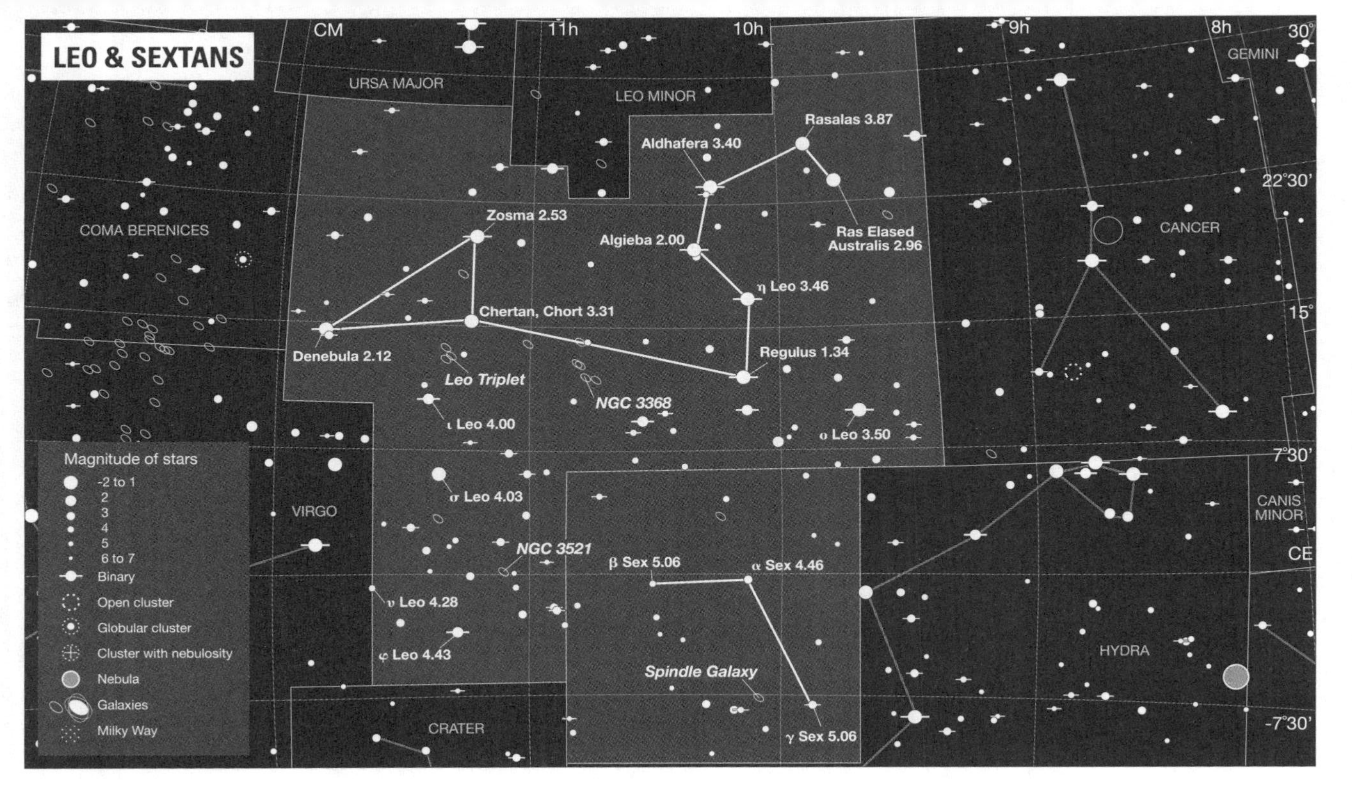
LEO & SEXTANS
CM
11h
10h
9h
8h
30°
22°30'
15°
7°30'
-7°30'
GEMINI
URSA MAJOR
LEO MINOR
COMA BERENICES
CANCER
VIRGO
CRATER
HYDRA
CANIS MINOR
CE
Rasalas 3.87
Aldhafera 3.40
Ras Elased Australis 2.96
Zosma 2.53
Algieba 2.00
η Leo 3.46
Chertan, Chort 3.31
Denebula 2.12
Regulus 1.34
Leo Triplet
NGC 3368
ι Leo 4.00
ο Leo 3.50
σ Leo 4.03
NGC 3521
β Sex 5.06
α Sex 4.46
υ Leo 4.28
φ Leo 4.43
Spindle Galaxy
γ Sex 5.06
Magnitude of stars
-2 to 1
2
3
4
5
6 to 7
Binary
Open cluster
Globular cluster
Cluster with nebulosity
Nebula
Galaxies
Milky Way

Lepus

Coordinates of Brightest Star Arneb (Mag 2.56): RA 5h 32m 44s, Dec -17° 49' 10"

Size of Constellation Boundaries 290 square degrees

Bounded by Orion, Eridanus, Caelum, Columba, Canis Major, Monoceros

Constellation Legend Lepus is Latin for "hare", the favourite prey of Orion the hunter. The constellation probably gained its name due to its position, lying at Orion's feet just ahead of Canis Major, who is often said to be chasing it. It was not consistently seen as a hare, however; the Arabs saw in its shape and position "the throne of the great one", or Orion's throne, while the ancient Egyptians saw it as the boat carrying Osiris through the sky.

Constellation Sights Gamma Leporis is a double star, both parts of which can be seen through binoculars. Lepus is also the home of NGC 1904 (also known as M79), a globular cluster viewable through a small scope, about forty thousand light years away.

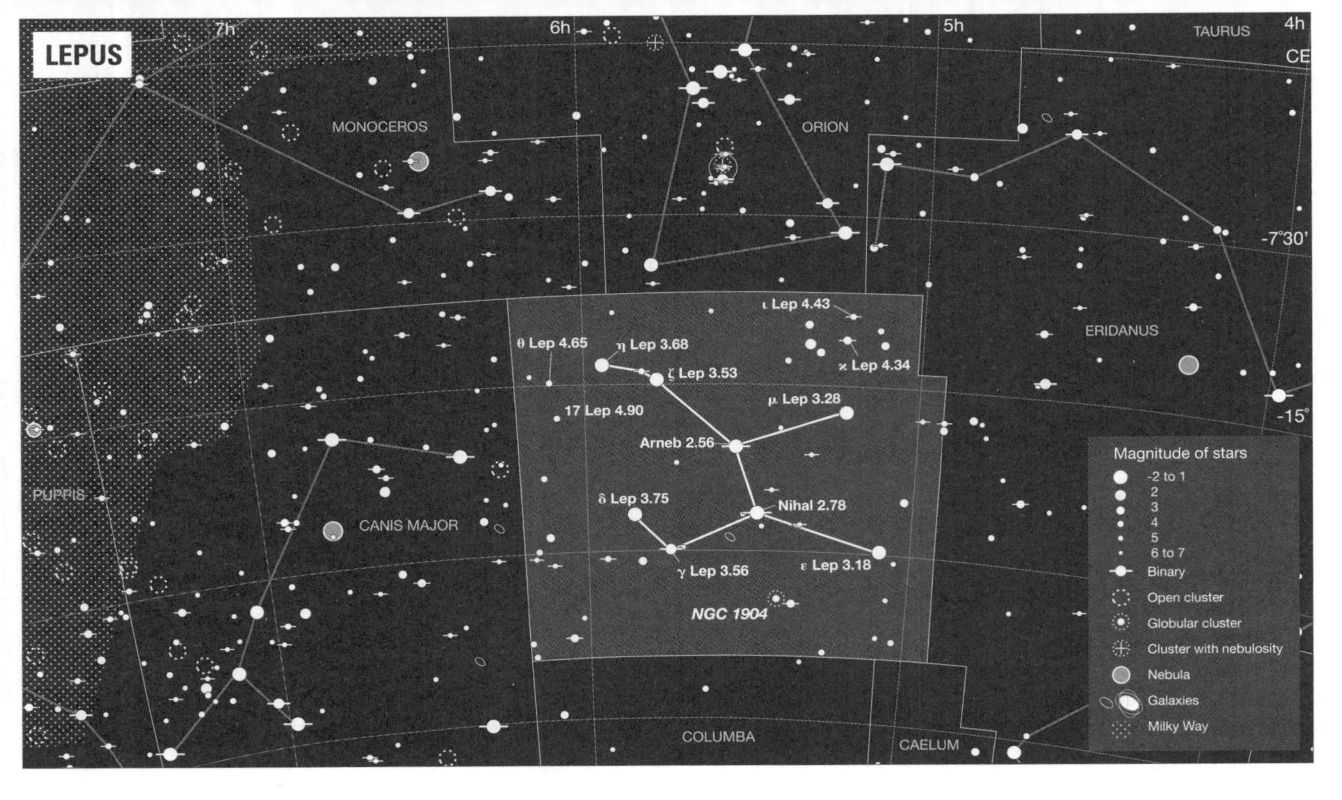
LEPUS
7h
6h
5h
4h
TAURUS
-7°30'
-15°
MONOCEROS
ORION
ERIDANUS
PUPPIS
CANIS MAJOR
COLUMBA
CAELUM
ι Lep 4.43
θ Lep 4.65
η Lep 3.68
ζ Lep 3.53
κ Lep 4.34
μ Lep 3.28
17 Lep 4.90
Arneb 2.56
δ Lep 3.75
Nihal 2.78
γ Lep 3.56
ε Lep 3.18
NGC 1904
Magnitude of stars
-2 to 1
2
3
4
5
6 to 7
Binary
Open cluster
Globular cluster
Cluster with nebulosity
Nebula
Galaxies
Milky Way

Libra

Coordinates of Brightest Star Zubeneschamali (Mag 2.59): RA 15h 17m 00s, Dec -9° 23' 33"

Size of Constellation Boundaries 538 square degrees

Bounded by Serpens Caput, Virgo, Hydra, Lupus, Scorpius, Ophiuchus

Constellation Legend Libra was seen as resembling a pair of scales by both the Egyptians and the Romans, who added it to the constellations which make up the signs of the Zodiac. The Roman astronomers saw in Libra the image of the scales of justice wielded by Caesar, a symbol of their empire's law and order. The Egyptians, on the other hand, pictured the mythical scales used to weigh the heart after death; if the deceased's heart was heavier than the feather of truth and justice, it meant that it was heavy with evil deeds and the deceased was condemned to eternal oblivion. Before the Romans gave it the name it has today, the constellation was part of Scorpius (where the Greeks had charted it) and to this day its brightest stars have names that translate as "northern claw" and "southern claw".

Constellation Sights First check out Zubeneschamali (Arabic for "northern claw"), which is the brightest star in the constellation. Many observers claim that the star shines with a green tint that's visible to the naked eye, but other observers see it as merely white. After you've decided on whether you can see Zubeneschamali's colouring, cruise down to Zubenelgenubi (the "southern claw"). It's a double star that you can separate out with binoculars or even the naked eye in favourable conditions. Iota Librae is another interesting multiple star. Binoculars will see it as a double star, while a small telescope will see the brighter star as yet another double star. And a larger telescope still will see the fainter of these stars as yet another double star. Zubenhakrabi (the "scorpion's claw") is a variable that fluctuates between magnitudes five and six every second and third day.

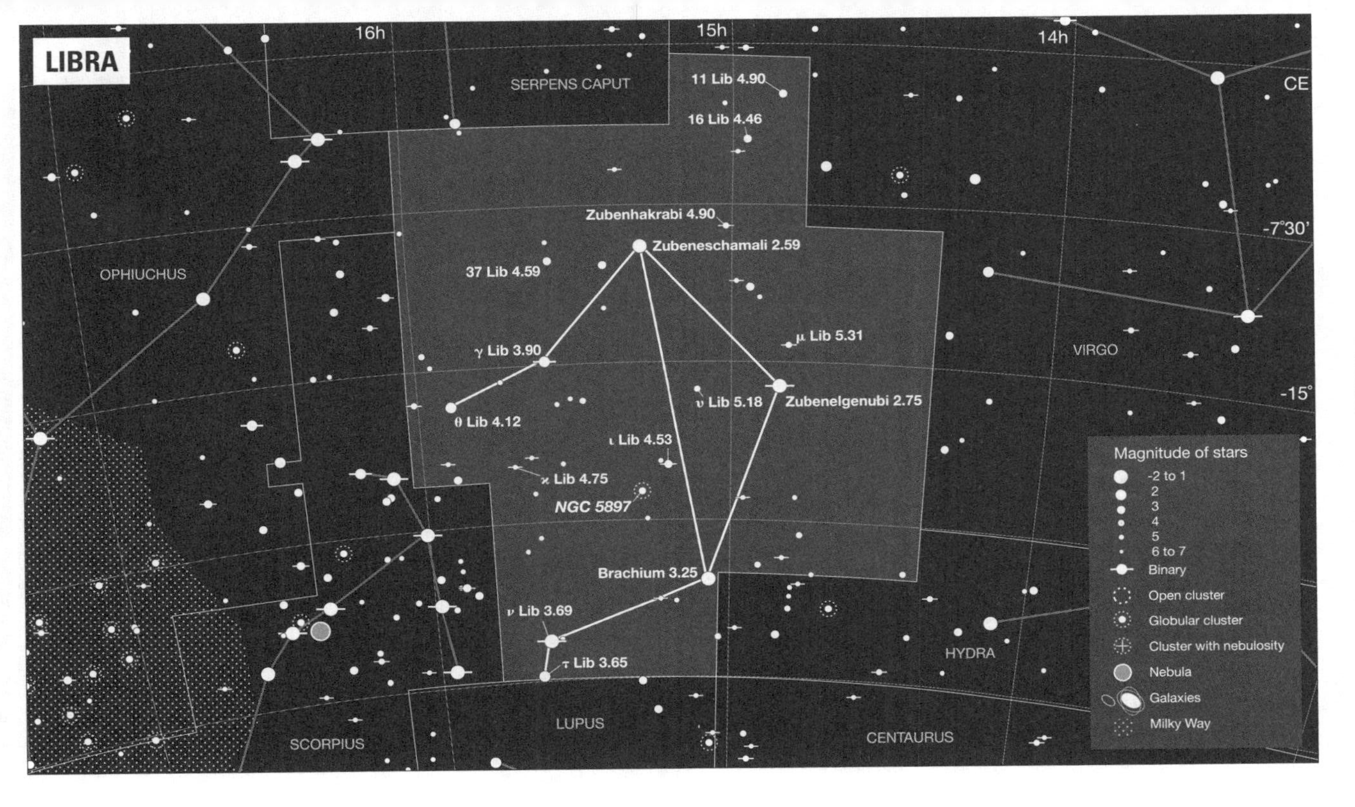

LIBRA
16h
15h
14h
SERPENS CAPUT
CE
11 Lib 4.90
16 Lib 4.46
Zubenhakrabi 4.90
Zubeneschamali 2.59
-7°30'
OPHIUCHUS
37 Lib 4.59
μ Lib 5.31
VIRGO
γ Lib 3.90
υ Lib 5.18
Zubenelgenubi 2.75
-15°
θ Lib 4.12
ι Lib 4.53
κ Lib 4.75
NGC 5897
Brachium 3.25
ν Lib 3.69
τ Lib 3.65
HYDRA
LUPUS
SCORPIUS
CENTAURUS
Magnitude of stars
-2 to 1
2
3
4
5
6 to 7
Binary
Open cluster
Globular cluster
Cluster with nebulosity
Nebula
Galaxies
Milky Way

Lupus & Norma

Coordinates of Brightest Star Alpha Lupi (Mag 2.28): RA 14h 41m 56s, Dec -47 23' 57"; Gamma-2 Normae (Mag 4.00): RA 16h 19m 50s, Dec -50° 09' 43"

Size of Constellation Boundaries 334 square degrees (Lupus); 165 square degrees (Norma)

Bounded by Libra, Centaurus, Circinus, Triangulum Australe, Ara, Scorpius

Constellation Legend Lupus, Latin for "wolf", was a constellation known to the Greeks and the Romans, although not necessarily as a wolf. The Greeks called it Therion, meaning "wild animal", while the Romans labelled it Bestia, which means roughly the same thing. Wolf or non-specified wild beast, the constellation's name is assumed to be related to its proximity to Centaurus and Ara – Lupus is a sacrifice to be "placed" by Chiron the centaur upon the nearby constellation Ara, whose name means "altar". Norma, which takes its name from a carpenter's square or level, is another of French astronomer Nicolas Louis de Lacaille's contributions to the night sky.

Constellation Sights Lupus is home to a number of double and multiple stars resolvable in small telescopes, including Kappa Lupi, Xi Lupi, Pi Lupi, Eta Lupi and Mu Lupi. Lupus also offers NGC 5822, an open cluster of over one hundred stars, visible in binoculars. Norma offers a binocular open cluster, NGC 6087, and a number of double stars, most obviously Gamma Normae, whose components can be separated by the naked eye. Epsilon Normae and Iota-1 Normae are also double stars, resolvable in a small telescope.

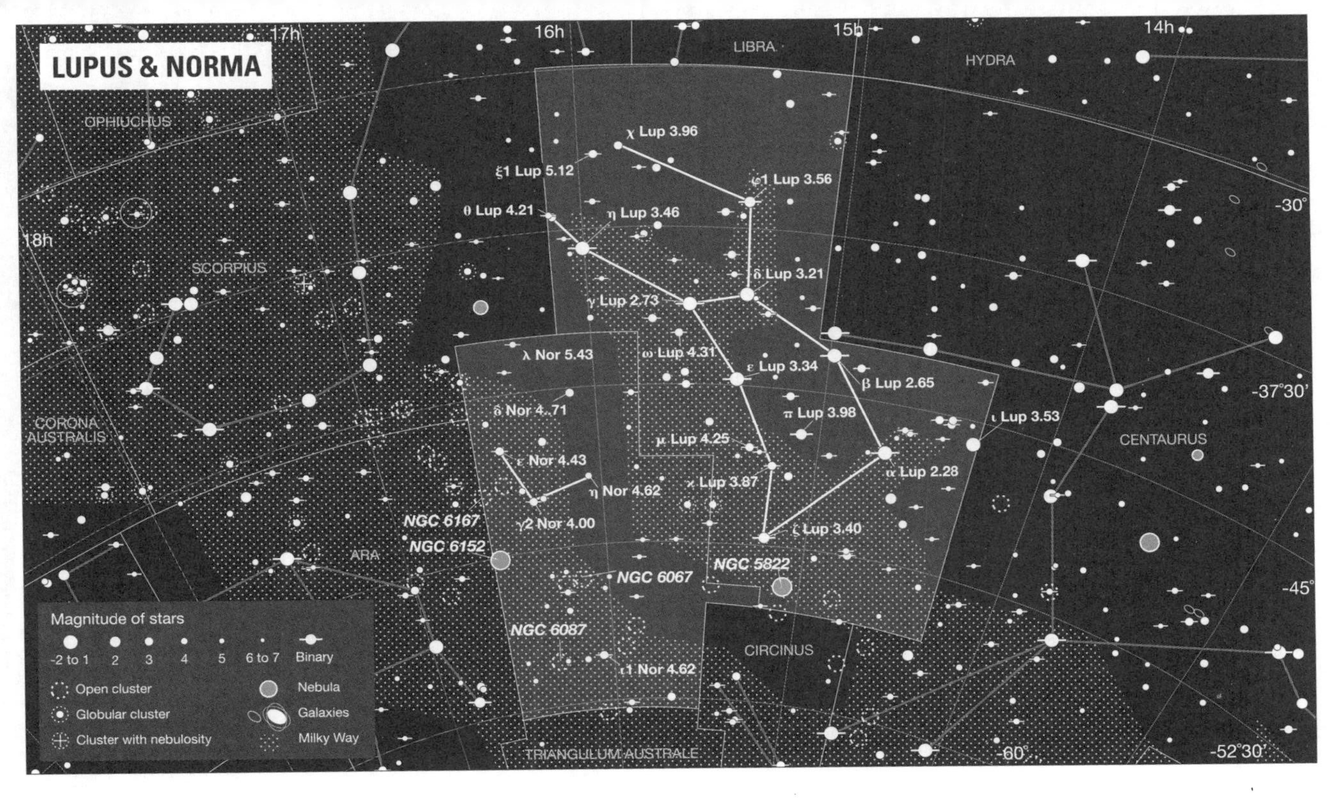
LUPUS & NORMA
14h
15h
16h
17h
18h
-30°
-37°30'
-45°
-52°30'
-60°
HYDRA
LIBRA
CENTAURUS
CIRCINUS
TRIANGULUM AUSTRALE
ARA
SCORPIUS
OPHIUCHUS
CORONA AUSTRALIS
χ Lup 3.96
ξ1 Lup 5.12
φ1 Lup 3.56
θ Lup 4.21
η Lup 3.46
δ Lup 3.21
γ Lup 2.73
λ Nor 5.43
ω Lup 4.31
ε Lup 3.34
β Lup 2.65
δ Nor 4.71
π Lup 3.98
ι Lup 3.53
μ Lup 4.25
ε Nor 4.43
κ Lup 3.87
α Lup 2.28
η Nor 4.62
γ2 Nor 4.00
ζ Lup 3.40
NGC 6167
NGC 6152
NGC 6067
NGC 5822
NGC 6087
ι1 Nor 4.62
Magnitude of stars
-2 to 1
2
3
4
5
6 to 7
Binary
Open cluster
Globular cluster
Cluster with nebulosity
Nebula
Galaxies
Milky Way

Lyra

Coordinates of Brightest Star Vega (Mag 0.03): RA 18h 36m 56s, Dec 38° 47' 06"

Size of Constellation Boundaries 286 square degrees

Bounded by Draco, Hercules, Vulpecula, Cygnus

Constellation Legend Lyra was visualized by the Greeks as a lyre, a musical instrument like a small harp, which in Greek mythology was invented by the god Hermes (Mercury). Hermes gave the lyre to Apollo who in turn passed it on to Orpheus – a man whose musical skill was so great that he could tame wild animals. Orpheus was eventually killed by frenzied followers of the god Dionysus. Other cultures saw the constellation as an eagle or vulture. Lyra's brightest star, Vega, is also at the heart of an ancient Chinese fable. It supposedly commemorates the daughter of a god separated from her husband (perceived as the star Altair, in Aquila) by the Milky Way. The pair were said to be reunited one day a year by the birds of China, who flew to the sky to make a bridge for the lovers to cross.

Constellation Sights Lyra has a number of striking sights, most notably Vega, the fifth brightest star in the sky. With its distinguishing blue tint, it's a grand sight either with the naked eye or through binoculars. Northeast of Lyra is Epsilon Lyrae, which sharp eyes will see as a double star and, via a telescope, a pair of double stars (which is why the star is known as the "Double-Double"). The Ring nebula (M57) also resides in Lyra and will show its ring-like structure in larger telescopes. NGC 6779 (M56) is a globular cluster that is less bright than many, but still fairly easily resolvable.

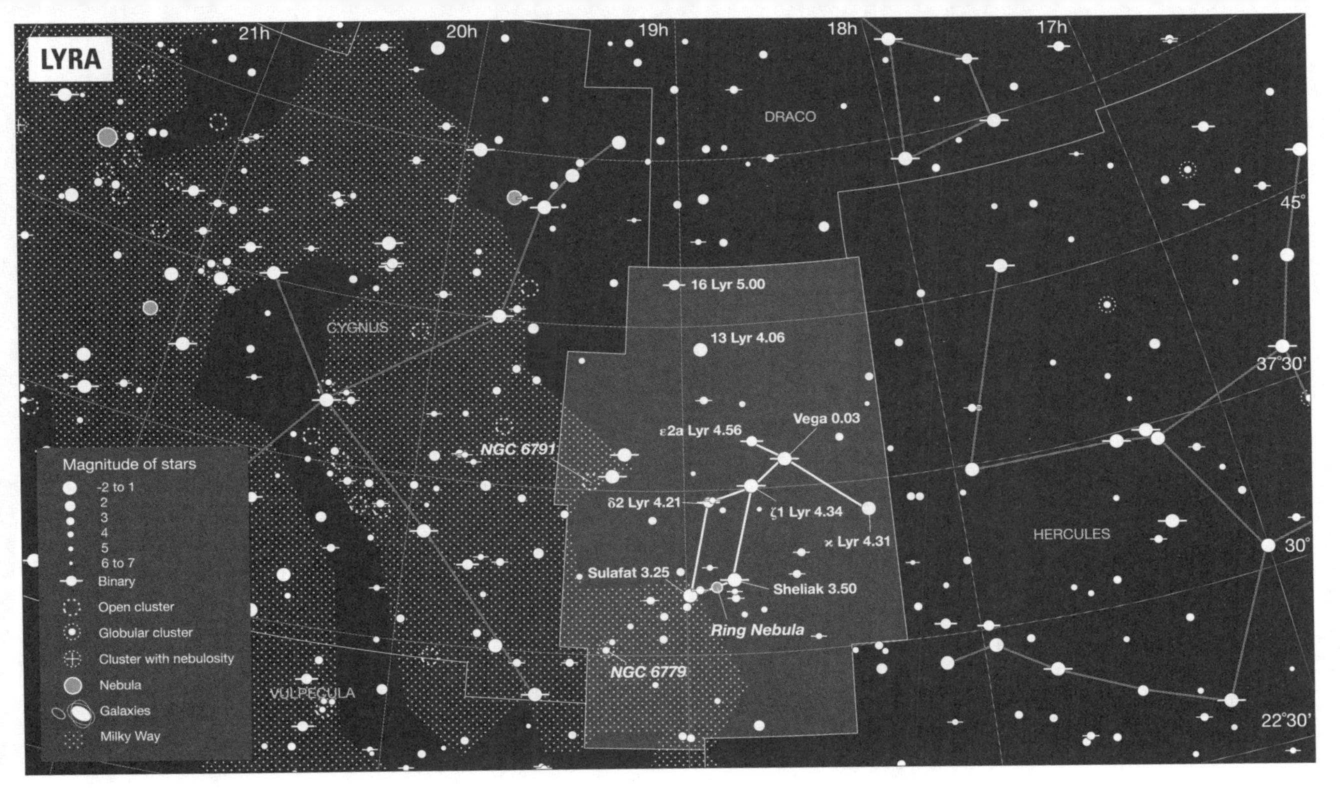

LYRA
21h
20h
19h
18h
17h
DRACO
CYGNUS
VULPECULA
HERCULES
45°
37°30'
30°
22°30'
16 Lyr 5.00
13 Lyr 4.06
Vega 0.03
ε2a Lyr 4.56
δ2 Lyr 4.21
ζ1 Lyr 4.34
κ Lyr 4.31
Sulafat 3.25
Sheliak 3.50
Ring Nebula
NGC 6791
NGC 6779
Magnitude of stars
-2 to 1
2
3
4
5
6 to 7
Binary
Open cluster
Globular cluster
Cluster with nebulosity
Nebula
Galaxies
Milky Way

Musca

Coordinates of Brightest Star Alpha Muscae (Mag 2.68): RA 12h 37m 11s, Dec -69° 08' 57"

Size of Constellation Boundaries 138 square degrees

Bounded by Crux, Centaurus, Carina, Chamaeleon, Apus, Circinus

Constellation Legend Musca (the Fly) is one of the twelve constellations added by Pieter Dirkszoon Keyser and Frederick de Houtman in the late sixteenth century. Its was known earlier as Musca Australis (the southern fly) to differentiate it from Musca Borealis (the northern fly). The latter is a small constellation directly above Aries that is no longer recognized, its stars having been assimilated into the Aries constellation. It was, incidentally, originally known as both Apis (the bee) and Vespa (the wasp).

Constellation Sights Musca features two interesting binaries, Theta Muscae and Beta Muscae. Theta can be resolved using a smaller telescope, and the fainter of its two stars is a Wolf-Rayet star (a rare type of extremely hot star). Beta Muscae, however, needs a medium-sized telescope to separate its two stars. Among the deep-sky objects to look out for are NGC 4372, a faint globular cluster, and NGC 5189, informally known as the Weird Planetary nebula on account of its unusual shape, which is also faint at tenth magnitude.

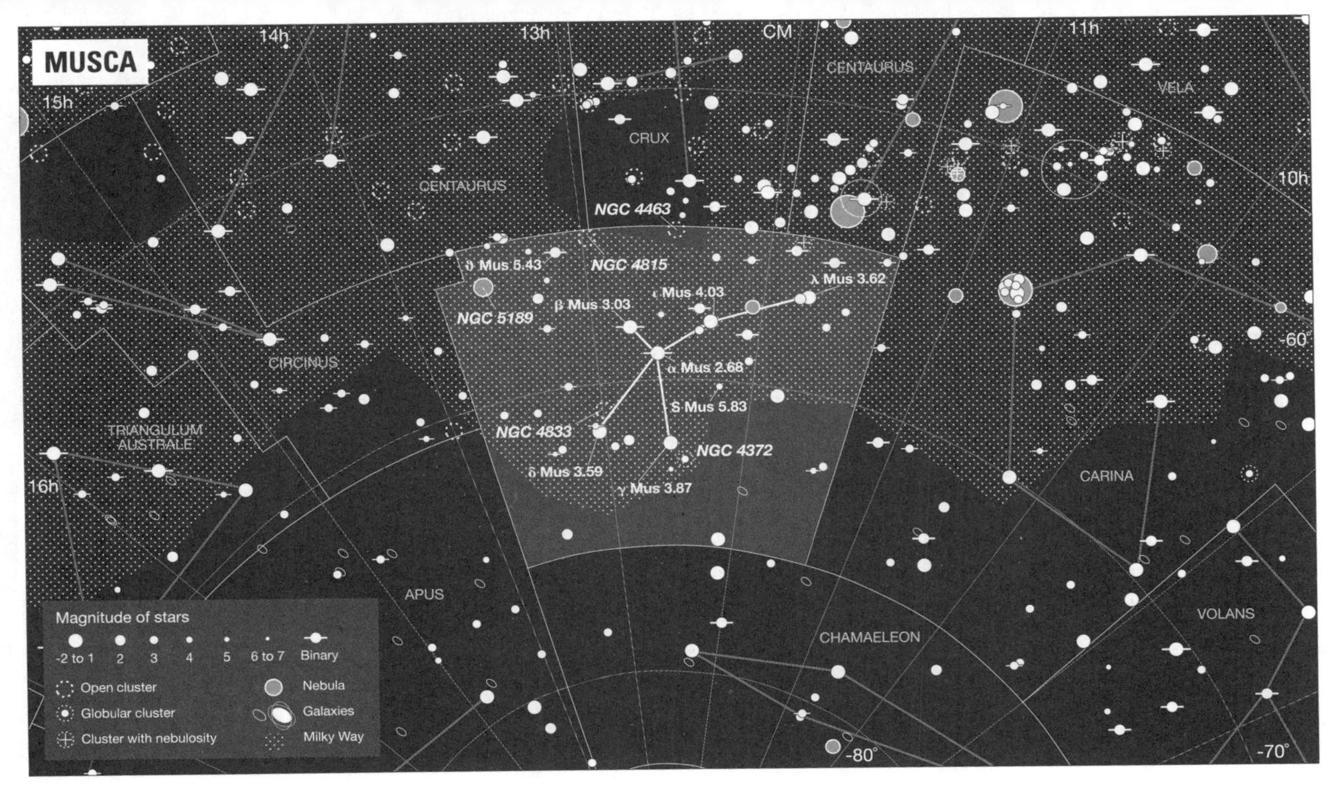
MUSCA
14h
13h
CM
11h
15h
10h
16h
-60°
-70°
-80°
VELA
CENTAURUS
CRUX
CENTAURUS
CIRCINUS
TRIANGULUM AUSTRALE
CARINA
APUS
CHAMAELEON
VOLANS
NGC 4463
NGC 4815
NGC 5189
NGC 4833
NGC 4372
ϑ Mus 5.43
ι Mus 4.03
λ Mus 3.62
β Mus 3.03
α Mus 2.68
S Mus 5.83
δ Mus 3.59
γ Mus 3.87
Magnitude of stars
-2 to 1
2
3
4
5
6 to 7
Binary
Open cluster
Globular cluster
Cluster with nebulosity
Nebula
Galaxies
Milky Way

Ophiuchus

Coordinates of Brightest Star Rasalhauge (Mag 2.06): RA 17h 34m 56s, Dec 12° 33' 26"

Size of Constellation Boundaries 948 square degrees

Bounded by Hercules, Serpens Caput, Libra, Scorpius, Sagittarius, Serpens Cauda, Aquila

Constellation Legend Ophiuchus, or the Serpent Holder, is a constellation identified with Aesculapius, the Greek god of healing and physician to the Argonauts. His father was Apollo and his mother Coronis, but he was brought up by the centaur Chiron. One day he saw a serpent use a herb it was carrying in its mouth to revive another, apparently dead, serpent. Aesculapius used that same herb to bring dead humans back to life. This angered Hades, god of the underworld, who asked Zeus to kill Aesculapius. Zeus did so, but placed him in the sky as a testament to his skill, honouring the serpent with its own position, which has its analogue today as the constellation Serpens. Although Ophiuchus intrudes into the plane of the Zodiac, between Scorpius and Sagittarius, it is not considered a sign of the Zodiac.

Constellation Sights Ophiuchus is home to over twenty globular clusters, including NGC 6254 (M10), NGC 6218 (M12), NGC 6333 (M9) and NGC 6273 (M19). 6254 and 6218 are the easiest to spot and are well worth viewing through binoculars. Ophiuchus also houses open clusters, including NGC 6633 and IC 4665, both of which are impressive sights. 36 Ophiuchi is a binary star whose two components resolve in a small telescope into orange stars that are twins of each other; 70 Ophiuchi is another good binary for small-to-moderate-sized telescopes.

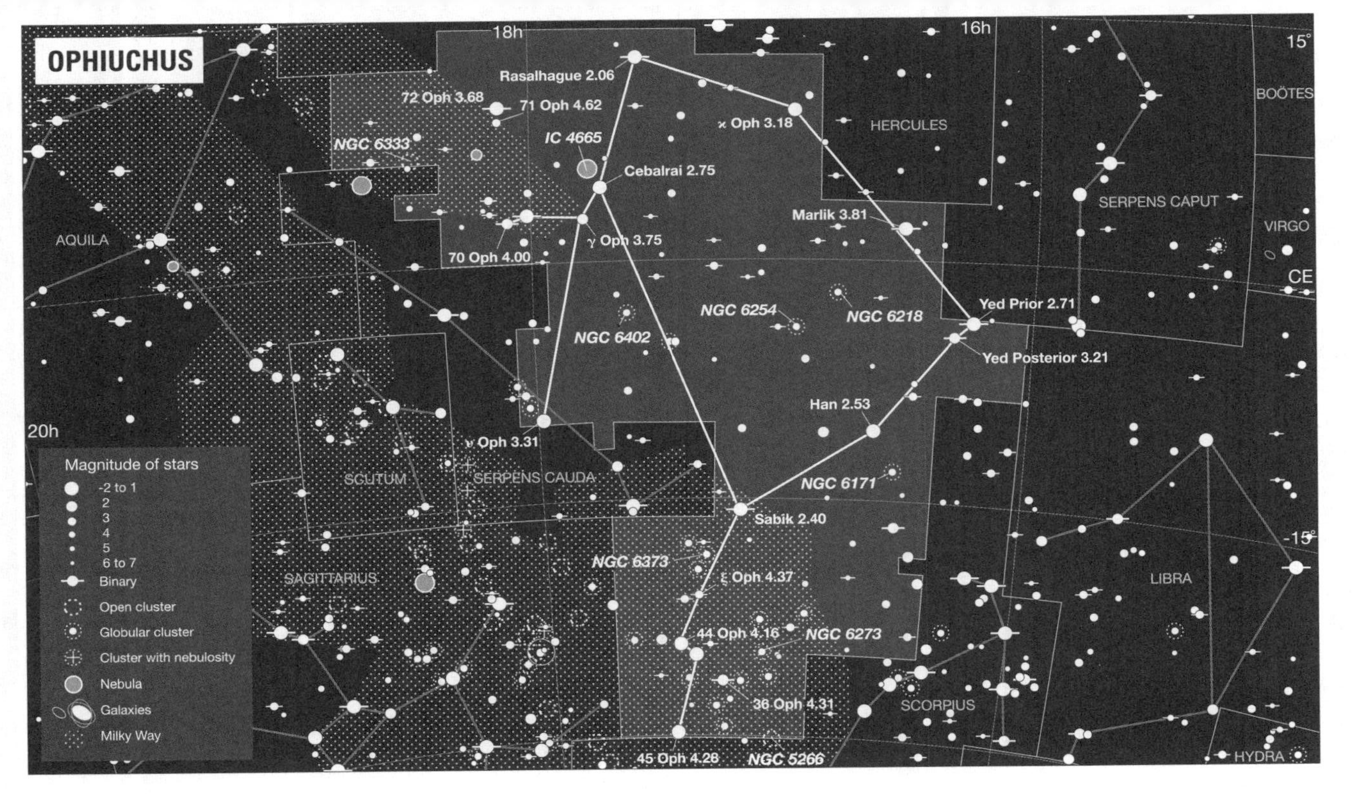
OPHIUCHUS
18h
16h
15°
20h
-15°
Rasalhague 2.06
72 Oph 3.68
71 Oph 4.62
κ Oph 3.18
HERCULES
BOÖTES
NGC 6333
IC 4665
Cebalrai 2.75
SERPENS CAPUT
Marlik 3.81
VIRGO
AQUILA
γ Oph 3.75
70 Oph 4.00
CE
Yed Prior 2.71
NGC 6254
NGC 6218
NGC 6402
Yed Posterior 3.21
Han 2.53
ν Oph 3.31
SCUTUM
SERPENS CAUDA
NGC 6171
Sabik 2.40
NGC 6373
ξ Oph 4.37
SAGITTARIUS
LIBRA
44 Oph 4.16
NGC 6273
36 Oph 4.31
SCORPIUS
45 Oph 4.28
NGC 5266
HYDRA
Magnitude of stars
-2 to 1
2
3
4
5
6 to 7
Binary
Open cluster
Globular cluster
Cluster with nebulosity
Nebula
Galaxies
Milky Way

Orion

Coordinates of Brightest Star Rigel (Mag 0.15): RA 5h 14m 32s, Dec -8° 11' 52"

Size of Constellation Boundaries 594 square degrees

Bounded by Taurus, Eridanus, Lepus, Monoceros, Gemini

Constellation Legend Orion was a great hunter in Greek mythology, the mortal son of Poseidon (Neptune), god of the sea. One myth tells of Orion's boastful threat to hunt down every animal on earth. This angered the gods, and Zeus sent a giant scorpion to kill him. After a long fight, Orion succumbed and was stung to death. At the request of Diana, goddess of the hunt, Zeus thereafter placed him in the sky; he also did the same for the scorpion, but at the opposite end of the sky to ensure that the two never fought again. Other ancient cultures saw the constellation as signifying a hunter or warrior, and the Egyptians identified it with the home of Osiris in the sky. Straddling the celestial equator, Orion is arguably the best known constellation, and the easiest one to recognize.

Constellation Sights Orion is positively jam-packed with sights, the most glorious of which is the Orion nebula (M42), perhaps the most famous one in the sky. The nebula is visible to the naked eye (it appears as a fuzzy star) and in binoculars it begins to reveal its shape, but to take in all its beauty you'll need a telescope. Around and above the Orion nebula are several other nebulae and deep-sky objects, including the nebula M43 (actually part of the same nebula as M42), and the open cluster NGC 1981. Directly in front of M42 lies the Trapezium, a multiple system which a small telescope will resolve into four separate stars arrayed in a trapezium shape. Orion is also home to a great number of bright stars, including Betelgeuse, which varies in brightness by more than a full degree in magnitude, and whose diameter is larger than the orbit of the Earth. Even brighter is Rigel, whose native luminosity is some sixty thousand times greater than our own Sun's.

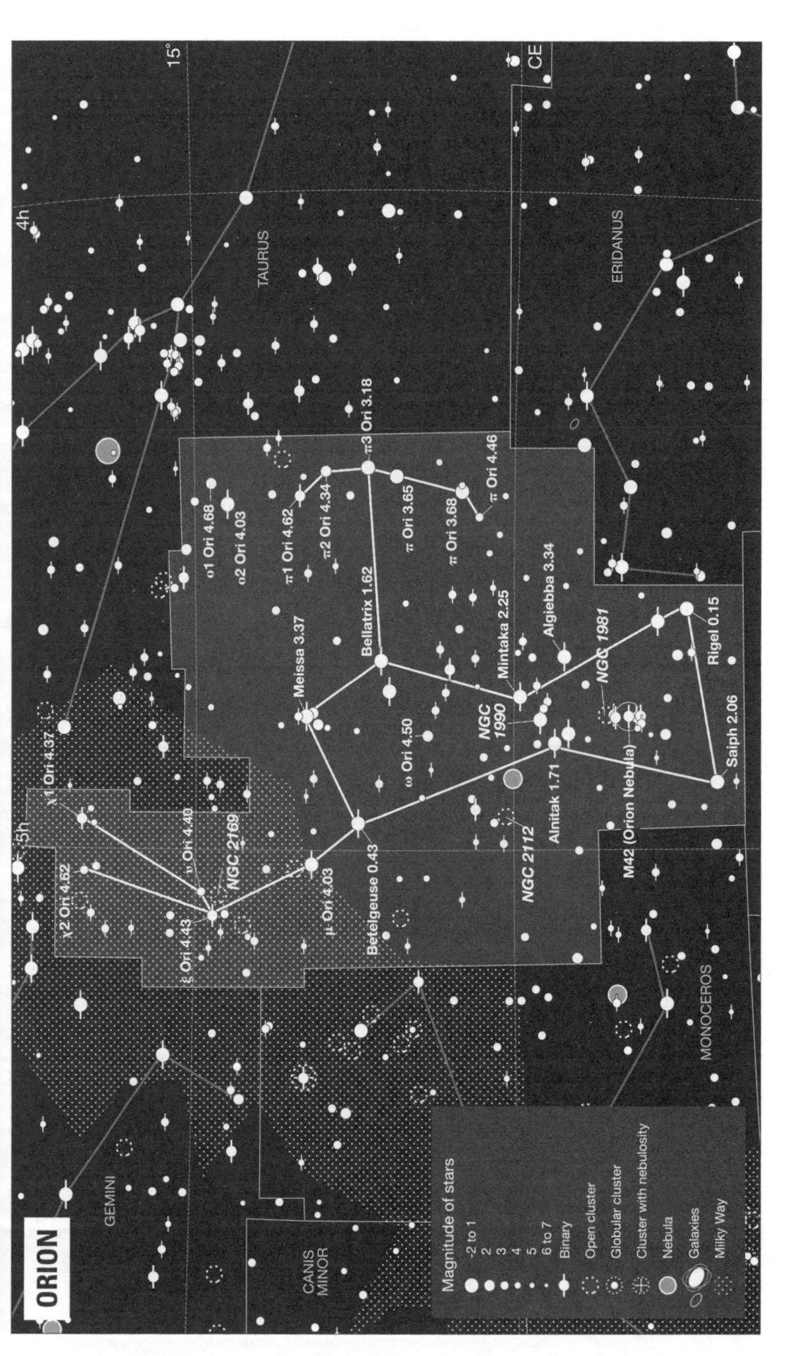
ORION
4h
5h
15°
TAURUS
ERIDANUS
GEMINI
CANIS MINOR
MONOCEROS
χ1 Ori 4.37
χ2 Ori 4.62
ν Ori 4.40
ξ Ori 4.43
NGC 2169
μ Ori 4.03
Betelgeuse 0.43
ο1 Ori 4.68
ο2 Ori 4.03
Meissa 3.37
ω Ori 4.50
π1 Ori 4.62
π2 Ori 4.34
π3 Ori 3.18
π Ori 3.65
π Ori 3.68
π Ori 4.46
Bellatrix 1.62
Mintaka 2.25
Algiebba 3.34
NGC 1990
NGC 2112
Alnitak 1.71
NGC 1981
M42 (Orion Nebula)
Rigel 0.15
Saiph 2.06
Magnitude of stars
-2 to 1
2
3
4
5
6 to 7
Binary
Open cluster
Globular cluster
Cluster with nebulosity
Nebula
Galaxies
Milky Way

Pegasus

Coordinates of Brightest Star Enif (Mag 2.37): RA 21h 44m 11s, Dec 9° 53' 08"

Size of Constellation Boundaries 1121 square degrees

Bounded by Andromeda, Lacerta, Cygnus, Vulpecula, Delphinus, Equuleus, Aquarius, Pisces

Constellation Legend In Greek mythology Pegasus was the winged horse created by the god Poseidon out of the blood of Medusa, the snake-haired Gorgon. He had many adventures, among which were helping Perseus rescue Andromeda, and Bellerophon to kill the Chimera. When Bellerophon attempted to ride Pegasus to the home of the gods, Zeus sent a gadfly to sting the horse, whose rider was sent crashing to earth. Zeus later employed Pegasus to carry his thunder and lightning. Like the constellation Hercules, Pegasus is oriented upside down, with its head and neck southernmost. The star Enif, pinpointed on Pegasus' nose, comes from an Arabic word meaning, predictably enough, "nose".

Constellation Sights Pegasus' most notable sight is an asterism known as the Great Square of Pegasus; it's formed with the stars Scheat, Markab and Algeneb, as well as Alpheratz, which technically lies in the constellation Andromeda. Scheat is also a variable which fluctuates by half a magnitude at unpredictable intervals. Enif, the constellation's brightest star, is a double star whose dimmer half (eighth magnitude) is resolvable in a small telescope. NGC 7078 (M15), a binocular globular cluster, lies to the northwest of Enif. NGC 7331 is a relatively bright spiral galaxy of the ninth magnitude.

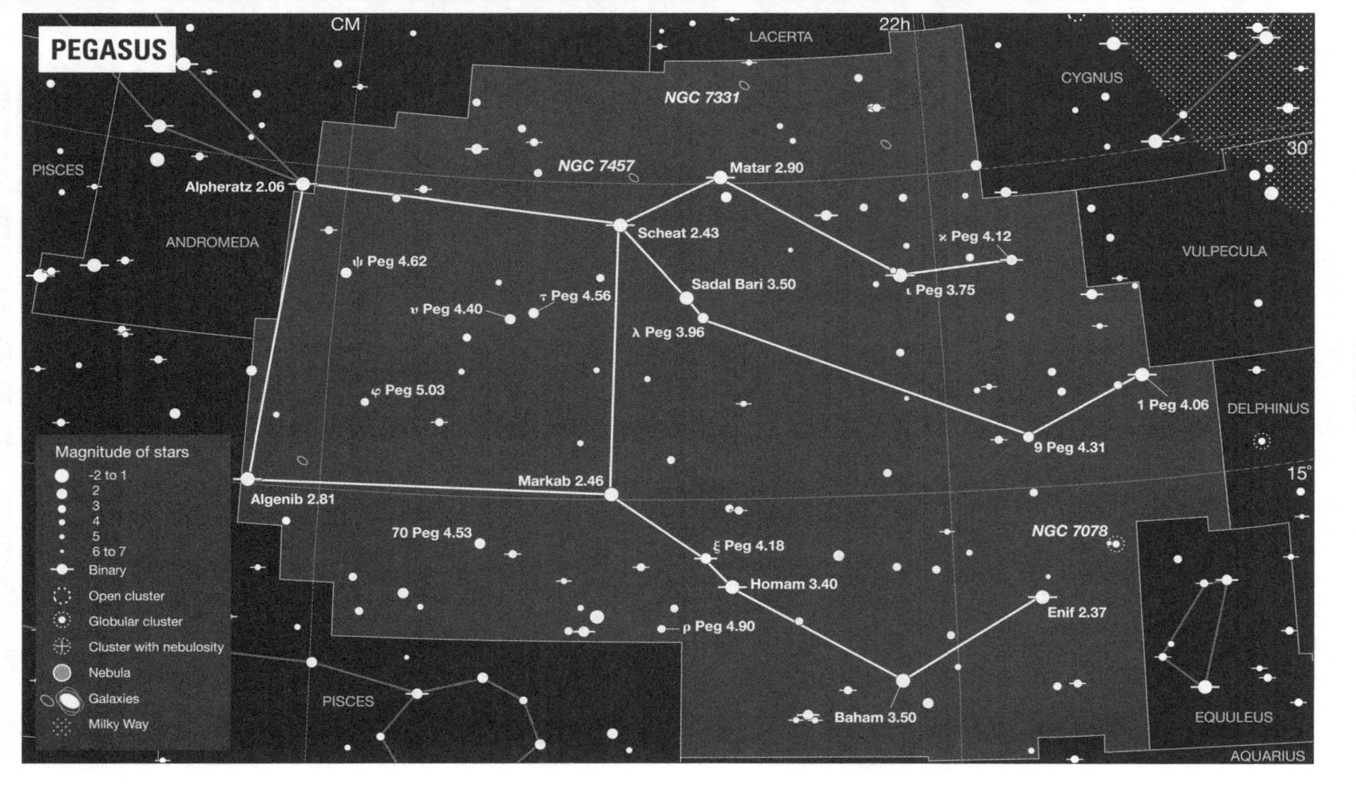
PEGASUS
CM
LACERTA
22h
CYGNUS
30°
PISCES
NGC 7331
NGC 7457
Alpheratz 2.06
Matar 2.90
Scheat 2.43
ANDROMEDA
ϰ Peg 4.12
VULPECULA
ψ Peg 4.62
Sadal Bari 3.50
ι Peg 3.75
τ Peg 4.56
υ Peg 4.40
λ Peg 3.96
φ Peg 5.03
1 Peg 4.06
DELPHINUS
9 Peg 4.31
Markab 2.46
15°
Algenib 2.81
70 Peg 4.53
NGC 7078
ξ Peg 4.18
Homam 3.40
Enif 2.37
ρ Peg 4.90
PISCES
Baham 3.50
EQUULEUS
AQUARIUS
Magnitude of stars
-2 to 1
2
3
4
5
6 to 7
Binary
Open cluster
Globular cluster
Cluster with nebulosity
Nebula
Galaxies
Milky Way

Perseus

Coordinates of Brightest Star Mirfak (Mag 1.78): RA 3h 24m 19s, Dec 49° 52' 13"

Size of Constellation Boundaries 615 square degrees

Bounded by Camelopardalis, Cassiopeia, Andromeda, Triangulum, Aries, Taurus, Auriga

Constellation Legend Perseus was a son of Zeus by Danae, daughter of King Acrisius of Argos. His most famous adventure involved beheading Medusa the Gorgon, a woman so hideous that looking at her turned the beholder to stone. Mounted on the flying horse Pegasus, Perseus later rescued the princess Andromeda from the sea monster which had been terrorizing her country due to the vanity of Andromeda's mother, Cassiopeia. The story is illustrated right across the sky in the constellations of Andromeda, Cassiopeia, Cetus, Pegasus and Cepheus. Among these constellations, Medusa herself is said to have her counterpart in the star Algol, depicted in many illustrations as the severed Gorgon's head carried by Perseus which he used to vanquish the sea monster.

Constellation Sights Mirfak, the constellation's brightest star, lies right in the middle of a loose cluster of stars known as Melotte 20, which can easily be seen via telescope. Algol is a famous variable, because it is an eclipsing binary, which means that when the dimmer star crosses in front of the brighter one (from our point of view), Algol's brightness drops by more than one magnitude for some ten hours. This happens every 2.7 days. There are several open clusters in Perseus, most notably the Double Cluster of Chi and Eta Persei, barely visible to the naked eye but a nice sight in binoculars or a small telescope. Also look for the Spiral Cluster (M34), so called because the stars appear to form three curving arms.

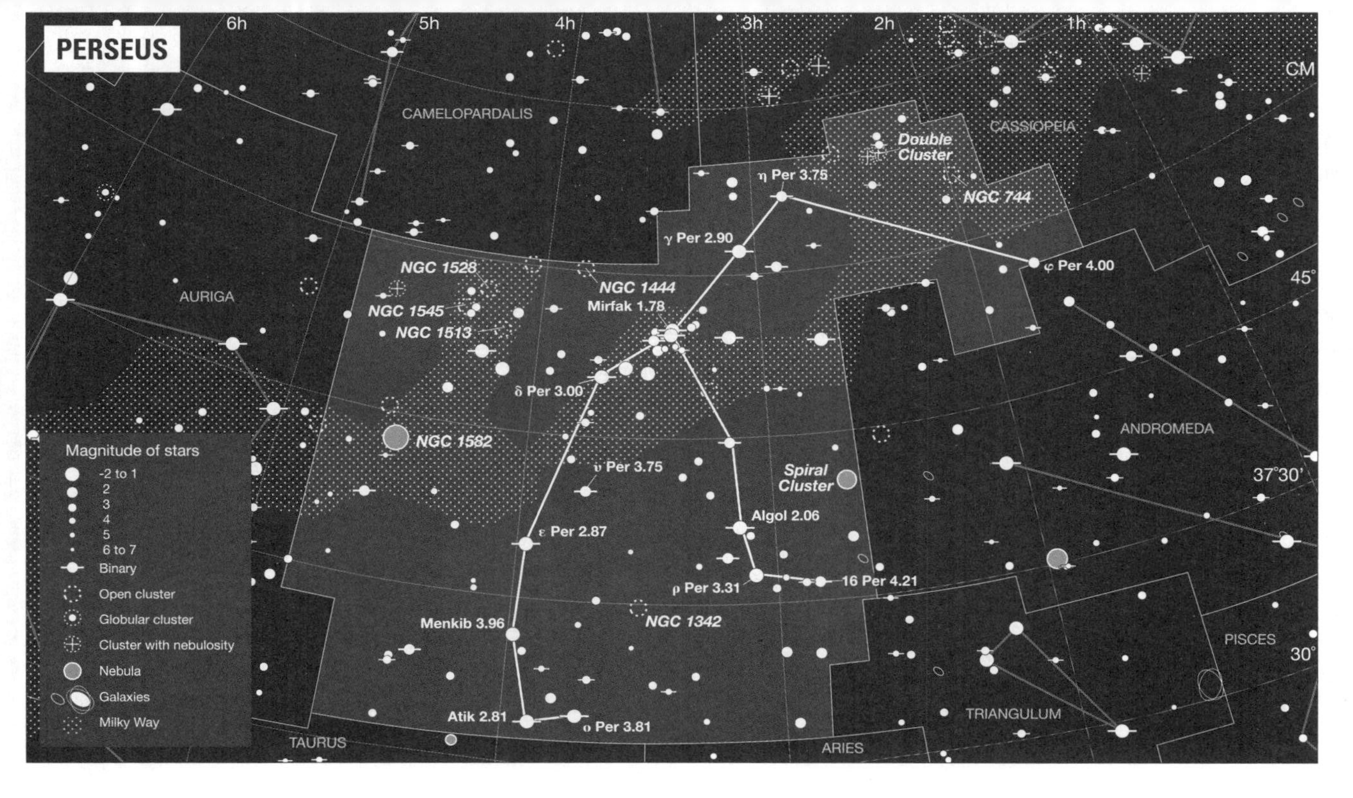
PERSEUS
6h
5h
4h
3h
2h
1h
CM
CAMELOPARDALIS
CASSIOPEIA
Double Cluster
η Per 3.75
NGC 744
γ Per 2.90
φ Per 4.00
45°
NGC 1528
NGC 1444
AURIGA
NGC 1545
Mirfak 1.78
NGC 1513
δ Per 3.00
ANDROMEDA
NGC 1582
υ Per 3.75
Spiral Cluster
37°30'
Algol 2.06
ε Per 2.87
ρ Per 3.31
16 Per 4.21
NGC 1342
Menkib 3.96
PISCES
30°
Atik 2.81
ο Per 3.81
TRIANGULUM
TAURUS
ARIES
Magnitude of stars
-2 to 1
2
3
4
5
6 to 7
Binary
Open cluster
Globular cluster
Cluster with nebulosity
Nebula
Galaxies
Milky Way

Phoenix & Sculptor

Coordinates of Brightest Stars Ankaa (Mag 2.37), RA 0h 26m 17s, Dec -42° 17' 33"; Alpha Sculptoris (Mag 4.28) RA 0h 58m 36s, Dec -29° 20' 38"

Size of Constellation Boundaries 469 square degrees (Phoenix); 475 square degrees (Sculptor)

Bounded by Cetus, Aquarius, Piscis Austrinus, Grus, Tucana, Hydra, Eridanus, Fornax

Constellation Legend According to tradition, the Phoenix was a red-and-gold bird about the size of an eagle. As it neared death, it would build itself a funeral pyre out of sweet spices on which it was burned to death by the rays of the Sun. From the ashes of its former self, a new Phoenix would emerge. The story of the Phoenix is an ancient one which appears in many different cultures, but the constellation itself is relatively recent, having been charted by Pieter Dirkszoon Keyser and Frederick de Houtman in the late sixteenth century. Sculptor is a constellation named in the eighteenth century by Nicolas Louis de Lacaille *L'Atelier du Sculpteur* (the Sculptor's Workshop). What exactly he saw in the sparse constellation that inspired the name is rather hard to fathom.

Constellation Sights Zeta Phoenicis is an eclipsing binary, similar to Algol in Perseus, which dips some half a magnitude every forty hours. In a larger telescope it also shows itself to be a triple star system. Beta Phoenicis is also a double star, whose individual components are best seen through a medium-to-large telescope. Sculptor is home to the Sculptor galaxy (NGC 253), a spiral galaxy visible in binoculars; larger telescopes will reveal the substantial clouds of dust within it. NGC 55 is another spiral galaxy, although it is only visible through a small telescope, through which you can see it edge-on.

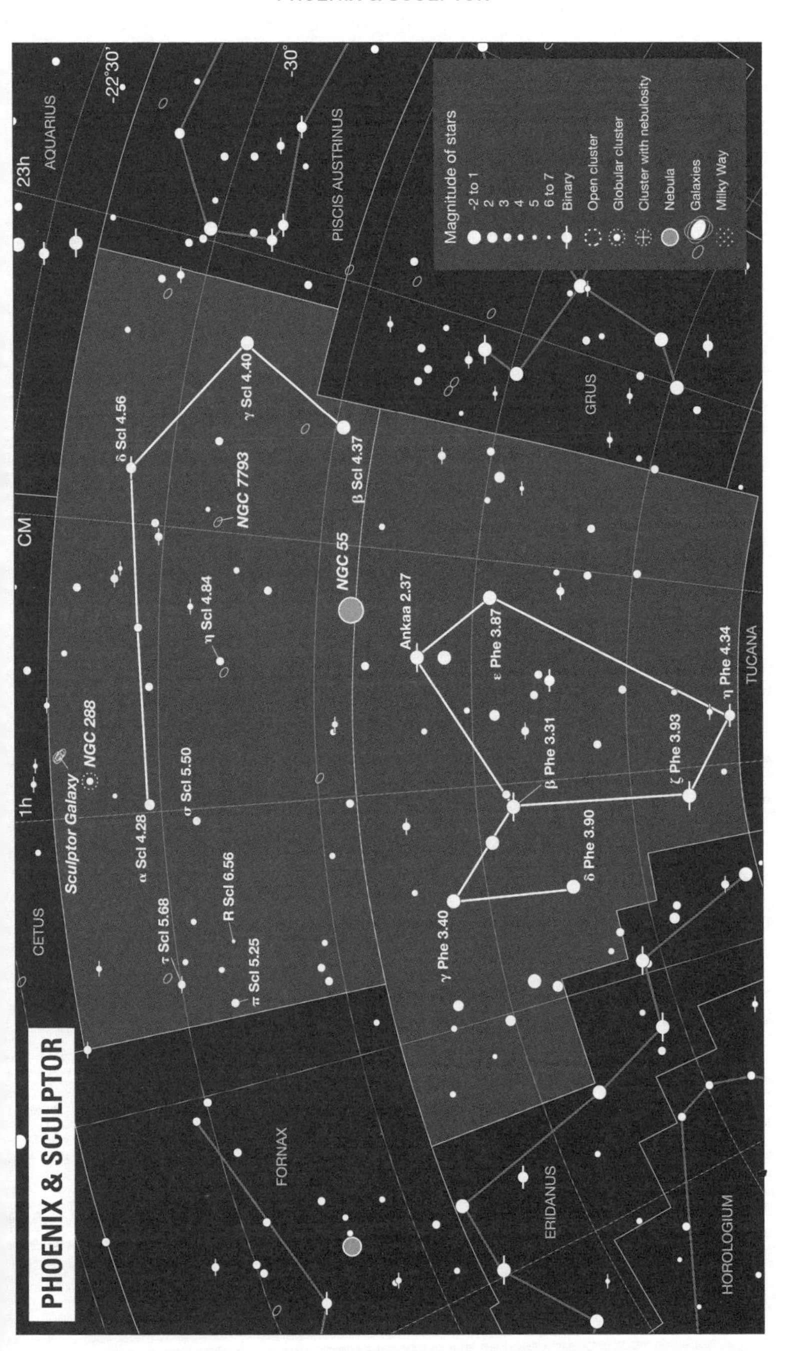
PHOENIX & SCULPTOR
23h
AQUARIUS
-22°30'
-30°
PISCIS AUSTRINUS
Magnitude of stars
-2 to 1
2
3
4
5
6 to 7
Binary
Open cluster
Globular cluster
Cluster with nebulosity
Nebula
Galaxies
Milky Way
δ Scl 4.56
γ Scl 4.40
NGC 7793
β Scl 4.37
GRUS
CM
NGC 55
η Scl 4.84
Ankaa 2.37
ε Phe 3.87
TUCANA
η Phe 4.34
NGC 288
Sculptor Galaxy
1h
σ Scl 5.50
β Phe 3.31
ζ Phe 3.93
α Scl 4.28
δ Phe 3.90
CETUS
R Scl 6.56
τ Scl 5.68
γ Phe 3.40
π Scl 5.25
FORNAX
ERIDANUS
HOROLOGIUM

Pictor

Coordinates of Brightest Star Alpha Pictoris (Mag 3.21): RA 6h 48m 11s, Dec -61° 56' 35"

Size of Constellation Boundaries 247 square degrees

Bounded by Columba, Caelum, Dorado, Volans, Carina, Puppis

Constellation Legend This dim constellation was originally named Equuleus Pictoris, or the Painter's Easel, and was added to our recognized constellations in the eighteenth century by Nicolas Louis de Lacaille. The word equuleus, which translates literally as "small horse", referred to the wooden horse-like stand that painters use as a kind of elongated stool on which an easel or drawing-board can be mounted. It seems as if de Lacaille never lost his predilection for naming his constellations after tools, gadgets and inventions, although perhaps Pictor and Sculptor show that he had some interest in the arts.

Constellation Sights There is almost nothing of interest in Pictor for the amateur astronomer. However, professional astronomers are particularly enamoured of Beta Pictoris, largely because it was one of the first stars observed to be surrounded by a disc of dust and gas with a warped inner area – evidence of a potential planetary system in a process of formation.

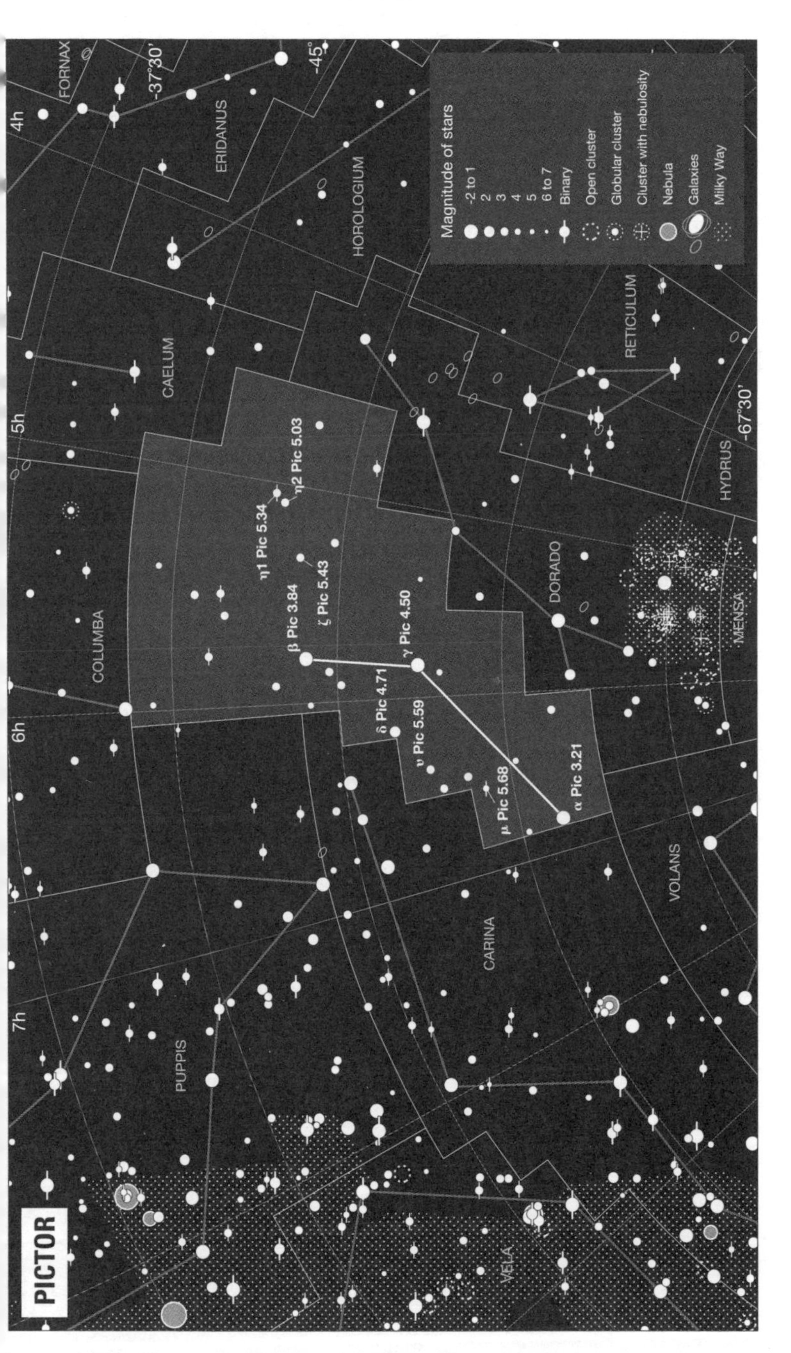
PICTOR
4h
5h
6h
7h
-37°30'
-45°
-67°30'
FORNAX
ERIDANUS
HOROLOGIUM
CAELUM
COLUMBA
PUPPIS
VELA
CARINA
VOLANS
DORADO
RETICULUM
HYDRUS
MENSA
η1 Pic 5.34
η2 Pic 5.03
ζ Pic 5.43
β Pic 3.84
γ Pic 4.50
δ Pic 4.71
υ Pic 5.59
μ Pic 5.68
α Pic 3.21
Magnitude of stars
-2 to 1
2
3
4
5
6 to 7
Binary
Open cluster
Globular cluster
Cluster with nebulosity
Nebula
Galaxies
Milky Way

Pisces

Coordinates of Brightest Star Eta Piscium (Mag 3.59): RA 1h 30m 55s, Dec 15° 21' 31"

Size of Constellation Boundaries 889 square degrees

Bounded by Andromeda, Pegasus, Aquarius, Cetus, Aries, Triangulum

Constellation Legend Pisces represents two fish tied by their tails, connected by a cord. The fish in question are no ordinary fish but are supposedly the Greek gods Aphrodite and her son Eros (Venus and Cupid in the Roman version), who turned themselves into fish to escape the wrath of the sea monster Typhon. The constellation was also seen as fish by the Persians and Babylonians. The Chinese, however, saw it very differently: sometimes as an emperor and sometimes as a pig – an interesting dichotomy! Pisces is on the celestial equator and contains the vernal equinox, so the First Point of Aries, which demarcates that event, actually resides in Pisces. This is due to precession of the seasons.

Constellation Sights The most notable sight in Pisces is the Circlet, an asterism composed of seven stars laid out in a rough circle. 19X Piscium (also known as TX Piscium) is one of these; it is also a variable star that fluctuates by nearly half a magnitude at irregular intervals. Pisces is home to a number of double and multiple stars, including Alrescha, Zeta Piscium and Psi Piscum, all of which resolve in telescopes. NGC 628 is a head-on spiral galaxy for which you'll need favourable viewing conditions, and at least a moderately powerful telescope if you want to see the arms.

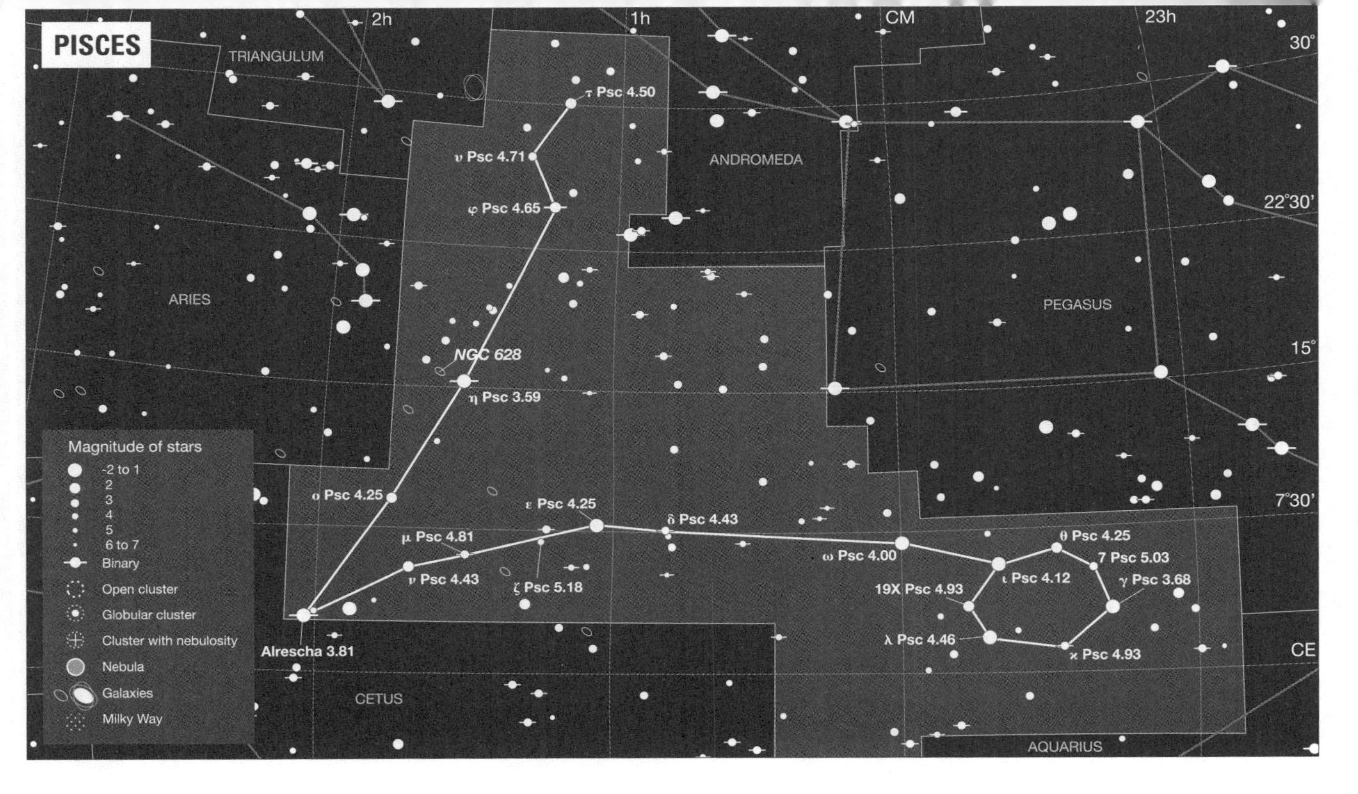
PISCES
TRIANGULUM
2h
1h
CM
23h
30°
22°30'
15°
7°30'
CE
ANDROMEDA
PEGASUS
ARIES
CETUS
AQUARIUS
τ Psc 4.50
υ Psc 4.71
φ Psc 4.65
NGC 628
η Psc 3.59
ο Psc 4.25
Alrescha 3.81
μ Psc 4.81
ν Psc 4.43
ζ Psc 5.18
ε Psc 4.25
δ Psc 4.43
ω Psc 4.00
θ Psc 4.25
7 Psc 5.03
ι Psc 4.12
γ Psc 3.68
19X Psc 4.93
λ Psc 4.46
κ Psc 4.93
Magnitude of stars
-2 to 1
2
3
4
5
6 to 7
Binary
Open cluster
Globular cluster
Cluster with nebulosity
Nebula
Galaxies
Milky Way

Puppis

Coordinates of Brightest Star Naos (Mag 2.18): RA 8h 3m 35s, Dec -40° 00' 33"

Size of Constellation Boundaries 673 square degrees

Bounded by Monoceros, Canis Major, Columba, Pictor, Carina, Vela, Pyxis, Hydra

Constellation Legend Puppis is Latin for the stern of a ship, and the constellation originally corresponded to the stern of the Argo (when Puppis was part of the now-obsolete constellation Argo Navis), the ship which carried the Greek hero Jason and his companions on the quest for the Golden Fleece. Argo Navis was informally broken up into three smaller constellations by de Lacaille in 1763, and officially divided into four parts in 1929. The other three are Carina (the Keel), Pyxis (the Compass) and Vela (the Sail). But the Argo also carried Hercules, Chiron, and the twins Castor and Pollux. According to one version of the story, Jason was killed in old age when one of the Argo's rotting beams fell on top of him.

Constellation Sights Puppis cruises through the Milky Way, and as such is home to an impressive number of deep-sky objects, particularly open clusters, viewable through binoculars. These include M93 (also known as the Butterfly Cluster), M46, a very rich cluster with well over 150 stars, and NGC 2422 (M47), which can be spotted by the naked eye in good conditions. The open cluster NGC 2451 is more easily discernible with the naked eye. Among the stars, V Puppis is an eclipsing binary that fluctuates by half a magnitude every 35 hours.

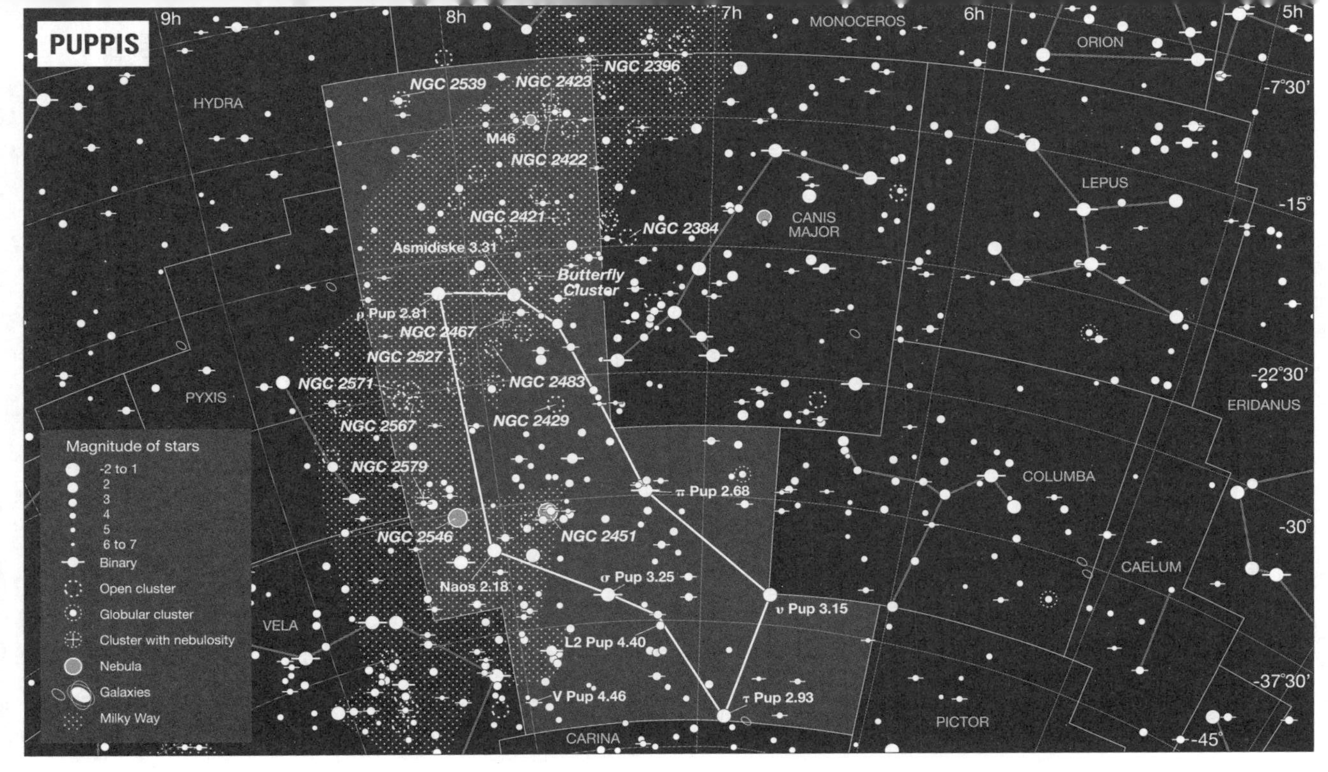
PUPPIS
9h
8h
7h
6h
5h
MONOCEROS
ORION
HYDRA
LEPUS
CANIS MAJOR
PYXIS
ERIDANUS
COLUMBA
CAELUM
VELA
PICTOR
CARINA
-7°30'
-15°
-22°30'
-30°
-37°30'
-45°
NGC 2396
NGC 2539
NGC 2423
M46
NGC 2422
NGC 2421
NGC 2384
Asmidiske 3.31
Butterfly Cluster
ρ Pup 2.81
NGC 2467
NGC 2527
NGC 2571
NGC 2483
NGC 2429
NGC 2567
NGC 2579
π Pup 2.68
NGC 2546
NGC 2451
Naos 2.18
σ Pup 3.25
υ Pup 3.15
L2 Pup 4.40
V Pup 4.46
τ Pup 2.93
Magnitude of stars
-2 to 1
2
3
4
5
6 to 7
Binary
Open cluster
Globular cluster
Cluster with nebulosity
Nebula
Galaxies
Milky Way

Sagittarius

Coordinates of Brightest Star Kaus Australis (Mag: 1.78): RA 18h 24m 10s, Dec 34° 23' 03"

Size of Constellation Boundaries 867 degrees

Bounded by Scutum, Serpens Cauda, Ophiuchus, Scorpius, Corona Australis, Telescopium, Microscopium, Capricornus, Aquila

Constellation Legend Sagittarius, the "Archer", is a name given to us by the Romans. The constellation represents a centaur, a mythical beast with the torso of a man and the body of a horse. There is some confusion as to which specific centaur Sagittarius was linked to. Crotus is the likeliest candidate, being the inventor of archery (Chiron, another celebrated centaur, is conventionally associated with the constellation Centaurus). Crotus was the offspring of the goat god Pan and the nymph Eupheme, nurse of the Muses (among whom he was raised). Crotus grew to be skilled in the arts as well as a great hunter, and at his death the Muses entreated Zeus to place him among the stars. Many of the stars in the constellation refer specifically to parts of the centaur or his bow, including the brightest, Kaus Australis, the Southern Bow.

Constellation Sights The Milky Way runs through Sagittarius and provides it with many of its most dazzling sights including several important nebulae. Primary among them is the Lagoon nebula (M8), which is visible to the naked eye and also makes a good sight in binoculars. Directly north is the Trifid nebula (M20), visible with a larger telescope, so named because of the trio of dust lanes that appear to radiate from its centre. The Horseshoe nebula (also known as the Omega nebula and M17) is in the far north of the constellation: you'll need a telescope to see its distinctive shape, although it is also viewable in binoculars. NGC 6626 (M22) is a brilliant globular cluster (the third brightest in the sky, with over seventy thousand stars) which also shows up in binoculars, while IC 4725 (M25) is a fine open cluster of more than eighty stars – perfect for small telescopes. Sagittarius also has an asterism known as the Teapot, with Kaus Australis and Ascella comprising the bottom, Kaus Borealis the top of the lid, and Alnasl at the tip of the spout.

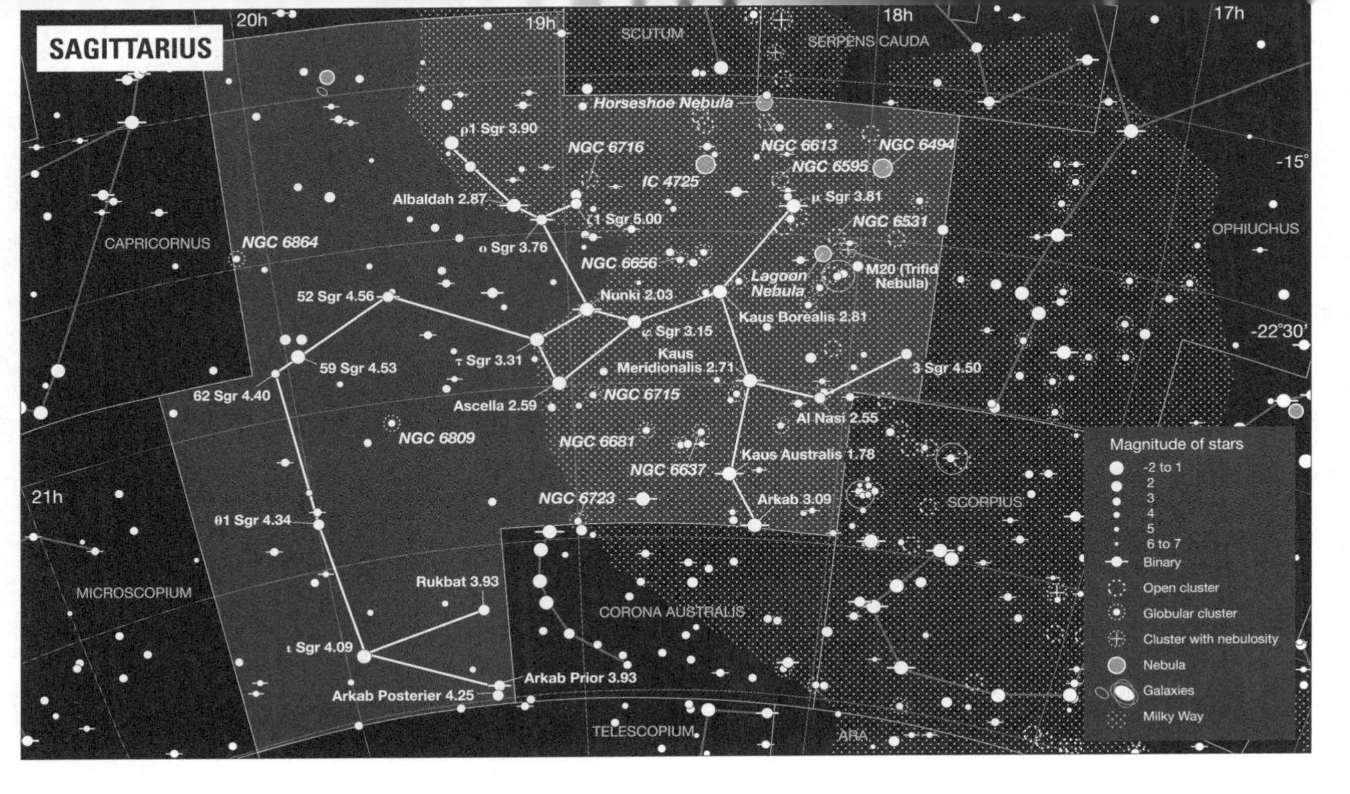
SAGITTARIUS
20h
19h
18h
17h
21h
-15°
-22°30'
SCUTUM
SERPENS CAUDA
OPHIUCHUS
CAPRICORNUS
MICROSCOPIUM
CORONA AUSTRALIS
TELESCOPIUM
ARA
SCORPIUS
Horseshoe Nebula
ρ1 Sgr 3.90
NGC 6716
NGC 6613
NGC 6494
NGC 6595
IC 4725
Albaldah 2.87
μ Sgr 3.81
ξ1 Sgr 5.00
NGC 6531
NGC 6864
ο Sgr 3.76
NGC 6656
Lagoon Nebula
M20 (Trifid Nebula)
52 Sgr 4.56
Nunki 2.03
Kaus Borealis 2.81
φ Sgr 3.15
τ Sgr 3.31
Kaus Meridionalis 2.71
59 Sgr 4.53
3 Sgr 4.50
62 Sgr 4.40
NGC 6715
Ascella 2.59
Al Nasl 2.55
NGC 6809
NGC 6681
Kaus Australis 1.78
NGC 6637
NGC 6723
Arkab 3.09
θ1 Sgr 4.34
Rukbat 3.93
ι Sgr 4.09
Arkab Prior 3.93
Arkab Posterier 4.25
Magnitude of stars
-2 to 1
2
3
4
5
6 to 7
Binary
Open cluster
Globular cluster
Cluster with nebulosity
Nebula
Galaxies
Milky Way

Sagitta & Vulpecula

Coordinates of Brightest Stars Gamma Sagittae (Mag 3.50): RA 19h 58m 45s, Dec 19° 29' 53"; Alpha Vulpeculae (Mag 4.43): RA 19h 28m 42s, Dec 24° 40' 08"

Size of Constellation Boundaries 80 square degrees (Sagitta); 268 square degrees (Vulpecula)

Bounded by Cygnus, Lyra, Hercules, Aquila, Delphinus, Pegasus

Constellation Legend Despite being the third-smallest constellation in the sky, Sagitta was known to the ancient Greeks. It represents an arrow, but there's some speculation as to who is shooting it. The archer of Sagittarius would seem to be the obvious culprit, only the bow is facing in the opposite direction. There's one Greek legend, however, in which the hero Hercules hunts two birds: Aquila (the Eagle) and Cygnus (the Swan). Both of these are constellations near Sagitta, so the Sagitta constellation could be a representation of Hercules' arrow flying towards Cygnus. Vulpecula, or the Ox, was a constellation added by Johannes Hevelius in the seventeenth century. It was originally called *Vulpecula cum Anser*, which means the "fox" and "goose", and illustrations of it often show the fox with a goose in its mouth. In 1967 Vulpecula was the site of the first pulsar to be discovered.

Constellation Sights Both Sagitta and Vulpecula offer faint stars which are of little interest. Sagitta does have a small globular cluster, NGC 6838, visible through binoculars or a small telescope, and which for many years was wrongly thought to be an open cluster. Vulpecula offers the Dumbbell nebula (M27), the first planetary nebula to be discovered, whose double-lobed appearance is viewable in larger telescopes (although the nebula can be seen in lesser detail in binoculars). Vulpecula also has the Coathanger, a small cluster of stars in a straight line, with a little "hook" of stars hanging down. 4 Vulpeculae, represented in the map, is one of the stars in the Coathanger.

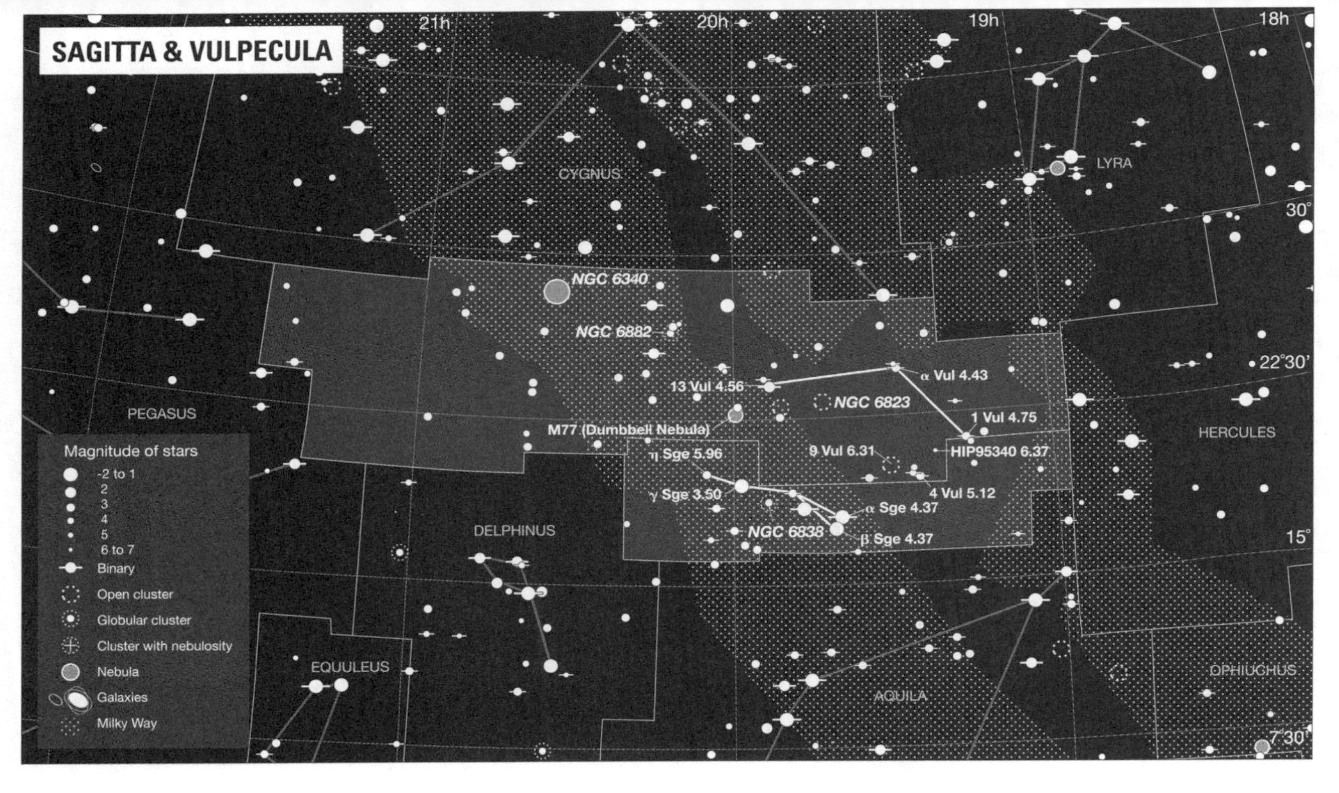
SAGITTA & VULPECULA
21h
20h
19h
18h
30°
22°30'
15°
7°30'
CYGNUS
LYRA
HERCULES
PEGASUS
DELPHINUS
EQUULEUS
AQUILA
OPHIUCHUS
NGC 6340
NGC 6882
13 Vul 4.56
α Vul 4.43
NGC 6823
1 Vul 4.75
M77 (Dumbbell Nebula)
η Sge 5.96
9 Vul 6.31
HIP95340 6.37
γ Sge 3.50
4 Vul 5.12
α Sge 4.37
NGC 6838
β Sge 4.37
Magnitude of stars
-2 to 1
2
3
4
5
6 to 7
Binary
Open cluster
Globular cluster
Cluster with nebulosity
Nebula
Galaxies
Milky Way

Scorpius

Coordinates of Brightest Star Antares (Mag 1.03): RA 16h 29m 24s, Dec -26° 26' 17"

Size of Constellation Boundaries 497 square degrees

Bounded by Ophiuchus, Libra, Lupus, Norma, Ara, Corona Australis, Sagittarius

Constellation Legend Scorpius represents the scorpion, sent by Zeus, which stung Orion the hunter to death, after he had boasted he could hunt down every animal on earth. Scorpius rises in the sky just as Orion sets, reflecting the mythological enmity between their two namesakes. It's one of the few constellations that convincingly resembles the thing it is named after. Mayan astronomers also associated the constellation with the scorpion, it being the form the god of hunting (or death, according to some) took in their culture. Polynesian astronomers, however, saw in it something much less sinister – the simple fish-hook.

Constellation Sights Antares means "rival of Mars" and is so called because of the star's reddish colour, which resembles that of our neighbouring planet. Scorpius features several double and multiple stars resolvable with binoculars, including Lesath, Sargas and Zeta Scorpii. It is also home to a number of prominent clusters, including the Butterfly Cluster (M6), not to be confused with the Butterfly Cluster in Puppis, and Ptolemy's Cluster (M7). Both of these are open clusters easily visible to the naked eye. M4, a globular cluster visible through binoculars, is one of the closest globular clusters to us, at 7200 light years away more or less. Despite its proximity, M4 is somewhat faint due to interstellar dust between it and the Earth, and it's best viewed on dark nights.

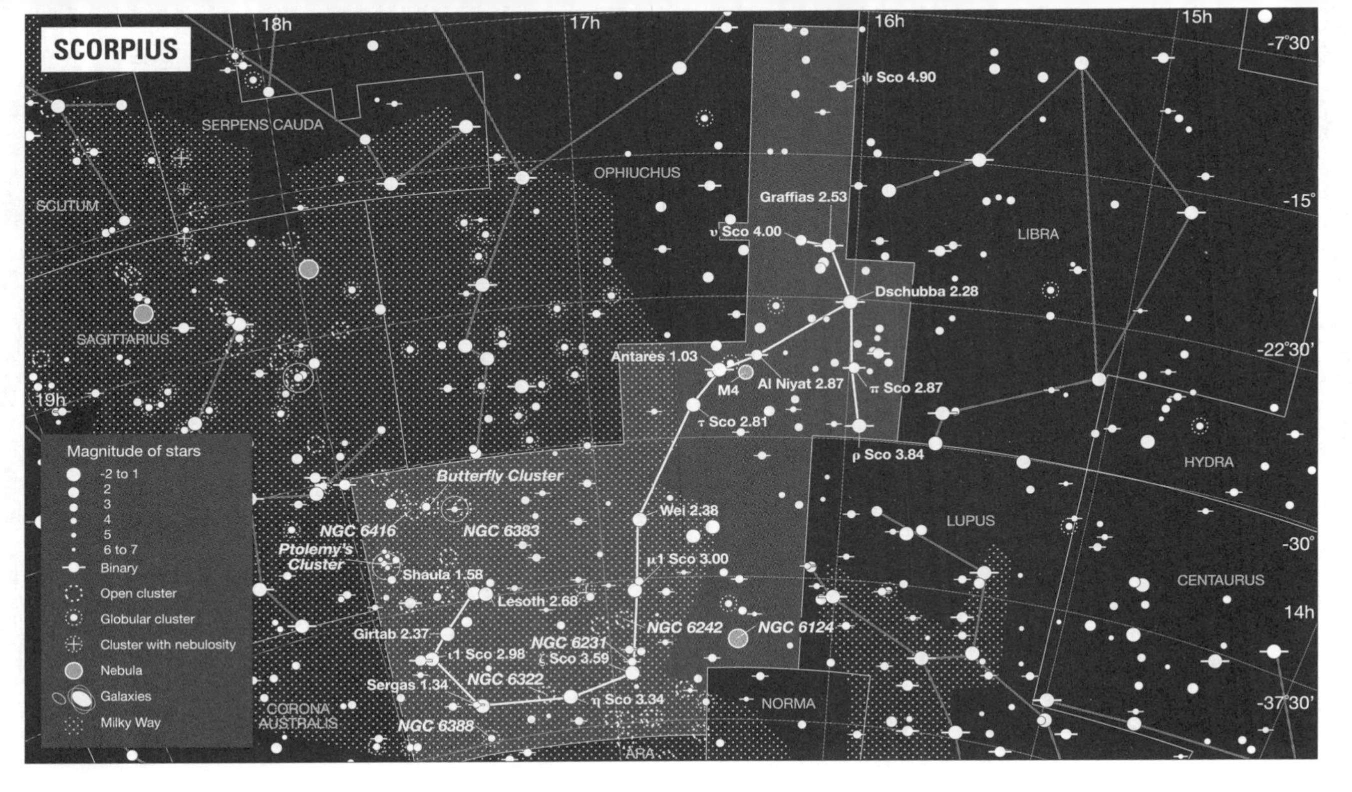
SCORPIUS
18h
17h
16h
15h
-7°30'
-15°
-22°30'
-30°
14h
-37°30'
19h
SERPENS CAUDA
OPHIUCHUS
SCUTUM
LIBRA
SAGITTARIUS
HYDRA
LUPUS
CENTAURUS
NORMA
ARA
CORONA AUSTRALIS
ψ Sco 4.90
Graffias 2.53
υ Sco 4.00
Dschubba 2.28
Antares 1.03
M4
Al Niyat 2.87
π Sco 2.87
τ Sco 2.81
ρ Sco 3.84
Butterfly Cluster
Wei 2.38
NGC 6416
NGC 6383
Ptolemy's Cluster
μ1 Sco 3.00
Shaula 1.58
Lesoth 2.68
Girtab 2.37
NGC 6242
NGC 6124
NGC 6231
ι1 Sco 2.98
ζ Sco 3.59
Sergas 1.34
NGC 6322
η Sco 3.34
NGC 6388
Magnitude of stars
-2 to 1
2
3
4
5
6 to 7
Binary
Open cluster
Globular cluster
Cluster with nebulosity
Nebula
Galaxies
Milky Way

Scutum & Serpens Cauda

Coordinates of Brightest Stars Alpha Scuti (3.84): RA 18h 35m 12s, Dec -8° 14' 35"; Eta Serpentis (3.21): RA 18h 21m 19s, Dec -2°, 53' 56"

Size of Constellation Boundaries 475 square degrees (Scutum); 637 square degrees (Serpens Cauda & Serpens Caput)

Bounded by Ophiuchus, Sagittarius, Aquila

Constellation Legend Scutum is an unusual constellation in that its name refers to an actual human being. John Sobieski was the king of Poland and a renowned military commander, who broke the Turkish siege of Vienna in 1683. Scutum was originally named Scutum Sobiescanium (Shield of Sobieski) and was added to the sky in honour of the king by the Polish astronomer Johannes Hevelius. Serpens Cauda (the "serpent's tail") is actually one portion of a single constellation known as Serpens; the other portion is called Serpens Caput (the "serpent's head"). The two halves are coiled around the constellation of Opiuchus.

Constellation Sights Scutum features two interesting variable stars: Delta Scuti, which fluctuates only a little (from magnitude 4.6 to 4.8), but does so very quickly and in just a few hours; R Scuti fluctuates rather more dramatically – over four orders of magnitude every 140 days or so. Scutum is also home to the Wild Duck Cluster (M11), an open cluster that features hundreds of stars. It's viewable in binoculars and shows up as a V-shape in small telescopes, hence its name. Serpens Cauda also features an interesting open cluster, NGC 6611, which is also known as M16. This cluster of stars is inside the Eagle nebula, and while it's possible to view the stars with binoculars, to get a good view of the nebula you'll need a large-aperture telescope.

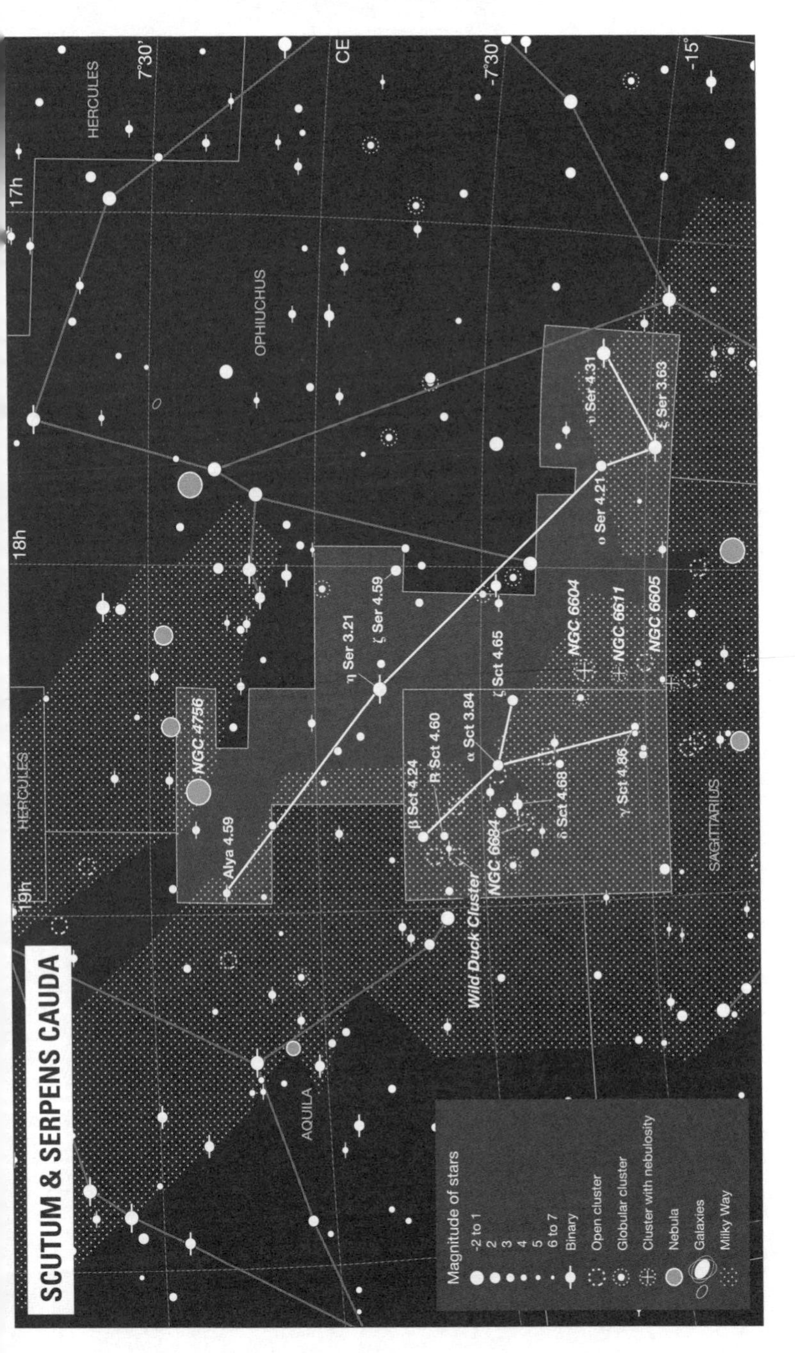
SCUTUM & SERPENS CAUDA
HERCULES
7°30'
17h
OPHIUCHUS
CE
-7°30'
-15°
18h
19h
HERCULES
NGC 4756
Alya 4.59
η Ser 3.21
ζ Ser 4.59
β Sct 4.24
R Sct 4.60
α Sct 3.84
ζ Sct 4.65
δ Sct 4.68
γ Sct 4.86
NGC 6604
NGC 6611
NGC 6605
ο Ser 4.21
ν Ser 4.31
ξ Ser 3.63
NGC 6684
Wild Duck Cluster
SAGITTARIUS
AQUILA
Magnitude of stars
-2 to 1
2
3
4
5
6 to 7
Binary
Open cluster
Globular cluster
Cluster with nebulosity
Nebula
Galaxies
Milky Way

Serpens Caput

Coordinates of Brightest Star Unukalhai (Mag 2.62): RA 15h 44m 16s, Dec 6° 25' 02"

Size of Constellation Boundaries 637 square degrees (Serpens Caput & Serpens Cauda)

Bounded by Corona Borealis, Boötes, Virgo, Libra, Ophiuchus, Hercules

Constellation Legend Serpens Caput (the Serpent's Head) is one half of the constellation Serpens (the other half being Serpens Cauda). Serpens is a unique constellation because it is the only one that is in two separate parts of the sky, lying on either side of Ophiuchus. Serpens was perceived as a snake held by Ophiuchus, and both constellations derive from Greek stories about Aesculepius, the god of healing. He had learned how to bring back men from the dead by watching a serpent revive a fellow snake using a special herb. Both Aesculepius and the snake were honoured with their own constellations.

Constellation Sights In the head portion of the serpent, the most notable sight is NGC 5904, also known as M5. It's a globular cluster that is just about visible to the eye in clear conditions, and makes for quite an impressive sight in a good telescope. Delta Serpentis is a binary star, resolvable in small telescopes.

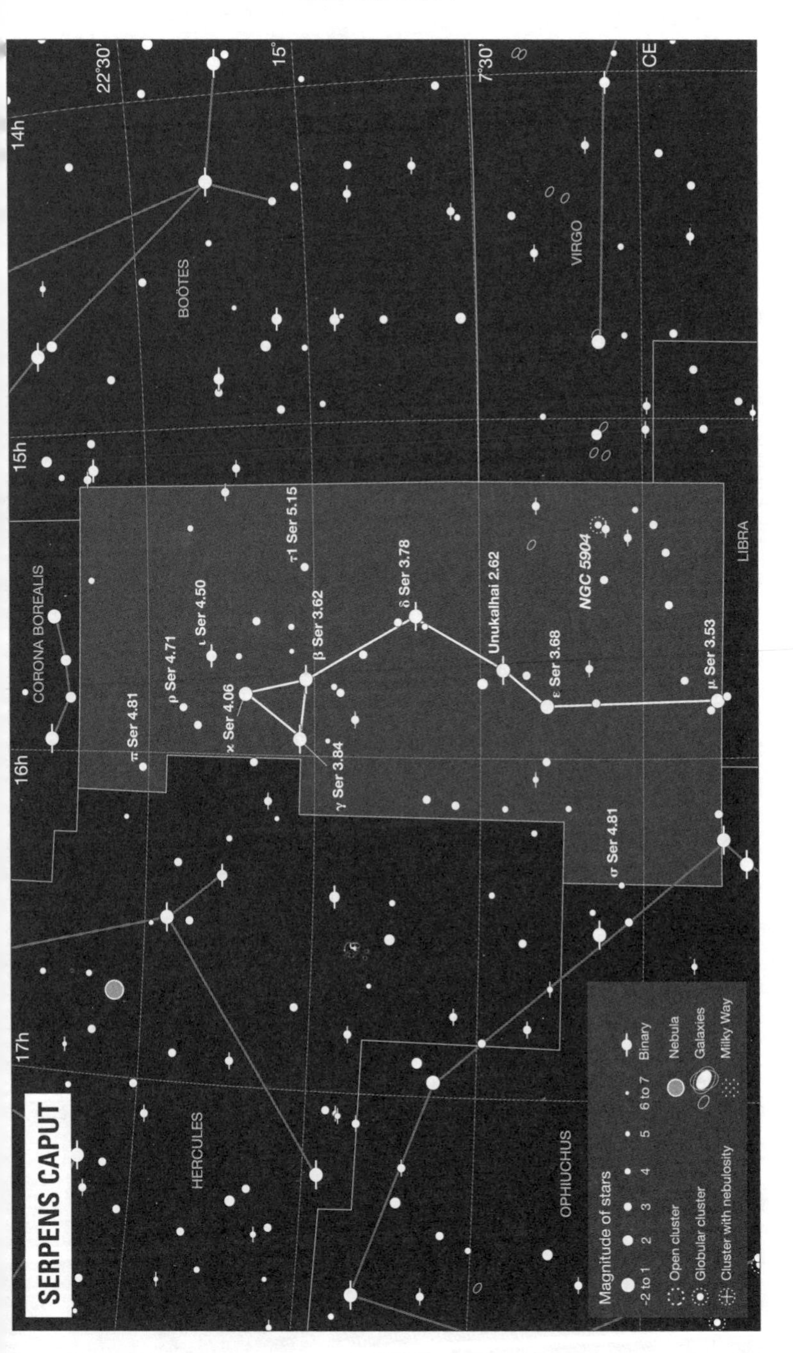
SERPENS CAPUT
14h
15h
16h
17h
22°30'
15°
7°30'
CE
BOÖTES
VIRGO
CORONA BOREALIS
LIBRA
HERCULES
OPHIUCHUS
τ1 Ser 5.15
δ Ser 3.78
Unukalhai 2.62
NGC 5904
ι Ser 4.50
β Ser 3.62
ε Ser 3.68
μ Ser 3.53
ρ Ser 4.71
κ Ser 4.06
π Ser 4.81
γ Ser 3.84
σ Ser 4.81
Magnitude of stars
-2 to 1
2
3
4
5
6 to 7
Binary
Open cluster
Globular cluster
Cluster with nebulosity
Nebula
Galaxies
Milky Way

Taurus

Coordinates of Brightest Star Aldebaran (Mag 0.84): RA 4h 35m 55s, Dec 16° 30' 53"

Size of Constellation Boundaries 797 square degrees

Bounded by Auriga, Perseus, Aries, Cetus, Eridanus, Orion, Gemini

Constellation Legend Taurus has been seen as a bull in many ancient cultures, including the Babylonians, Sumerians and Greeks. For the Greeks it referred to the myth of the princess Europa, whom Zeus seduced by taking on the form of a bull and carrying her across the sea to Crete. There Zeus returned to his proper form and Europa bore him three children. The constellation only shows half a bull, as its bottom half is supposedly under water. Within the constellation, the Pleiades and the Hyades also have their own affiliated stories. The Pleiades were seven daughters of the Titan Atlas who, while being pursued by Orion the hunter, were transformed by the gods into doves. The Hyades, also daughters of Atlas, were turned into stars in pity at their weeping for their dead brother Hyas; their appearance in the sky was said to bring rain.

Constellation Sights Taurus has several sights worthy of view. The Pleiades are easily the most famous open cluster in the sky. Sharp-eyed sky-watchers will be able to spot the "Seven Sisters", although binoculars and telescopes will reveal dozens more. The Hyades help to form the "V" in the constellation, and this entire cluster is best seen in binoculars. Aldebaran, the constellation's brightest star, appears to be in the Hyades but is actually some sixty light years closer to us. Taurus is also home to the Crab nebula, which has the honour of being M1 – the first object in the Messier catalogue. The Crab nebula is the remnant of a star that went supernova in 1054; it's discernible in binoculars or small telescopes but larger scopes are needed to properly appreciate it. Lambda Tauri is an eclipsing binary star that fluctuates by half a magnitude over the space of four days.

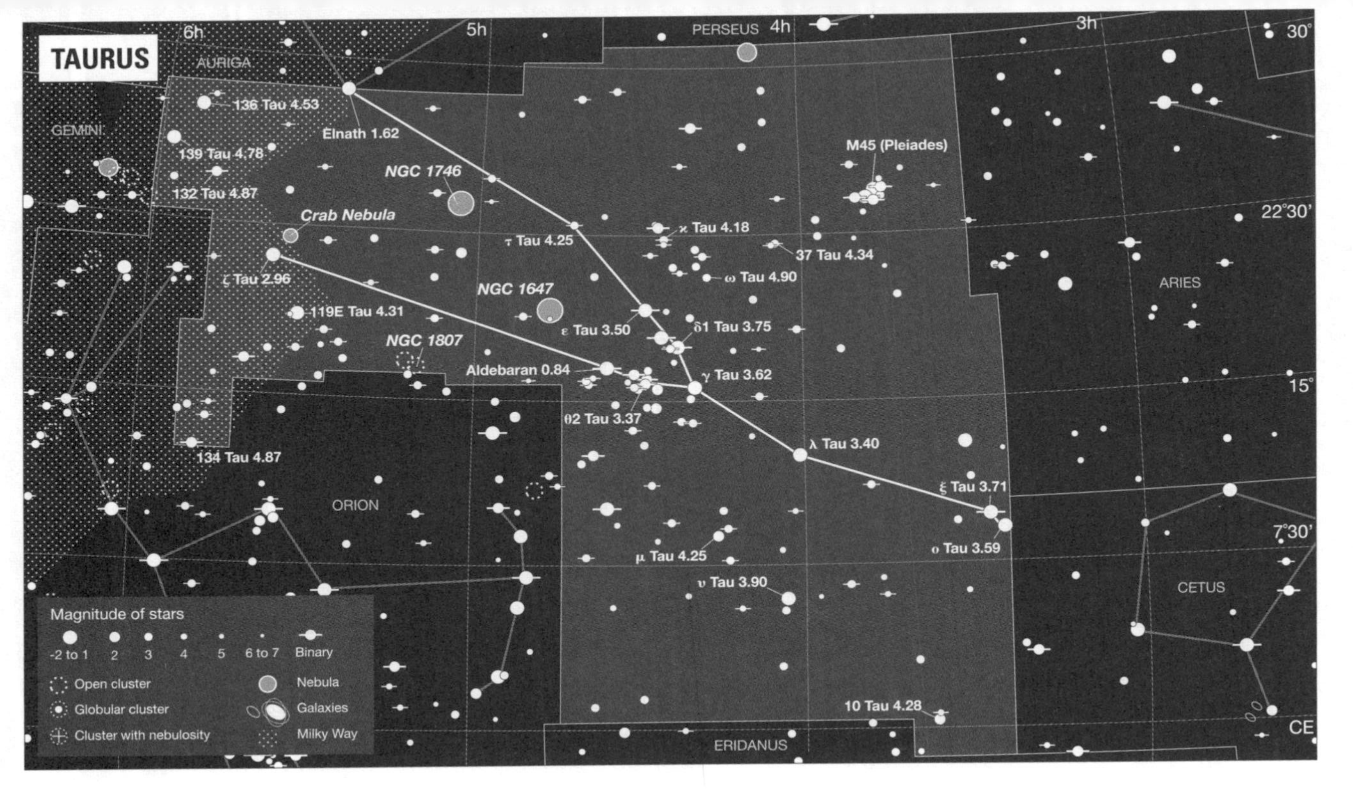
TAURUS
6h
5h
4h
3h
30°
22°30'
15°
7°30'
CE
PERSEUS
AURIGA
GEMINI
ARIES
ORION
CETUS
ERIDANUS
136 Tau 4.53
Elnath 1.62
139 Tau 4.78
132 Tau 4.87
NGC 1746
M45 (Pleiades)
Crab Nebula
τ Tau 4.25
κ Tau 4.18
37 Tau 4.34
ζ Tau 2.96
ω Tau 4.90
NGC 1647
119E Tau 4.31
ε Tau 3.50
δ1 Tau 3.75
NGC 1807
Aldebaran 0.84
γ Tau 3.62
θ2 Tau 3.37
λ Tau 3.40
134 Tau 4.87
ξ Tau 3.71
ο Tau 3.59
μ Tau 4.25
υ Tau 3.90
10 Tau 4.28
Magnitude of stars
-2 to 1
2
3
4
5
6 to 7
Binary
Open cluster
Nebula
Globular cluster
Galaxies
Cluster with nebulosity
Milky Way

Telescopium & Pavo

Coordinates of Brightest Stars Alpha Telescopii (Mag 3.46): RA 18h 26m 58s, Dec -45° 58' 04"; The Peacock Star (Mag 1.93): RA 20h 25m 39s, Dec -56° 43' 40"

Size of Constellation Boundaries 252 square degrees (Telescopium); 378 square degrees (Pavo)

Bounded by Sagittarius, Corona Austrinus, Ara, Apus, Octans, Indus, Microscopium

Constellation Legend Pavo represents a peacock, and was added to the sky by the sixteenth-century Dutch explorers Pieter Dirkszoon Keyser and Frederick de Houtman. It may be that they were following the tradition of mythic reference with the name, since there is a Greek legend attached to the peacock (although it is not specific to the Pavo constellation). Argus, who had one hundred eyes, was charged by the goddess Hera to watch over a white heifer that she suspected was really the nymph Io, transformed by Io's lover (and Hera's husband) Zeus. At Zeus's request, the god Hermes lulled Argus to sleep by playing sweet music on his lyre, and then decapitated him. Hera then placed Argus's hundred eyes upon the tail of the peacock, her sacred bird. Telescopium is another of Nicolas Louis de Lacaille's southern constellations, which he named after scientific instruments with monotonous regularity. Inevitably, he finally got round to the telescope.

Constellation Sights Telescopium has few sights of any real interest. Delta Telescopii is a double star sharp eyesight resolvable with, but that's about all there is that's worth seeing. Pavo is slightly more interesting, and features NGC 6752, a bright globular cluster. In good conditions, it's visible to the naked eye, and it's a very nice sight in telescopes.

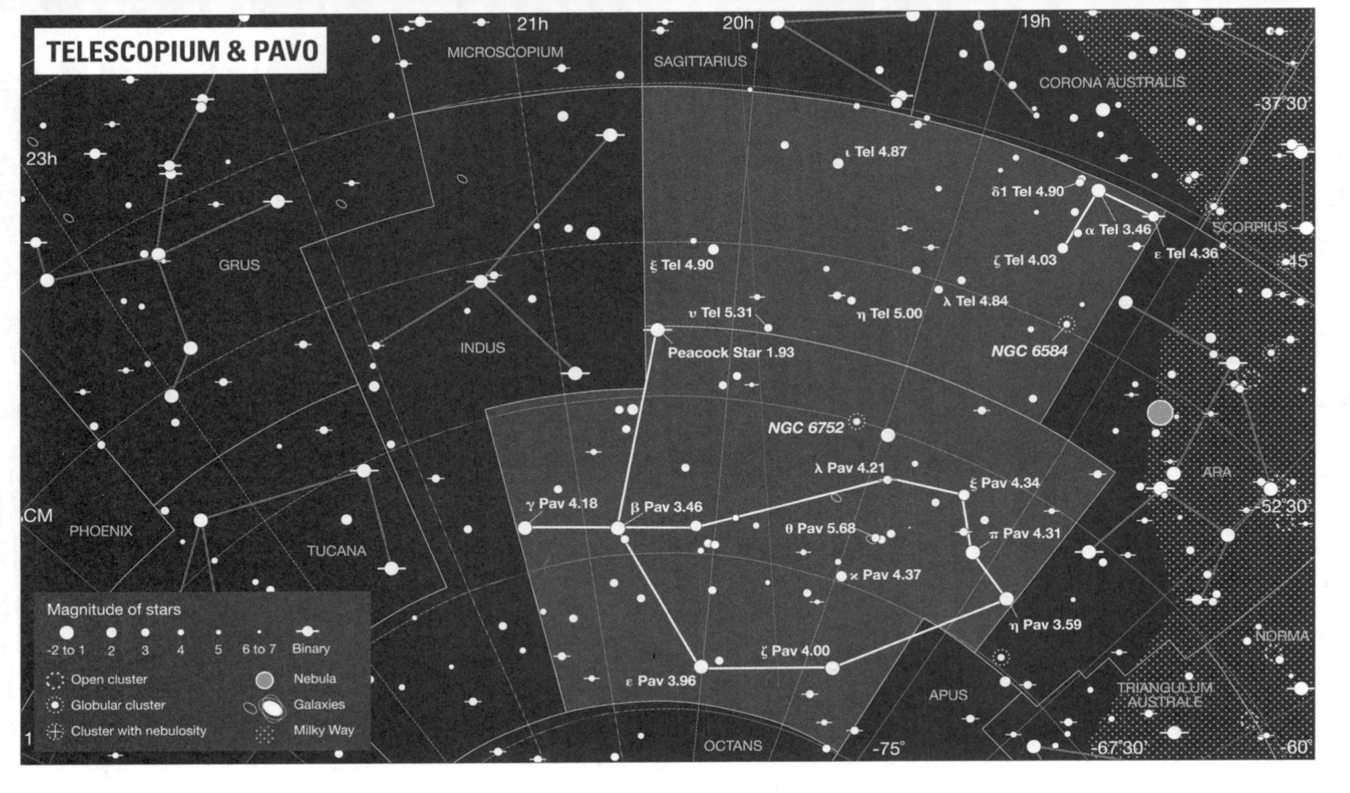
TELESCOPIUM & PAVO
21h
20h
19h
23h
MICROSCOPIUM
SAGITTARIUS
CORONA AUSTRALIS
GRUS
INDUS
PHOENIX
TUCANA
SCORPIUS
ARA
NORMA
TRIANGULUM AUSTRALE
APUS
OCTANS
CM
-37°30'
-45°
-52°30'
-60°
-67°30'
-75°
ι Tel 4.87
δ1 Tel 4.90
α Tel 3.46
ζ Tel 4.03
ε Tel 4.36
ξ Tel 4.90
λ Tel 4.84
υ Tel 5.31
η Tel 5.00
NGC 6584
Peacock Star 1.93
NGC 6752
λ Pav 4.21
ξ Pav 4.34
γ Pav 4.18
β Pav 3.46
θ Pav 5.68
π Pav 4.31
κ Pav 4.37
η Pav 3.59
ζ Pav 4.00
ε Pav 3.96
Magnitude of stars
-2 to 1
2
3
4
5
6 to 7
Binary
Open cluster
Nebula
Globular cluster
Galaxies
Cluster with nebulosity
Milky Way

Triangulum

Coordinates of Brightest Star Beta Trianguli (Mag 3.0): RA 2h 9m 44s, Dec 34° 59' 57"

Size of Constellation Boundaries 132 square degrees

Bounded by Andromeda, Pisces, Aries, Perseus

Constellation Legend This is an old constellation, and is commonly associated with Sicily, the triangular island off the tip of Italy. According to one Roman myth, the goddess Ceres, guardian divinity of Sicily and goddess of the harvest, entreated her brother Jupiter to honour the island with a place in the sky. To the Greeks the constellation was Deltoton because its three main stars form a semblance of the Greek capital letter delta. Egyptian astronomers also thought of these stars as a delta, but in the sense of the delta of the River Nile, and they accorded it the title "Home of the Nile". The first known asteroid, Ceres, was discovered within this constellation in 1801.

Constellation Sights Triangulum houses the Triangulum Pinwheel galaxy (M33), a member of the Local Group of galaxies, which also includes the Milky Way and Andromeda galaxies. The galaxy can be spotted by the naked eye only in exceptionally good viewing conditions, making it the most distant object that can be seen unaided. To all intents and purposes, it's best viewed in binoculars or a small telescope, which can handle its low overall brightness.

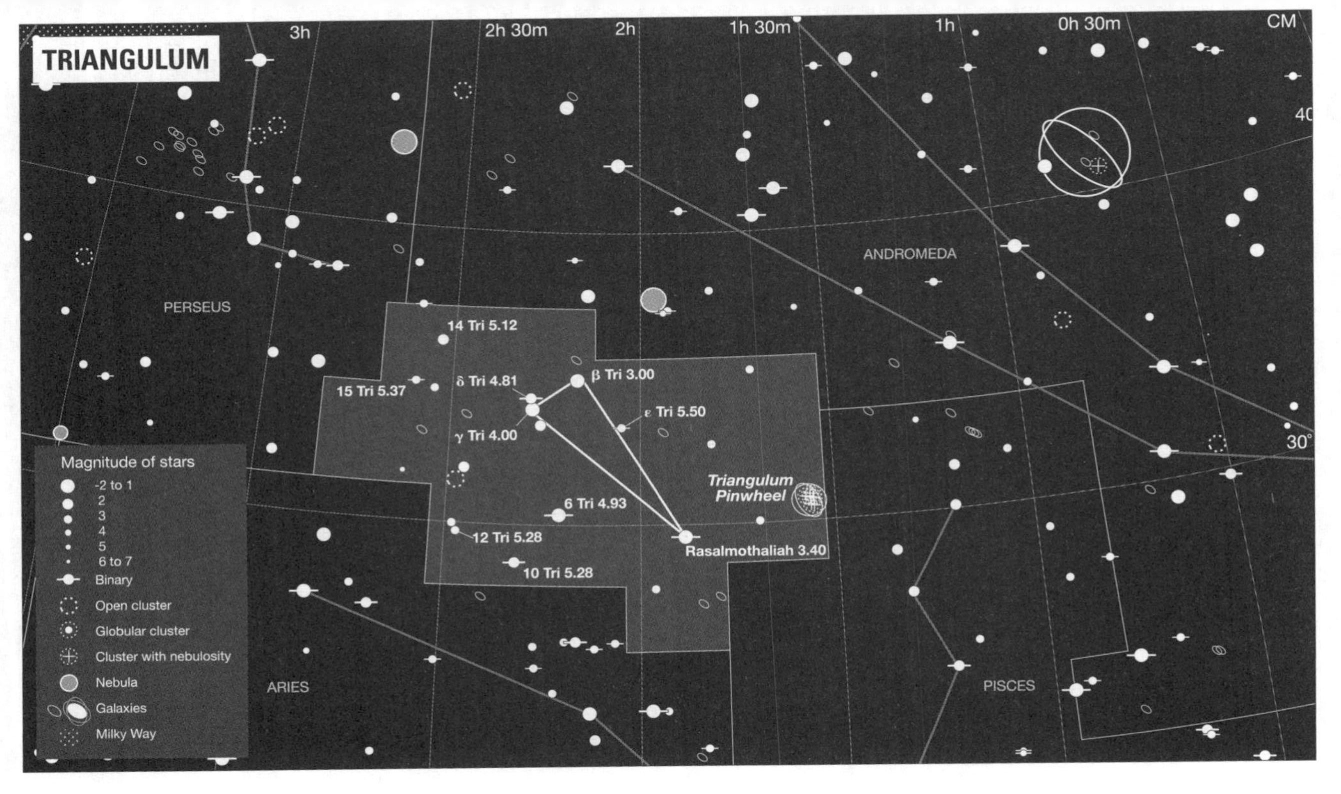

TRIANGULUM
3h
2h 30m
2h
1h 30m
1h
0h 30m
CM
40
30°
PERSEUS
ANDROMEDA
ARIES
PISCES
14 Tri 5.12
15 Tri 5.37
δ Tri 4.81
β Tri 3.00
ε Tri 5.50
γ Tri 4.00
Triangulum Pinwheel
6 Tri 4.93
12 Tri 5.28
Rasalmothaliah 3.40
10 Tri 5.28
Magnitude of stars
-2 to 1
2
3
4
5
6 to 7
Binary
Open cluster
Globular cluster
Cluster with nebulosity
Nebula
Galaxies
Milky Way

Ursa Major

Coordinates of Brightest Star Dubhe (Mag 1.78): RA 11h 3m 44s, Dec 61° 44' 16"

Size of Constellation Boundaries 1280 square degrees

Bounded by Camelopardalis, Lynx, Leo Minor, Leo, Canes Venatici, Boötes, Draco

Constellation Legend Ursa Major (the Great Bear) relates to the story of Callisto, the daughter of King Lycaon of Arcadia. She was one of the attendants of Artemis (Diana), goddess of the hunt and of chastity, and was one of Zeus's many lovers. Hera, Zeus's jealous wife, transformed Callisto into a bear (although in some versions of the story, it's Artemis who does this). Later when their son, Arcas, started to hunt (not knowing who she was), Zeus placed her among the stars where she would forever be safe. Ursa Major has a long tail because Zeus had to tug on it to swing her into the sky. Interestingly, Native Americans saw the constellation as seven hunters (corresponding to The Plough) pursuing a bear. It is said that many of their tribes used The Plough to test whether a young man was fit to become a warrior: if he could describe the double star in the centre of its "handle", then his eyes were sharp enough to make him a worthy hunter.

Constellation Sights The most prominent feature of Ursa Major is The Plough, an asterism that's better known than the constellation itself. Over the years, across different cultures it's been thought of as a saucepan, a ladle, a wagon and a chariot, while Americans know it as The Big Dipper. Mizar, the second star in The Plough's handle, is a visual double with a smaller star known as Alcor, but these two aren't related. A small telescope will spot another star twinned with Mizar. Bode's galaxy (M81), a spiral, is one of the brightest galaxies in the northern sky at seventh magnitude, and is worth viewing through binoculars and telescopes. M82, an irregular galaxy, is a close companion to Bode's galaxy. The Pinwheel galaxy (M101) is a face-on spiral galaxy, which you can spot with a small telescope in favourable conditions.

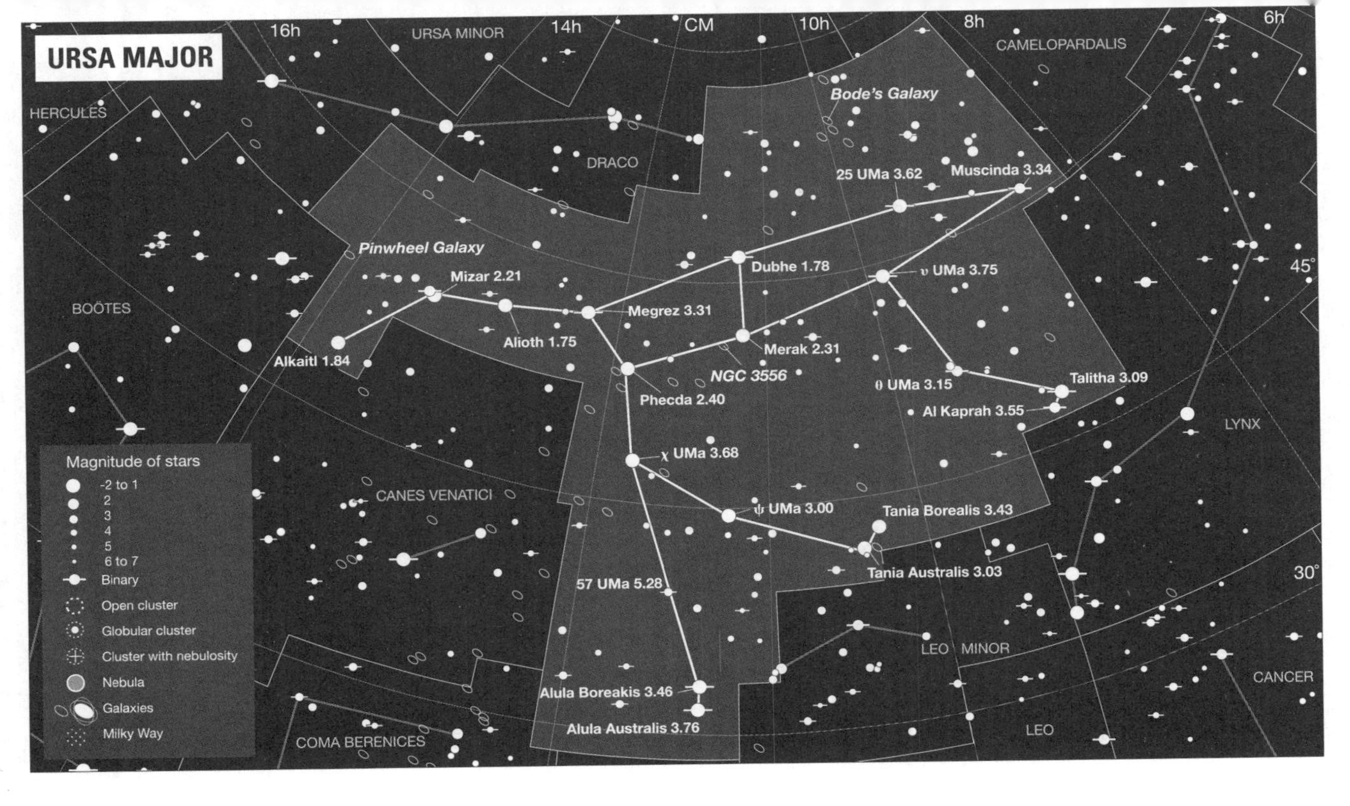
URSA MAJOR
16h
14h
CM
10h
8h
6h
URSA MINOR
CAMELOPARDALIS
HERCULES
DRACO
Bode's Galaxy
25 UMa 3.62
Muscinda 3.34
Pinwheel Galaxy
Dubhe 1.78
υ UMa 3.75
Mizar 2.21
BOÖTES
Megrez 3.31
Alioth 1.75
Merak 2.31
Alkaitl 1.84
NGC 3556
θ UMa 3.15
Talitha 3.09
Phecda 2.40
Al Kaprah 3.55
45°
LYNX
χ UMa 3.68
Magnitude of stars
-2 to 1
2
3
4
5
6 to 7
Binary
Open cluster
Globular cluster
Cluster with nebulosity
Nebula
Galaxies
Milky Way
CANES VENATICI
ψ UMa 3.00
Tania Borealis 3.43
Tania Australis 3.03
57 UMa 5.28
30°
LEO MINOR
CANCER
Alula Boreakis 3.46
Alula Australis 3.76
COMA BERENICES
LEO

Ursa Minor

Coordinates of Brightest Star Polaris (Mag 1.96): RA 2h 31m 49s, Dec 89° 16' 31"

Size of Constellation Boundaries 256 square degrees

Bounded by Cepheus, Camelopardalis, Draco

Constellation Legend Ursa Minor translates as the "little bear" and there are several stories connected to it. The Greeks held that Ursa Minor was originally Arcas, the son of Callisto, a nymph of Artemis who had been turned into a bear by the jealous goddess Hera. As Callisto roamed the forests in her new form she came across her son and, overwhelmed with joy, bounded over to embrace him. Arcas, having no idea that this bear was his mother, was about to kill her when Zeus intervened, turning him into a bear as well, and setting them both in the sky where they could be properly reunited. The only problem with this interpretation is that the constellation Boötes is also strongly associated with Zeus and Callisto's son. Ursa Minor is the northernmost constellation in the sky and the home of the current Pole Star, Polaris.

Constellation Sights Polaris is the big draw here, a reasonably bright star (second magnitude) located about a degree from the celestial north pole. Its close proximity is merely a coincidence. Polaris is currently inching ever closer to the pole and will reach its closest point in 2102. It will then – over a period of thousands of years – drift away from the pole, as a result of precession. Aside from Polaris, other attractions include Pherkad and Eta Ursae Minoris, both of which are optical double stars.

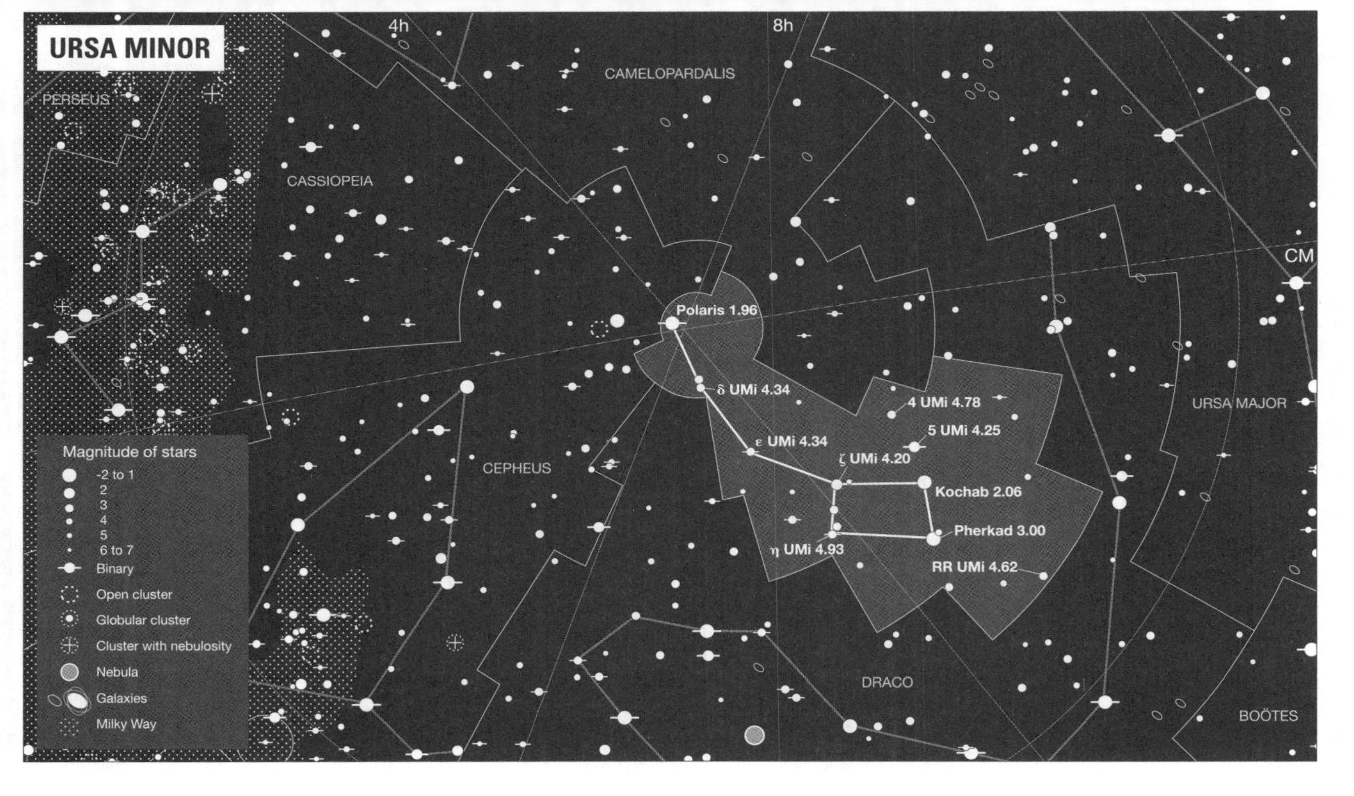

URSA MINOR
4h
8h
CAMELOPARDALIS
PERSEUS
CASSIOPEIA
CM
Polaris 1.96
δ UMi 4.34
4 UMi 4.78
5 UMi 4.25
ε UMi 4.34
ζ UMi 4.20
Kochab 2.06
Pherkad 3.00
η UMi 4.93
RR UMi 4.62
URSA MAJOR
CEPHEUS
DRACO
BOÖTES
Magnitude of stars
-2 to 1
2
3
4
5
6 to 7
Binary
Open cluster
Globular cluster
Cluster with nebulosity
Nebula
Galaxies
Milky Way

Vela

Coordinates of Brightest Star Suhail al Muhlif (Mag 1.75): RA 8h 9m 32s, Dec -47° 20' 35"

Size of Constellation Boundaries 500 square degrees

Bounded by Antlia, Pyxis, Puppis, Carina, Centaurus

Constellation Legend Vela represents the sail in the obsolete constellation Argo Navis, the famous ship in which the Greek hero Jason sailed with his companions the Argonauts during his many adventures. Argo Navis was informally divided up into three separate constellations by the French astronomer Nicolas Louis de Lacaille in his sky chart of 1763, and was officially split into four in 1929. The other constellations that once comprised Argo Navis are Carina (the Keel), Pyxis (the Compass) and Puppis (the Stern).

Constellation Sights The Omicron Velorum Cluster (IC 2391) is an easy sight for the naked eye and is also a good sight through binoculars. NGC 2547 is likewise an impressive open cluster when viewed with binoculars. The Eight Burst nebula (NGC 3132) is a planetary nebula that will appear as a disc in a small scope. Among the stars, Suhail al Muhlif is a multiple star that resolves into four stars in a small telescope. The brightest star in the constellation, Suhail al Muhlif is yet another binary, one of whose stars is a Wolf-Rayet star which is ejecting its outer layers into space, and one day may go supernova.

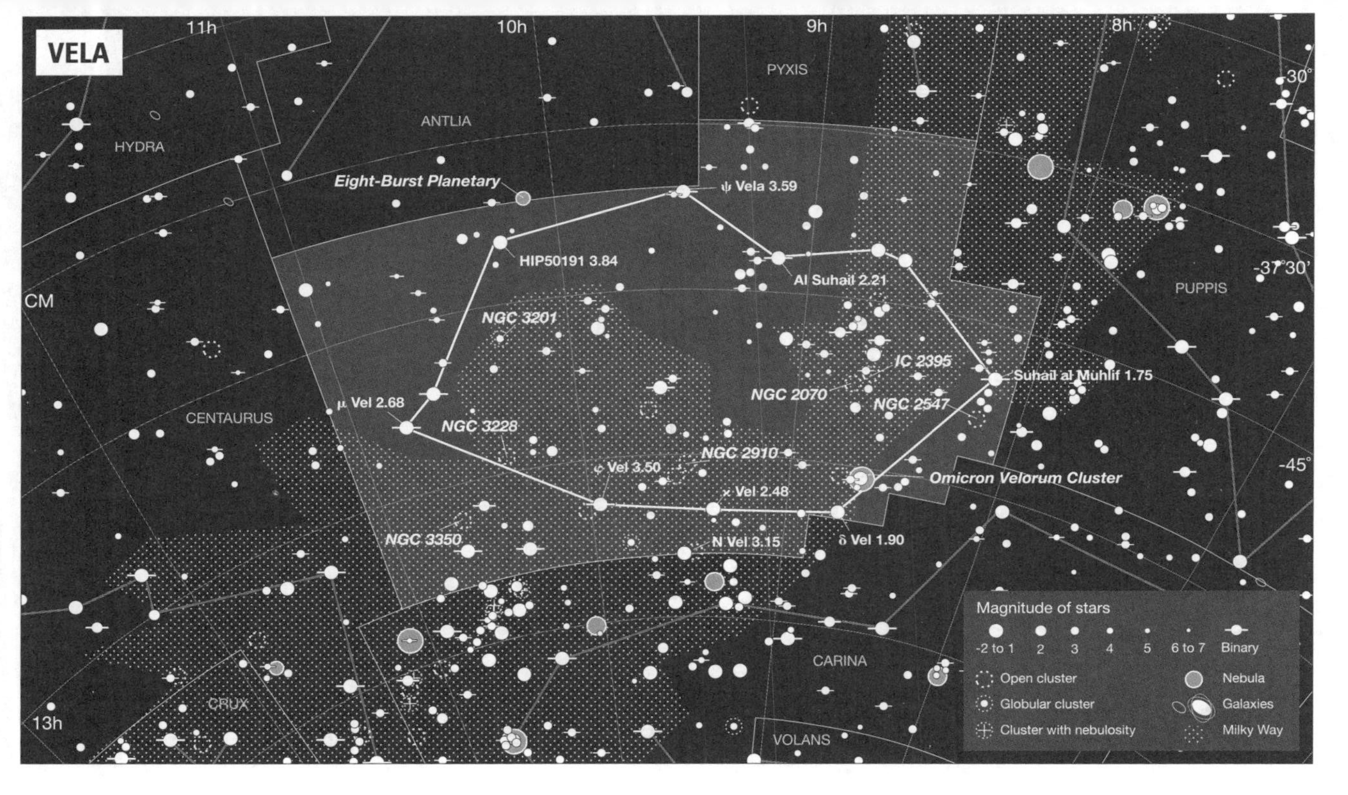
VELA
11h
10h
9h
8h
HYDRA
ANTLIA
PYXIS
-30°
Eight-Burst Planetary
ψ Vela 3.59
HIP50191 3.84
Al Suhail 2.21
-37°30'
PUPPIS
CM
NGC 3201
IC 2395
Suhail al Muhlif 1.75
NGC 2070
NGC 2547
μ Vel 2.68
CENTAURUS
NGC 3228
NGC 2910
φ Vel 3.50
-45°
Omicron Velorum Cluster
κ Vel 2.48
NGC 3350
N Vel 3.15
δ Vel 1.90
Magnitude of stars
-2 to 1
2
3
4
5
6 to 7
Binary
CARINA
Open cluster
Nebula
Globular cluster
Galaxies
Cluster with nebulosity
Milky Way
CRUX
13h
VOLANS

Virgo

Coordinates of Brightest Star Spica (Mag 0.96): RA 13h 37m 13s, Dec -11° 10' 27"

Size of Constellation Boundaries 1294 square degrees

Bounded by Coma Berenices, Leo, Crater, Corvus, Hydra, Libra, Serpens Caput, Boötes

Constellation Legend The constellation Virgo (the Virgin) is associated with several female deities. The fact that it is sometimes depicted as a figure holding wheat, and that Spica (Virgo's brightest star) means "ear of wheat" suggests a strong connection with Demeter (Ceres), the Greek goddess of the harvest. This is reinforced by the related connection to Atargatis, the Phoenician goddess of fertility. Another interpretation links the constellation to Astraea, the goddess of justice (and connects Virgo with Libra, the scales of justice). Virgo is the largest constellation of the Zodiac, and the sun spends over six weeks within its boundaries, the most time spent in any sign.

Constellation Sights Virgo is the home of the Virgo Cluster, a vast grouping of galaxies in the northwest of the constellation, and the centre of the Virgo Supercluster, of which our own galaxy is a part. As such, there are several prominent galaxies to view here, including The Smoking Gun (M87), a huge elliptical galaxy that is also a giant radio source. The galaxy also features huge gas jets, which unfortunately you won't be able to view with home telescopes. Another famous galaxy is the Sombrero galaxy (M104), an edge-on spiral whose name is derived from its unusual flattened shape. Spica, Virgo's brightest star, is the sixteenth brightest star in the night sky.

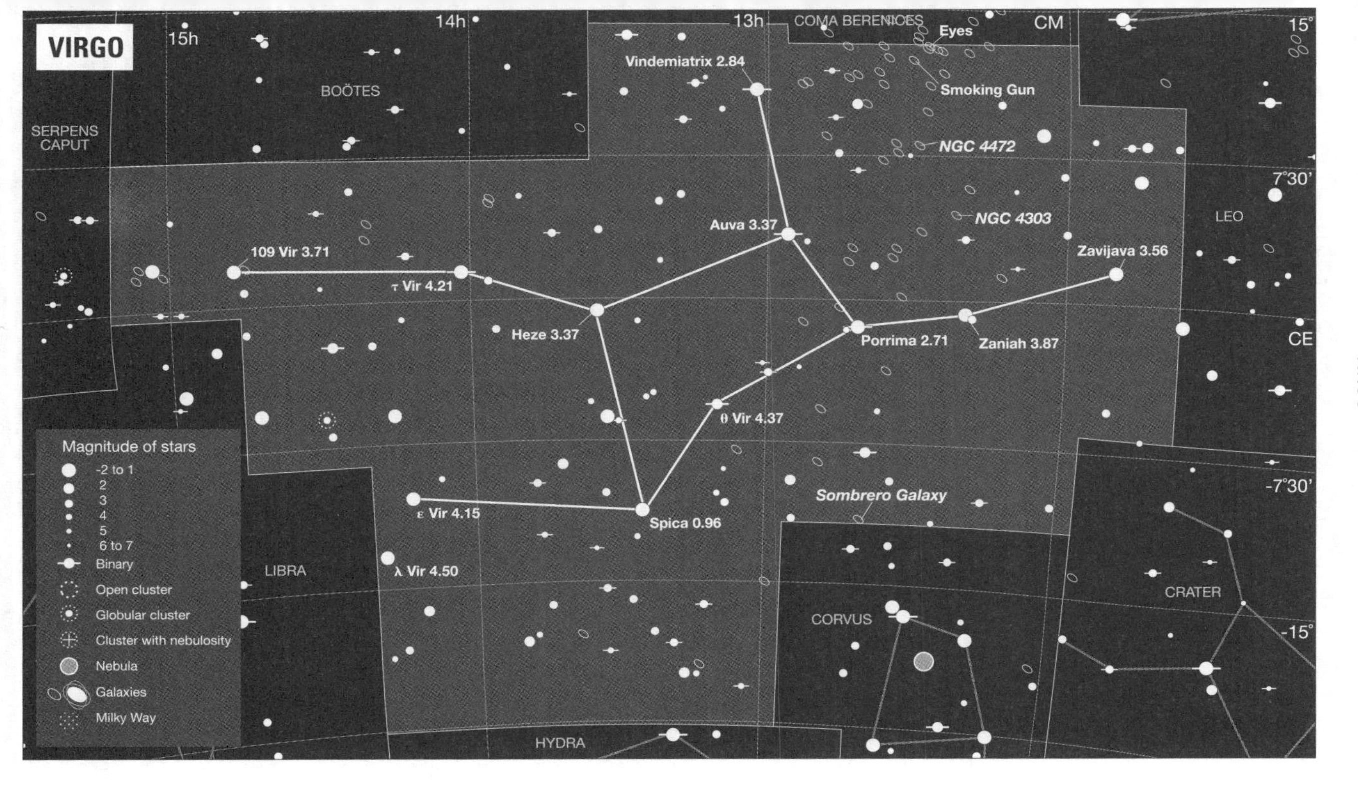
VIRGO
15h
14h
13h
COMA BERENICES
Eyes
CM
15°
BOÖTES
SERPENS CAPUT
Vindemiatrix 2.84
Smoking Gun
NGC 4472
7°30'
LEO
NGC 4303
Auva 3.37
109 Vir 3.71
τ Vir 4.21
Zavijava 3.56
Heze 3.37
Porrima 2.71
Zaniah 3.87
CE
θ Vir 4.37
-7°30'
Sombrero Galaxy
ε Vir 4.15
Spica 0.96
λ Vir 4.50
LIBRA
CRATER
CORVUS
-15°
HYDRA
Magnitude of stars
-2 to 1
2
3
4
5
6 to 7
Binary
Open cluster
Globular cluster
Cluster with nebulosity
Nebula
Galaxies
Milky Way

RESOURCES

Resources

Books

The *Rough Guide to the Universe* is intended to help you get started in the world of astronomy and the observation of the night sky. When you're ready to head out into deeper waters, here are ten books – from the thousands available – you should find helpful.

General astronomy

Universe: The Definitive Visual Guide (Dorling Kindersley, 2005). In addition to good basic information about the universe we live in, this large, multi-author coffee-table book is lavishly illustrated with the visual wonders of the cosmos, which makes it a good choice to leave out, allowing casual readers to get sucked into the beauty of the cosmos.

Brian Greene The Elegant Universe: Superstrings, Hidden Dimensions, and the Quest for the Ultimate Theory (Vintage, 2000). For those who want to dive into some of the concepts surrounding string theory (which posits that everything in the universe is created from tiny, vibrating 11-dimensional strings), this companion book to an acclaimed television series is an excellent place to start. It's heavy stuff, but explained well.

Stephen Hawking The Illustrated Brief History of Time, Updated and Expanded Edition (Bantam, 1996). The world's most famous living astrophysicist compresses fantastically strange concepts into writing that's (mostly) easy to follow. This edition updates the text and adds hundreds of interesting illustrations. Hawking's recent sequel to *Time, The Universe in a Nutshell*, is also well worth reading.

Philip Plait Bad Astronomy: Misconceptions and Misuses Revealed, from Astrology to the Moon Landing "Hoax" (John Wiley & Sons, 2002). A companion book to the well-regarded website, this book handily dismantles some of the common conspiracy theories and myths surrounding astronomy and cosmology. A fun read but also a fine reference book for the next time you get into a debate down the pub about life, the universe and everything.

Carl Sagan Cosmos (Ballantine, 1980). The science in this classic book is now a bit dated (it was originally published more than twenty years ago), but Sagan's enthusiasm for astronomy comes through in clear, compelling and engaging prose. A great "starter" book for people looking to start a love affair with the night sky and everything in it.

Advanced amateur sky-watching

Patricia L. Barnes-Svarney and Michael R. Porcellino Through the Telescope: A Guide for the Amateur Astronomer (McGraw-Hill, 1999). This comprehensive book for amateur stargazing includes quite a bit of detail on telescopes and how to optimize their use, and also presents detailed information about each of the planets and other objects you can spot in the night sky.

Michael A. Covington Astrophotography for the Amateur (Cambridge University Press, 1999). This book offers an excellent overview of astrophotography – taking pictures of the objects you see in your telescopes – covering the media from traditional film to the latest computer imaging. Very useful if you're wishing to document what you see in the sky, rather than just observe it.

David H. Levy The Sky: A User's Guide (Cambridge University Press, 1991). The author is one of the most famous amateur astronomers around (he was the co-discoverer of Shoemaker-Levy 9, the comet that impacted on Jupiter), and this book of sky watching is based on decades of personal experience. Useful, complete, and written with an enthusiast's sensibility.

Jay M. Pasachoff Stars and Planets (Houghton Mifflin, 1990). In terms of pure "field guides" (astronomy books intended specifically for use while you're out observing), this is one of the most complete, with detailed star charts, extensive maps of the Moon, charts of planetary positions in the night sky and more.

Fred Schaaf 40 Nights to Knowing the Sky (Owl Books, 1998). This book takes a systematic approach to learning about and enjoying the night sky, giving its reader 40 different projects to undertake with telescopes, binoculars and their own eyes. These include spotting artificial satellites, tracking the planets, and observing deep-sky objects.

Websites

The Web is arguably a more useful tool to the amateur astronomer than even a good telescope – but there is so much astronomical information available out there that it can be hard to know where to begin. So to get you started, here are some of the best sites available on and about astronomy and observing. This list is by no means complete, so your own explorations starting off from these sites will undoubtedly lead you to new discoveries both online and in the night sky.

News, magazines and information

These are the sites that will keep you up to date on the latest events in astronomy and space exploration news, and act as quick resources for your basic astronomical questions. Many of these sites also act as "portals" to point you to other information on the Web.

Astronomy.com www.astronomy.com Home to the US magazine *Astronomy*, this is a good astronomy portal featuring news, articles from the magazine, and features and links for hobbyists and amateur astronomers.

Astronomy Now www.astronomynow.com *Astronomy Now* is the UK's top-selling astronomy magazine, and its website puts top space and astronomy stories front and centre. Other areas of the site seem less frequently updated.

BBCi Space www.bbc.co.uk/science/space This flashy site is rich in top-flight content, with news, polls, features and interactive galleries. It's a little overdesigned – it looks cool at the expense of easy navigation – but it's well worth digging through.

CNN Science & Space www.cnn.com/TECH/space/archive The CNN site isn't pretty (it's designed to be blandly utilitarian), but as a quick-hit stop for the latest news in space and science in general, it's hard to beat. In addition to news stories the site also includes NASA Webcams and the dates of important space launches.

Sci.Astro FAQs sciastro.astronomy.net "Sci.Astro" is an Internet newsgroup frequented by both amateur and professional astronomers alike, and its "FAQs" ("frequently asked questions") on astronomical issues are well researched and very useful. Topics range from basic information on the solar system to deeper discussions of cosmology and extraterrestrial life.

Science@NASA science.nasa.gov This site keeps you current with the latest news and accomplishments from NASA, and presents it in an engaging format that doesn't smell like just another press release.

Sky and Telescope www.skyandtelescope.com The online home to the US magazine of the same name. The design is nothing to shout about but content is good, and focuses on what's out there to see in the night sky and how best to see it.

SkyNews www.skynewsmagazine.com Canada has its own stargazing magazine, and its accompanying website is primarily concerned with promoting it; there are a number of useful links, including news updates and resources.

Space.com www.space.com The big boy on the online block, this site was designed from the ground up to be all about space. The current design of the site is not especially good – it emphasizes a few "hot" stories over the site's wealth of features – but wander site a bit and you'll find just about everything you need. It also provides links to a number of other astronomy and science sites.

SpaceDaily www.spacedaily.com Another news site designed with utility in mind, SpaceDaily lists dozens of space-related news stories on its front page every day. It also tracks some interesting topics in subsections, particularly its "Dragon Watch", which keeps tabs on Chinese space efforts.

Telescopes and instruments

If you haven't got kitted up yet, here are some of the leading telescope manufacturers in the US and UK. Note that their appearance here is not an endorsement of one brand over another. It's a good idea to shop around to find the telescope that fits your needs and budget.

Celestron www.celestron.com Prominent US maker of telescopes and binoculars whose site features product information and telescope and binocular basics.

Meade www.meade.com Meade positions itself as the world's leading manufacturer of amateur telescopes. This primary site features links to the company's international sites as well as product information.

Orion Telescopes (US) www.telescope.com This nicely designed site features easily accessible astronomical information as well as the usual information on its telescopes, binoculars and accessories.

Orion Optics (UK) www.orionoptics.co.uk This telescope manufacturer is unrelated to the similarly-named US firm; the site offers basic information on its products and accessories.

Konus www.konus.com This Italian telescope-maker has recently begun selling its products in the US. The site is unwieldy but offers information on its telescopes and other scientific products.

Stellarvue Telescopes www.stellarvue.com Stellarvue prides itself on the fact that it makes each of its telescopes by hand. The site features products and a brief essay for people who are new to sky watching.

AST Optics www.astoptics.co.uk AST Optics offers free delivery within the UK on many of its products. This very basic site features price information on telescopes and accessories.

ATM Resource List home.cfl.rr.com/bobnpam If you want to put your telescope together with your own bare hands, this is the site for you.

Astronomical software

Today's astronomical software can be extremely useful in planning a night's observing activities – and indeed, can be a night of astronomical fun and learning in itself. Here is some of the best commercial software available, and some of the best freeware (free software).

Starry Night www.starrynightstore.com Starry Night was used to help make the star charts in this book. The software comes in three varieties, each offering varying levels of information on the night sky, with the "Pro" version packed with an almost ridiculous amount of features. Very complete and highly recommended.

AstroTips www.stargazing.net/AstroTips/english For those of you who'd prefer your astronomical software to be free, this site lists a couple of dozen freeware, shareware and demonstration model programs. Perhaps unsurprisingly, the quality of these packages varies.

Google Earth earth.google.com What does Google's planet-mapping program have to do with astronomy? Well recent versions of the software also include a comprehensive planetarium option that does an excellent job covering the basics of the stars, planets and other observable phenomena in the night sky. And, as with most things Google, it's free for anyone to download and fiddle with.

Home Planet www.fourmilab.ch/homeplanet One of the best-known freeware planetarium packages out there.

Sky View Café www.skyviewcafe.com When you enter details of your geographical location, this free online brower-based site tells you what's happening in your local night sky. It also offers star charts, a 3-D orrery (model of the solar system), displays of the moons of Jupiter and Saturn, and an astronomical event calendar.

Software Bisque www.bisque.com Software Bisque offers the well-regarded The Sky software, which comes in three editions; depending which version you get, the database includes information on a billion stars and deep-sky objects, thousands of "thumbnail" images of deep-sky objects, and computerized telescope support.

Southern Star Systems www.southernstars.com This company offers SkyChart, which has many of the features of high-end astronomical software, including a large star and deep-sky database, and computerized telescope support, all for a very competitive price. It also offers a downloadable demo version for free.

The night sky

What is out there when you look up? This collection of sites samples it all, from near-Earth encounters with asteroids to the possibility of contact with intelligent alien life.

American Meteor Society amsmeteors.org This site keeps you in the loop for which comets are travelling through the sky and which meteor showers you should plan for.

AstronomyDaily.com www.astronomydaily.com Another site that provides you with stargazing information for your specific area (including a weather forecast) when you enter your geographical location. A very useful site.

Campaign for Dark Skies www.darkskies.org/introq A UK organization is which is battling to keep the skies of Britain dark enough for good stargazing.

The Constellations www.astronomical.org Detailed information on constellations, their stars and deep-sky objects, from the Peoria Astronomical Society.

The Extra-Solar Planets Encyclopedia vo.obspm.fr/exoplanetes/encyclo More than 260 planets orbiting other stars have been discovered. This is the site to find out where they are; it also provides information on how the search for more planets is happening.

International Dark Sky Association www.darksky.org This organization is fearlessly trying to convince people all over the world to point their lights down towards the ground, not uselessly up into the sky.

The Minor Planet Center cfa-www.harvard.edu/iau/mpc This site tracks the orbits of minor planets, and provides useful astronomical information for skygazers hoping to catch one in their telescopes.

Near-Earth Object Project neo.jpl.nasa.gov If you're worried about an asteroid ploughing into the Earth, this site keeps track of such "near-Earth" objects so you can sleep soundly.

NGC 2000.0 Catalog seds.org/~spider/ngc/ngc Detailed information on the deep-sky objects covered in the *New General Catalogue* and the *Index Catalogues* (which means, most of the deep-sky objects you'd want to look at).

SETI www.seti.org The people who will be on the other end of the line when the aliens give us a call. This site speaks to their mission and features a screensaver you can download to help identify signals from intelligent life.

What's Up Tonight www.space.com/spacewatch/sky_calendar A quick shot of information as to what astronomical goodies await you when you go stargazing tonight.

The solar system

Before you tour the rest of the cosmos, it's worth spending a little time getting to know your own neighbourhood, the solar system. These sites give you the inside information about everything from Mercury to Pluto.

The Nine Planets www.seds.org/nineplanets An unflashy but highly informative site (now cheekily with the "nine" of the title scribbled out and the number "8" spraypainted above, to reflect Pluto's demotion) with extensive information on the Sun and every planet, as well as most of the major moons. Each entry contains links to other relevant sites.

The Virtual Solar System www.solarsystem.org.uk/index.php This site offers brief information on the planets and a virtual flyby of the solar system. Visitors are also encouraged to create a "10th planet" with the aim of creating life.

Lunar and Planetary Institute www.lpi.usra.edu This helpful site offers information on the Moon and planets for the casual visitor and also lesson plans on astronomy for teachers and educators.

Solar System Simulator space.jpl.nasa.gov Want to see how the Earth looks from Mars? Or Jupiter from the Cassini spacecraft? This simulator lets you do it, and creates images that show how every major object in the solar system is viewed from every other major object.

Solar System Live www.fourmilab.ch/solar/solar This basic orrery shows you where each of the planets is in its orbit, and provides you with all the right ascension and declination information for finding the planets in the sky.

Associations and clubs

Stargazing doesn't have to be a solitary pursuit: there are thousands of stargazing clubs and associations around the world. You can locate one that's near you, wherever you are.

Astronomical Societies www.universetoday.com/html/directory/astronomicalsocieties A collection of hundreds of links to astronomical societies all over the globe, arranged alphabetically by association name.

Australian Astronomical Societies www.quasarastronomy.com.au/society A list of Australian astronomical societies maintained by Quasar publishing. Includes information on locations, meetings and membership fees.

Canadian Astronomical Societies www.skynewsmagazine.com/pages/astronomy_clubs This page lists and links to national and regional Canadian astronomical societies and also other Canadian space-related sites.

New Zealand Astronomical Societies homepages.paradise.net.nz/~zog/ds/socks/nzsocs Provides contact information and Web links to astronomical associations in New Zealand.

UK Astronomical Societies www.r-clarke.org.uk/astrolinks_amateur A collection of local astronomical sites affiliated with the Federation of Astronomical Societies. It features clubs and associations in England, Scotland, Wales, Northern Ireland and other UK territories. Listings from the Republic of Ireland are also included.

US Astronomical Societies www.astroleague.org/al/general/society Astronomical club and associations in the US affiliated with The Astronomical League. Includes contact addresses and phone numbers as well as Web pages where applicable. Clubs in every state (apart from Rhode Island) are represented.

The Astronomical League www.astroleague.org The largest association of amateur astronomers. The site contains information on local branches, as well as information on becoming an "at large" member if no club is near you.

International Astronomical Union www.iau.org The worldwide professional organization for astronomers. Among other things, it gets to decide what are "official" constellations and planets (they're the ones that demoted Pluto). The site features the latest news and activities of the IAU, plus member directories and other information.

Royal Astronomical Societies These associations are open to astronomical professionals and amateurs and work to advance the cause of astronomy in their home countries. UK: **www.ras.org.uk** Canada: **www.rasc.ca** New Zealand: **www.rasnz.org.nz**

The Planetary Society www.planetary.org Society co-founded by the late Carl Sagan to promote solar-system exploration and the search for intelligent life in the universe. Site includes society news and information, updates on projects, and a learning centre.

The Society for Popular Astronomy www.popastro.com The name gives it away. This UK-based site wants to bring the sky to the masses. Their site certainly helps; it's a fun and well-designed collection of news and events, and provides useful information for beginners.

Space agencies

The United States doesn't have a monopoly on space – far from it. Space agencies all over the globe are planning missions for scientific and exploration purposes, manned and unmanned. Here's where you can find out what's being planned.

NASA www.nasa.gov NASA's massive site has just about everything on it, ranging from information on the latest space launches to Hubble Telescope photos to NASA administrative quality guidelines. For the amateur astronomer, it's like manna from heaven.

European Space Agency www.esa.int Europe's primary space agency has a nicely designed site featuring all the latest on ESA missions, science and activities. There's also an extensive multimedia gallery.

Canadian Space Agency www.space.gc.ca A well-designed but modest site which gets the visitor up to date on the latest in Canadian space endeavours. The multimedia site features pictures of Canadian astronauts.

National Space Development Agency of Japan www.jaxa.jp This English version of the Japanese agency's site fills you in on Japan's most recent launches and upcoming space-oriented projects. There's a reasonably large selection of English information.

Federal Space Agency (Russia) www.rosaviakosmos.ru/english/eindex The fabled Russian space programme may have seen better days than these, but still maintains an active role in space science. This site is utilitarian at best, containing basic information on recent Russian space projects and activities.

China National Space Administration www.cnsa.gov.cn/n615709/cindex China has ambitious plans for its space programme, but its English-version website sports very poorly translated text and a cheesy design.

British National Space Centre www.bnsc.gov.uk A catch-all site for information about the UK's space-related projects. Research, rocket launching and space policy are all reported and linked to here.

Photo and multimedia galleries

The Web is full of thousands of dazzling pictures of space. These picture archives will get you up close and personal with galaxies, nebulae, stars and planets.

Anglo-Australian Observatory Astronomical Images www.aao.gov.au/images A good selection of images taken from the AAO's telescopes, with an easy-to-use index right on the front page.

Astronomy Digital Image Library imagelib.ncsa.uiuc.edu/imagelib A comprehensive collection of research-grade images. This is a site primarily designed by scientists, so it's not a breeze to navigate. But for astro geeks, it's a wonderful resource.

Astronomy Picture of the Day antwrp.gsfc.nasa.gov/apod/astropix One of the most popular astronomy features on the Web. NASA reaches into its voluminous picture collection and finds something interesting, or posts pictures fresh from Hubble or other telescopes. A great, quick hit.

The Galaxy Catalog www.astro.princeton.edu/~frei/galaxy_catalog 113 different nearby galaxies, ready for your visual inspection (and download). Fun in itself and as an illustration of the varieties of galaxies that are out there.

Hubblesite Gallery hubblesite.org/gallery A well-designed site featuring the many stunning pictures that have come out of the famous space telescope. The site also features information on the Hubble and a selection of "wallpaper" images.

The Messier Catalog seds.lpl.arizona.edu/messier/Messier Tour of the deep-sky objects identified by Messier, most of which are easily viewable in binoculars and small telescopes. Each entry includes pictures and detailed information on the object.

Mr. Eclipse.com www.mreclipse.com Online home of Fred Espenak, noted eclipse expert. The site features a gallery of eclipse photos, including several "time lapse" and false colour images to go with the more conventional pictures.

NASA Multimedia Gallery www.nasa.gov/gallery/index The mother of all astronomy picture sites. Thousands of pictures from NASA's decades of exploration are here, from the moonwalks to Hubble, as well as additional galleries of movies, sound files and artwork.

Planetary Photojournal photojournal.jpl.nasa.gov For those interested in pictures of the solar system, the Planetary Photojournal catalogue provides NASA pictures from various planetary missions and also provides ample background information.

The SOHO Gallery sohowww.nascom.nasa.gov/gallery SOHO stands for "Solar and Heliospheric Observatory", so what you get are lots of Sun pictures, often as it's erupting in a dramatic and visually startling fashion.

Star Journey www.nationalgeographic.com/features/97/stars *National Geographic* here let the viewer to cruise through the galaxy and click on interesting objects – clicking brings up a larger picture and explanatory text.

Observatories and museums

Ideally, observatories need the clearest possible skies, which is why so many of them are sited in relatively remote and often mountainous spots. Despite this, several are open to the public and have good education and visitor facilities that are well worth taking advantage of. Separate listings for planetaria are to be found on p.387.

UK

Royal Observatory, Greenwich www.rog.nmm.ac.uk Part of a complex that includes The National Maritime Museum, the Royal Observatory offers a wealth of visitor programmes and services (and is free). The website has plenty of information, plus several online-only exhibits.

The Observatory Science Centre www.the-observatory.org The Observatory Centre at Hurstmonceaux in Sussex is a public-oriented site that offers an extensive range of visitor services, including tours, hands-on demonstrations and exhibits (you can even host a birthday party there).

Mills Observatory, Dundee www.dundeecity.gov.uk/mills This observatory, in the Scottish city of Dundee, is open free to the public all year round and offers regular lecture series and events. Its refracting telescopes are used for public displays during winter evenings.

Armagh Observatory www. arm.ac.uk The Armagh Observatory in Northern Ireland is a leading scientific research establishment. It has regular educational programmes and lectures, plus the Astropark – an outdoor model of the solar system.

USA

Kitt Peak National Observatory www.noao.edu/kpno Kitt Peak is located outside Tucson, Arizona, and has a visitor centre open daily to the public as well as nightly viewing programmes. The website has a virtual tour of the observatory.

Lowell Observatory www.lowell.edu Based at Flagstaff, Arizona, this is the observatory at which Pluto was discovered. The website has information on the observatory's extensive public programmes, which includes daily tours and an interactive exhibition hall.

McDonald Observatory mcdonaldobservatory.org The observatory of the University of Texas at Austin claims to have the darkest skies in the continental US. Its visitor centre offers daily tours and night-time viewing programmes three evenings a week.

Mauna Kea Observatories www. ifa.hawaii.edu/mko Mauna Kea is in Hawaii and its observatories are not that suitable for visitors (its Web page all but begs you not to bother); nevertheless, for the determined, the site's visitor centre offers a range of programmes and sky-watching opportunities.

The US Naval Observatory www.usno.navy.mil The USNO is the official timekeeper for the US, housing the country's Master Clock. Situated at Washington DC, the observatory offers limited public tours and opportunities for sightseeing with tightened security following the 9/11 attacks.

Mount Wilson Observatory www.mtwilson.edu This website for the Pasadena, California-based observatory has a virtual tour, and links to the Mount Wilson Observatory Society, which sponsors stargazing events at the observatory site itself.

Neil Armstrong Air & Space Museum www.ohiohistory.org/places/armstron The home town of the first man on the Moon has a museum in his honour, which features artifacts from the Gemini and Apollo missions. So if you're ever breezing through Wapakoneta, Ohio, you know where to go.

Smithsonian National Air and Space Museum www.nasm.si.edu This famous Washington DC museum is the home to numerous famous air-and spacecraft, and the website offers extensive information on all the offerings of the museum, including its planetarium and IMAX theatre.

Space Center Houston www.spacecenter.org The visitor centre for the Johnson Space Center, at Houston in Texas, has numerous educational and group programmes commemorating the US space missions, plus other exhibits on science and space.

CANADA

The Dominion Astrophysical Observatory www.hia-iha.nrc-cnrc.gc.ca This Victoria-based observatory has a new interpretive centre, "Centre of the Universe", which offers programmes for the public; these include tours and stargazing sessions using the observatory's 1.8-metre Plaskett telescope.

AUSTRALIA

The Parkes Observatory www.parkes.atnf.csiro.au This radio telescope observatory, also in New South Wales, features a recently expanded visitor's centre with free admission (some attractions require a small fee) and is open every day except Christmas and Boxing Day.

Sydney Observatory www.phm.gov.au/observe Part of the Museum of Applied Arts and Sciences, Sydney Observatory has a regular programme of exhibitions, films and talks, as well as night-time viewings seven evenings a week.

NEW ZEALAND

Stardome Observatory www.stardome.org.nz At the Stardome Observatory at Auckland you can learn about telescope technology, and view the stars through outdoor telescopes. There are also regular family days.

Carter Observatory www.carterobs.ac.nz The national observatory of New Zealand at Wellington offers a wide range of visitor facilities, courses, daily solar viewing and the use of its historic refractor telescope.

SOUTH AFRICA

Boyden Observatory www.assabfn.co.za/friendsofboyden/boyden Located 26km east of the city of Bloemfontein, Boyden has the second largest telescope in South Africa and is committed to public outreach programmes, including "open evenings" with lectures and stargazing sessions, group tours, and meetings of amateur astronomers.

South African Astronomical Observatory www.saao.ac.za/ index The SAAO has its headquarters in Cape Town and its main telescopes in Sutherland; the Sutherland site can only be visited for day-trips but Cape Town features monthly stargazing parties.

Planetaria

Planetaria come in all shapes and sizes, from humble converted rooms in schools to state-of-the-art multimedia experiences intended to entertain just as much as educate. All of them can open your eyes to the wonder of the night sky, and depending on the weather conditions and the degree of light pollution, may actually give a clearer view of the "night sky" than you would get from the real thing.

If none of the planetaria listed below are near your home, or if you are travelling abroad, the following two websites provide worldwide listings.

International Planetarium Society www.ibiblio.org/ips This is mainly a professional site for those involved with planetaria, but it also features a comprehensive list of planetaria all around the world.

European Planetarium Network www.artofsky.com/epn Lists locations in 25 countries, including the UK and Ireland.

UK

Armagh, Northern Ireland Armagh Planetarium **www.armaghplanet.com** 028/3752 3689

Bognor Regis, England South Downs Planetarium **www.southdowns.org.uk/sdpt** 01243/829868

Bristol, England Orange Imaginarium **www.at-bristol.org.uk** 0845/345 1235

Cardiff, Wales Techniquest **www.techniquest.org** 02920/475475

Dundee, Scotland Mills Observatory Planetarium **www.dundeecity.gov.uk/mills/displays** 01382/435846

Leicester, England National Space Centre **www.spacecentre.co.uk** 0870/607 7223

Liverpool, England Liverpool Museum Planetarium **www. nmgm.org.uk/livmus/planetariumframeset** 0151/478 4283

London, England London Planetarium **www.london-planetarium.com** 020 7486 1121

Macclesfield, England Jodrell Bank Planetarium **www.jb.man.ac.uk/scicen** 01477/571339

USA

Arkadelphia, Arkansas Reynolds Science Center Planetarium **www.hsu.edu/dept/phy/planetarium** 870/230-5170

Baltimore, Maryland Davis Planetarium **www.mdsci.org** 410/685-5225

Bozeman, Montana Taylor Planetarium **www.montana.edu/wwwmor/planetarium/index** 406/994-2251

Boston, Massachusetts Charles Hayden Planetarium **www.mos.orgmos/planet** 617/723-2500

Casper, Wyoming Casper Planetarium **www.trib.com wyoming/ncsd /planetarium** 307/577-0310

Chicago, Illinois Adler Planetarium **www.adlerplanetarium.org** 312/922-STAR

Cleveland, Ohio Shafran Planetarium **www.cmnh.org/exhibits/shafran** 216/ 231-4600

Columbus, Georgia Coca-Cola Space Science Center **www.ccssc.org** 706/649-1470

Columbus, Ohio The Dimon R. McFerson Planetarium & Theater **www.cosi.org** 888/819.COSI

Concord, New Hampshire Christa McAuliffe Planetarium **www. starhop .com** 603/271-STAR

Dallas, Texas The Science Place **www .scienceplace.org** 214/428-5555

Des Moines, Iowa Sargent Space Theater **www.sciowa.org** 515/274-4138

Evansville, Indiana Koch Science Center and Planetarium **www .emuseum.org/shows** (812)/425-2406

French Camp, Mississippi Rainwater Observatory & Planetarium **rainwater .astronomers.org** 662/547-6865

Gastonia, North Carolina James H. Lynn Planetarium **www. schielemuseum.org/start.asp** 704/866 6900

Honolulu, Hawaii Bishop Museum Planetarium **www.bishop museum.org/planetarium** 808/848-4162

Houston, Texas Burke Baker Planetarium **www.hmns.org** 713/639-4629

Indianapolis, Indiana SpaceQuest Planetarium **www.childrens museum.org/opendoors/sqprogs** 317/334-3322

Kansas City, Missouri Kansas City Museum Planetarium **www . kcmuseum.com** 816/483-8300

Lincoln, Nebraska Mueller Planetarium **www.spacelaser. com** 402/472-2641

Los Angeles, California Griffith Observatory **www.griffithobs. org** 323/664-1181

Memphis, Tennessee Sharpe Planetarium **www.memphis museums.org/planet** 901/320-6320

Milwaukee, Wisconsin Manfred Olsen Planetarium **www.uwm.edu/Dept /Planetarium** 414/229-4961

Minneapolis, Minnesota Minneapolis Planetarium **www.mplanetarium.org /planet_home**

Mystic, Connecticut Mystic Seaport Planetarium **www.mysticseaport.org /participate/po-planetarium**

New Orleans, Louisiana Judith W. Freeman Planetarium **www .audubоninstitute.org/lnc/tour _planetarium** 504/861-2537

Newark, New Jersey Dreyfuss Planetarium **www.newarkmuseum .org/planetarium** 973/596-6550

Newport News, Virginia Virginia Living Museum Planetarium **www .valivingmuseum.org/planetarium**

New York, New York Hayden Planetarium Rose Center for Earth and Space **www.amnh.org** 212/769-5200

Oklahoma City, Oklahoma Kirkpatrick Planetarium **www .omniplex.org/html/planetarium** 405/602-OMNI

Orlando, Florida Dr. Phillips CineDom **www.osc.org** 407/514-2000

Philadelphia, Pennsylvania Fels Planetarium (Franklin Institute) **sln .fi.edu/tfi/info/fels** 215/448-1200

Phoenix, Arizona Dorrance Planetarium **www.azscience.org** 602/716-2000

Pittsburgh, Pennsylvania Buhl Planetarium **www.carnegiescience center.org/exhibits/planet.asp** (412)/237-3397

Portland, Oregon Murdock Planetarium (Oregon Museum of Science & Industry) **www.omsi.edu / explore/planetarium** 503/797-4610

Reno, Nevada Fleischmann Planetarium **planetarium.unr.nevada .edu** 775/784-4811

Richmond, Virginia Ethyl IMAXDome and Planetarium **www. smv.org/ethyl** 804/864-1400

Rochester, New York Strasenburgh Planetarium **www.rmsc .orgplanetarium/planetframeset** 585/271-4320

St Louis, Missouri McDonnell Planetarium **www.slsc.org** 314/289-4400

Salt Lake City, Utah Hansen Planetarium **www.hansen planetarium.net** 801/358-2104

San Francisco, California Morrison Planetarium **www.calacademy.org /planetarium** 415/750-7127

Seattle, Washington Willard W. Smith Planetarium **www.pacsci.org /planetarium** 206/443-3648

Tucson, Arizona Flandrau Planetarium **www.flandrau.org** 520/621-STAR

Washington, District of Columbia Albert Einstein Planetarium **www .nasm.si.edu/nasm/planetarium /Einstein** 202/357-2700

Wichita, Kansas Exploration Place Cyberdome **www.exploration.org /html/cyberdome** (316)/263-3373

CANADA

Calgary, Alberta Calgary Science Centre **www.calgaryscience.ca** 403/268-8300

Edmonton, Alberta Margaret Zeidler Star Theatre **www.odyssium.com/mzt** 780/452-9100

Hamilton, Ontario W.J. McCallion Planetarium **www.physics.mcmaster .ca/Planetarium/Planetarium** 905/525-9140 ext 27777

Montreal, Quebec Montréal Planetarium **www.planetarium .montreal.qc.ca/index_a** 514/872-4530

Winnipeg, Manitoba Manitoba Planetarium **www.manitobamuseum . mb.ca/pl_info** 204/956-2830

AUSTRALIA

Adelaide, South Australia Adelaide Planetarium **ching.apana.org. au/~oliri/planet** (+61 8) 8302 3138

Brisbane, Queensland Sir Thomas Brisbane Planetarium **www.brisbane .qld.gov.au/community_facilities /leisure/planetarium** 07 3403 2578

Canberra, Australian Capital Territory Canberra Space Dome and Observatory **www.ctuc.asn.au/planetarium** 02 6248 5333

Launceston, Tasmania Launceston Planetarium **www.vision.net .au/~peter/AST/launplan/launplan** +61 3 6323 3777

Melbourne, Victoria Melbourne Planetarium **www.mov.vic.gov.au /planetarium/index** (03) 9392 4800

Sydney, New South Wales Sydney Observatory and Planetarium **www .phm.gov.au/observe** (02) 9217 0485

NEW ZEALAND

Auckland, North Island Stardome Observatory **www.stardome.org.nz** 64 9 624 1246

Wellington, North Island Carter Observatory **www.carterobs.ac.nz** 64 04 472 8167

SOUTH AFRICA

Cape Town, Western Cape Province South African Museum Planetarium **www.museums.org.za/planetarium** 021 424 3330

Johannesburg, Gauteng Johannesburg Planetarium (University of the Witwatersrand) **www.wits.ac.za/planetarium** 011 717 1392

Sites for kids

A love of astronomy can start at an early age. These sites introduce children to the wonders of space, and all the exciting things that exist beyond our planet – and make the learning process fun at the same time.

Astro for Kids www.astronomy.com/asy/default.aspx?c=a&id=1091 A cute and basic primer on the solar system for young-to-intermediate readers.

Hands on Universe www.handsonuniverse.org This site is associated with the Lawrence Hall of Science at Berkeley, California, and offers astronomical activities and projects kids can try on-and offline.

KidsAstronomy.com www.kidsastronomy.com This site is a bit dated but offers a fair amount of information on the universe, aimed at younger kids. It also offers printable colouring activities and video games.

KidSpace www.space.gc.ca/asc/eng/kidspace/kidspace Sponsored by the Canadian Space Agency, this site features quizzes and activities aimed at intermediate students, with an understandable bias towards things Canadian.

Laugh and Learn www.jpl.nasa.gov/kids NASA has a number of excellent science sites geared towards children of varying ages. "Laugh and Learn" collects them into a single portal that makes it easy to find the right site for your child.

Observatorium's Fun and Games observe.arc.nasa.gov/nasa/fun/fun_index These online games are related to astronomy and space science (more or less), and while they're designed for kids, some are also fun for grown-ups too.

The Sky www.seasky.org/sky A comprehensive site for young adults and teens that covers a wide range of news, activities and games. The site also has an area devoted to undersea exploration.

Space Place spaceplace.nasa.gov/en/kids NASA's kid-oriented site features quite a collection of material and information on space exploration and science aimed at younger children.

Other interesting sites

These sites don't fit in any other category, but are fun, informative and well worth exploring.

Ask the Experts www.sciam.com/askexpert_directory.cfm?category=space No astronomy question is too tough for *Scientific American*'s experts, who answer reader-generated queries with sound but easy-to-follow responses. An index lets you catch up on previously answered questions.

Phil Plait's Bad Astronomy www.badastronomy.com Phil Plait, an astronomer himself, takes on all the poorly understood (and just plain wrong) astronomy found in movies, conspiracy theories and Internet rumours.

Index

A

B

C

G

H

I

N

O

P

Q–R

S

T

U

V

W

X–Z

ROUGH GUIDES Complete Listing

UK & Ireland
Britain
Devon & Cornwall
Dublin **D**
Edinburgh **D**
England
Ireland
The Lake District
London
London **D**
London Mini Guide
Scotland
Scottish Highlands & Islands
Wales

Europe
Algarve **D**
Amsterdam
Amsterdam **D**
Andalucía
Athens **D**
Austria
Baltic States
Barcelona
Barcelona **D**
Belgium & Luxembourg
Berlin
Brittany & Normandy
Bruges **D**
Brussels
Budapest
Bulgaria
Copenhagen
Corfu
Corsica
Costa Brava **D**
Crete
Croatia
Cyprus
Czech & Slovak Republics
Denmark
Dodecanese & East Aegean Islands
Dordogne & The Lot
Europe
Florence & Siena
Florence **D**
France
Germany
Gran Canaria **D**
Greece
Greek Islands
Hungary
Ibiza & Formentera **D**
Iceland
Ionian Islands
Italy
The Italian Lakes
Languedoc & Roussillon
Lanzarote & Fuerteventura **D**
Lisbon **D**
The Loire Valley
Madeira **D**
Madrid **D**
Mallorca **D**
Mallorca & Menorca
Malta & Gozo **D**
Menorca
Moscow
The Netherlands
Norway
Paris
Paris **D**
Paris Mini Guide
Poland
Portugal
Prague
Prague **D**
Provence & the Côte D'Azur
Pyrenees
Romania
Rome
Rome **D**
Sardinia
Scandinavia
Sicily
Slovenia
Spain
St Petersburg
Sweden
Switzerland
Tenerife & La Gomera **D**
Turkey
Tuscany & Umbria
Venice & The Veneto
Venice **D**
Vienna

Asia
Bali & Lombok
Bangkok
Beijing
Cambodia
China
Goa
Hong Kong & Macau
Hong Kong & Macau **D**
India
Indonesia
Japan
Kerala
Laos
Malaysia, Singapore & Brunei
Nepal
The Philippines
Rajasthan, Dehli & Agra
Singapore
Singapore **D**
South India
Southeast Asia
Sri Lanka
Taiwan
Thailand
Thailand's Beaches & Islands
Tokyo
Vietnam

Australasia
Australia
Melbourne
New Zealand
Sydney

North America
Alaska
Baja California
Boston
California
Canada
Chicago
Colorado
Florida
The Grand Canyon
Hawaii
Honolulu **D**
Las Vegas **D**
Los Angeles
Maui **D**
Miami & South Florida
Montréal
New England
New Orleans **D**
New York City
New York City **D**
New York City Mini Guide
Orlando & Walt Disney World® **D**
Pacific Northwest
San Francisco
San Francisco **D**
Seattle
Southwest USA
Toronto
USA
Vancouver
Washington DC
Washington DC **D**
Yellowstone & The Grand Tetons
Yosemite

Caribbean & Latin America
Antigua & Barbuda **D**
Argentina
Bahamas
Barbados **D**
Belize
Bolivia
Brazil
Cancùn & Cozumel **D**
Caribbean
Central America
Chile
Costa Rica
Cuba
Dominican Republic
Dominican Republic **D**
Ecuador
Guatemala
Jamaica
Mexico
Peru
St Lucia **D**
South America
Trinidad & Tobago
Yúcatan

Africa & Middle East
Cape Town & the Garden Route
Dubai **D**
Egypt
Gambia
Jordan

D: Rough Guide **DIRECTIONS** for short breaks

Available from all good bookstores

enya
arrakesh D
orocco
outh Africa, Lesotho
& Swaziland
yria
anzania
unisia
est Africa
anzibar

ravel Specials
irst-Time Africa
irst-Time Around
the World
irst-Time Asia
irst-Time Europe
irst-Time Latin
America
ravel Health
ravel Online
ravel Survival
alks in London
& SE England
omen Travel
orld Party

aps
lgarve
msterdam
ndalucia
& Costa del Sol
rgentina
thens
ustralia
arcelona
erlin
oston & Cambridge
rittany
russels
alifornia
hicago
hile
orsica
osta Rica
& Panama
rete
roatia
uba
yprus
zech Republic
ominican Republic
ubai & UAE
ublin
gypt

Florence & Siena
Florida
France
Frankfurt
Germany
Greece
Guatemala & Belize
Iceland
India
Ireland
Italy
Kenya & Northern
Tanzania
Lisbon
London
Los Angeles
Madrid
Malaysia
Mallorca
Marrakesh
Mexico
Miami & Key West
Morocco
New England
New York City
New Zealand
Northern Spain
Paris
Peru
Portugal
Prague
Pyrenees & Andorra
Rome
San Francisco
Sicily
South Africa
South India
Spain & Portugal
Sri Lanka
Tenerife
Thailand
Toronto
Trinidad & Tobago
Tunisia
Turkey
Tuscany
Venice
Vietnam, Laos
& Cambodia
Washington DC
Yucatán Peninsula

Dictionary Phrasebooks
Croatian
Czech
Dutch
Egyptian Arabic
French
German
Greek
Hindi & Urdu
Italian
Japanese
Latin American
Spanish
Mandarin Chinese
Mexican Spanish
Polish
Portuguese
Russian
Spanish
Swahili
Thai
Turkish
Vietnamese

Computers
Blogging
eBay
iPhone
iPods, iTunes
& music online
The Internet
Macs & OS X
MySpace
PCs and Windows
PlayStation Portable
Website Directory

Film & TV
American
Independent Film
British Cult Comedy
Chick Flicks
Comedy Movies
Cult Movies
Film
Film Musicals
Film Noir
Gangster Movies
Horror Movies
Kids' Movies
Sci-Fi Movies
Westerns

Lifestyle
Babies
Ethical Living
Pregnancy & Birth
Running

Music Guides
The Beatles
Blues
Bob Dylan
Book of Playlists
Classical Music
Elvis
Frank Sinatra
Heavy Metal
Hip-Hop
Jazz
Led Zeppelin
Opera
Pink Floyd
Punk
Reggae
Rock
The Rolling Stones
Soul and R&B
Velvet Underground
World Music
(2 vols)

Popular Culture
Books for Teenagers
Children's Books,
5-11
Conspiracy Theories
Crime Fiction
Cult Fiction
The Da Vinci Code
His Dark Materials
Lord of the Rings
Shakespeare
Superheroes
The Templars
Unexplained
Phenomena

Science
The Brain
Climate Change
The Earth
Genes & Cloning
The Universe
Weather

For more information go to www.roughguides.com

ROUGH GUIDES

"The brilliance of the Rough Guide concept is breathtaking"

BBC Music Magazine

BROADEN YOUR HORIZONS

COMPUTERS Blogging • The Internet The iPhone • iPods, iTunes & music online • Macs & OSX • PCs & Windows PlayStation Portable (PSP) Website Directory: Shopping Online & Surfing the Net

FILM & TV American Independent Film British Cult Comedy • Chick Flicks Comedy Movies • Cult Movies • eBay Film • Film Musicals • Film Noir Gangster Movies • Horror Movies Kids' Movies • Sci-Fi Movies • Westerns

LIFESTYLE Babies • Ethical Living Pregnancy & Birth • Running

MUSIC The Beatles • Bob Dylan Classical Music • Blues • Elvis • Frank Sinatra • Heavy Metal • Hip-Hop • Jazz Opera • Book of Playlists • Led Zeppelin Pink Floyd • Punk • Reggae • Rock The Rolling Stones • Soul and R&B Velvet Underground • World Music (2 vols)

POPULAR CULTURE Books for Teenagers • Children's Books 5-11 Crime Fiction • Cult Fiction • The Da Vinci Code • His Dark Materials • Lord of the Rings Poker • Shakespeare • Superheroes Conspiracy Theories • The Templars Unexplained Phenomena

SCIENCE The Brain • Climate Change The Earth • Genes & Cloning The Universe • Weather

www.roughguides.com

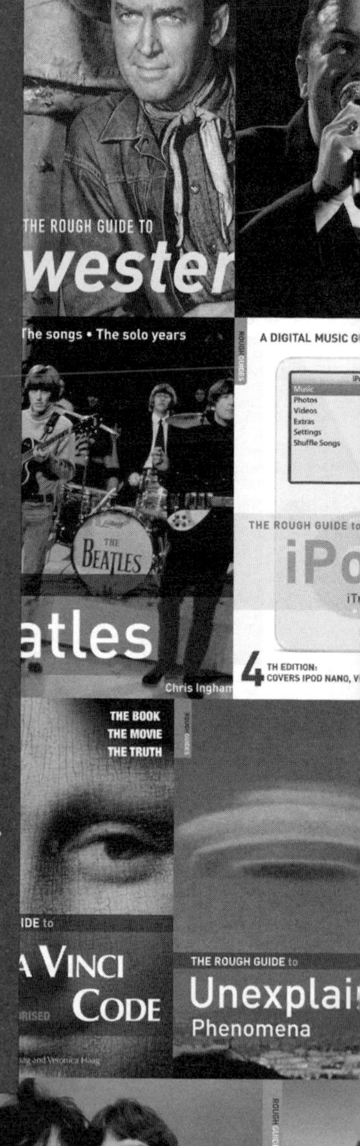

THE ROUGH GUIDE PHRASEBOOK
Japanese
Mexican Spanish
French
Hindi & Urdu
Greek
BROADEN YOUR HORIZONS
'he new Rough Guides are a different beast, they have evolved into something
SLEEK, SEXY and VERY STYLISH"
ndrew Steed, Stanfords Map and Travel Bookshop
ROUGH GUIDES
ore than 300 travel titles • guides • maps • phrasebooks • www.roughguides.com
THE ROUGH GUIDE to
Australia
Britain
New York City
THE ROUGH GUIDE First-Time Asia
ROUGH GUIDE MAP
Dublin
Hong Kong
London
Paris
Ken
Oxford Stre
Piccadilly
GREEN
Rough Guide DIRECTIONS
San Francisco
Marrakesh
Madrid
Singapore